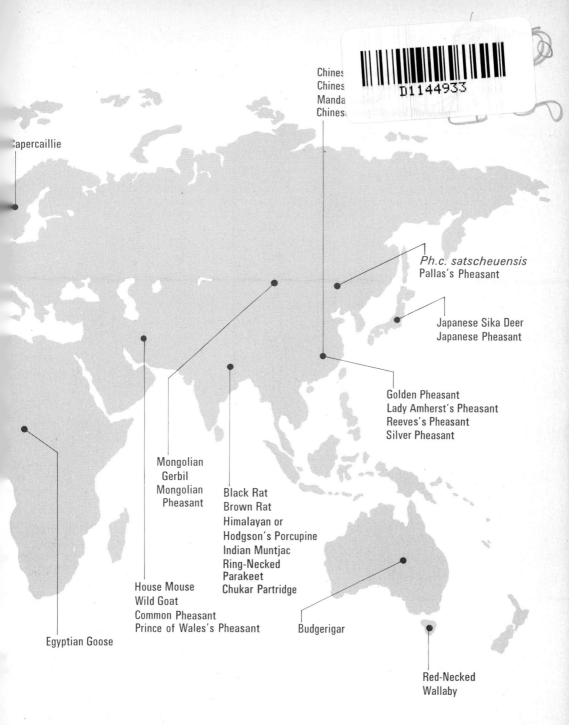

Capercaillie

Chines
Chines
Manda
Chines

Ph.c. satscheuensis
Pallas's Pheasant

Japanese Sika Deer
Japanese Pheasant

Golden Pheasant
Lady Amherst's Pheasant
Reeves's Pheasant
Silver Pheasant

Mongolian
Gerbil
Mongolian
Pheasant

Black Rat
Brown Rat
Himalayan or
Hodgson's Porcupine
Indian Muntjac
Ring-Necked
Parakeet
Chukar Partridge

House Mouse
Wild Goat
Common Pheasant
Prince of Wales's Pheasant

Budgerigar

Egyptian Goose

Red-Necked
Wallaby

The Naturalized Animals of the British Isles

Also by Christopher Lever

Goldsmiths and Silversmiths of England

The Naturalized Animals
of the British Isles

Christopher Lever

Foreword by Peter Scott

Drawings by Ann Thomson

Hutchinson of London

Hutchinson & Co (Publishers) Ltd
3 Fitzroy Square, London W1

London Melbourne Sydney Auckland
Wellington Johannesburg and agencies
throughout the world

Produced by Hutchinson Benham Ltd

First published 1977
© text and illustrations Christopher Lever 1977

Set in Monotype Times

Printed in Great Britain by The Anchor Press Ltd
and bound by Wm Brendon & Son Ltd
both of Tiptree, Essex

ISBN 0 09 127790 6

Dedicated to the Mandarin Duck – most beautiful of birds – for providing me with the inspiration for this book

Contents

Foreword

I first met Christopher Lever during the International Fourteen Foot Dinghy Championship at Torquay in 1956. I little thought at that time that twenty years later I should be writing a foreword to his history of the naturalized animals of the British Isles.

As the author points out, in the opening words of his prologue, there is a paucity of native wild animals in Britain compared to mainland Europe. It therefore follows that our naturalized wildlife is of proportionately greater importance to the general ecology of these islands.

Christopher Lever, though not a professional zoologist, has made a special study of our alien fauna for some nineteen years: he has researched his subject with a thoroughness which enables him to write with authority on a most interesting branch of natural history which has hitherto been sadly neglected. His book is the first comprehensive account of the exotic vertebrates of the British Isles, and thus fills an empty niche on the zoological library shelf: it must surely be the definitive work on the subject for many years to come.

The author may feel that being so close to his subject makes it difficult for him to pronounce a verdict on the environmental and ecological pros and cons of transferring wild animals to areas outside their natural range. Perhaps we should be guided by current ecological opinion in these matters. All too often the introduction of an exotic species threatens the survival of a native one through direct predation or competition for food and living space. Thus the responsible view seems to suggest that there is no reasonable excuse for the deliberate introduction of exotics. But it is certainly arguable that this conclusion might be modified in the case of a settler arriving in a countryside devoid of, for example, song birds. Must he, for the sake of possible ecological side effects, do without the bird song that he loved as a child? Who is to say that he has no right to introduce the song thrush or the skylark? Admittedly great damage has been caused especially to island fauna and flora, by such introductions, although in other cases there has been a nett enrichment.

Many introductions have been based on commercial rather than aesthetic considerations and for the most part these are the ones that have wrought the worst havoc. Attempts to control rats by introducing the mongoose in some tropical islands have been wholly disastrous to the endemic fauna. The escape of fur-bearing animals like the mink has been serious in many places including our own country.

Having been carelessly responsible myself for allowing the North American Ruddy Duck to escape and build up to what seems to be a small but viable population in England, I am in no position to pass judgement on others. To be sure the Ruddy Duck is decorative and apparently harmless but no one can know what insidious effect it may have on the ecological web. I really should not have allowed them to fly out into the countryside – although they look delightful in flight.

Christopher Lever has produced a most readable and cohesive survey of the naturalized wild animals in Britain, in which each chapter also forms a self-contained monograph on an individual species. It will appeal to all – professional and amateur alike – who are interested in natural history and are concerned with the conservation and preservation of our natural environment.

Peter Scott.

Acknowledgements

First and foremost my sincere thanks are due to Dr G. B. Corbet (Deputy Keeper, Department of Zoology, British Museum); Mr Robert Hudson (Research Officer and Honorary Librarian, British Trust for Ornithology); Dr E. N. Arnold (Senior Scientific Officer, Department of Zoology, British Museum); and Mr A. C. Wheeler (Senior Scientific Officer, Department of Zoology, British Museum). I am grateful to them all for the interest which they showed in my project when I approached them for their help, which they gave unstintingly; they pointed me in the right direction for my researches; they drew my attention to relevant published and unpublished material; and finally they made time from their own busy programmes to read through my manuscript chapters on, respectively, mammals; birds; amphibians and reptiles; and fish, and pointed out my various errors and omissions. To them all I am deeply indebted.

My thanks are also due to Mr Harry V. Thompson of the Pest Infestation Control Laboratory and to Dr L. M. Gosling of the Coypu Research Laboratory of the Ministry of Agriculture, Fisheries and Food: the former kindly read and annotated the relevant mammal chapters: the latter read and annotated the chapters on the coypu and porcupines, and generously supplied much of the information on the latter species.

To the librarians and staff of the various libraries where I have carried out my researches I extend my sincere thanks for their patience and understanding in searching out for me innumerable books and documents; I am particularly grateful to the staff of the various libraries in the British Museum (Natural History); Mr M. J. Rowlands (Museum Librarian); Mr M. R. Halliday (Reference Librarian, General Library); Mrs A. Datta (Zoology Librarian); Mrs M. Anthony; Miss G. Cornelius; Miss D. M. Hills; Miss M. Levitt;

Miss D. Norman; Miss M. Seely; and Miss E. Taylor. My gratitude is also due to the librarians of the British Trust for Ornithology; the Edward Grey Institute of Field Ornithology; the Ministry of Agriculture, Fisheries and Food; the Science Museum; the University of Reading; the London Library and the Zoological Society of London.

For kindly permitting me to include in the chapter on the mandarin duck material which first appeared in *Country Life* in 1957, I am grateful to the editor, Mr Michael Wright.

My frequent requests for help and information have met with great kindness and patience from a host of correspondents; for their assistance and generosity in various ways I am grateful to the following:

Mr I. R. H. Allan (Ministry of Agriculture, Fisheries and Food); Major C. J. Armour (M.A.F.F.); Mr H. R. Arnold (Institute of Terrestrial Ecology); Mr W. J. Ayton (Principal Fisheries Officer, Welsh National Water Development Authority); Mr H. E. Axell (Warden, Minsmere Bird Reserve); Mr R. Bagot; Mr R. Baker (Norfolk and Norwich Naturalists' Society); Mr F. R. S. Balfour; Mrs R. M. Barnes (Keeper of Natural History, Castle Museum, Norwich); the Duke of Bedford; Dr T. Beebee; Dr J. Berry (Director, Nature Conservancy); Mr D. Bird; Mr F. W. Black (M.A.F.F.); Mr F. Boyce; Dr J. M. Boyd and Mr R. N. Campbell (Nature Conservancy); Mr F. R. Cann and Mrs P. Carver (M.A.F.F.); Mr C. P. Carpenter; Viscount Chaplin; Mr R. E. Chaplin (Educational Department, Edinburgh Zoo); Mr A. S. Churchward (Severn–Trent Water Authority); Mr D. J. Clarke (Assistant Curator, Carlisle Museum); Mr W. L. Coleridge; Miss M. L. Connolly and Mr O. A. Crimmen (Department of Zoology, British Museum); the Earl of Cranbrook; Dr O. Dansie; Mr B. Dean; Mr J. M. de Bunsen; Mr J. O. D'Eath; Mrs M. G. de Udy; Mrs T. Dorrien-Smith; Mr P. Dumbill; Mr E. V. Evans (M.A.F.F.); Mr P. D. Ferguson; Mr P. Fetherston-Godley (Avon and Airlie Game Farms); Mr R. S. R. Fitter (Fauna Preservation Society); Dr J. Flegg (British Trust for Ornithology); Capt. E. A. M. Fox-Pitt; Mr A. Fraser (Middle Thames Natural History Society); Dr J. F. D. Frazer (Nature Conservancy); Mr O. Frazer; Dr W. Frost (Freshwater Biological Association); Mr J. G. Goldsmith (Assistant, Natural History, Castle Museum, Norwich); Mr R. P. D. Goodwin (Principal

Senior Scientific Officer, Department of Zoology, British Museum); Mr I. J. Grainger (Warden, Landmark Trust, Lundy Island); Miss E. Green (Fur Breeders' Association of the U.K.); Mr and Mrs J. J. Gurney; Miss P. Gurney; Sir Hildebrand Harmsworth, Bt; Mr B. Hawkes; Mr J. Hawkings; Dr B. J. Hill (M.A.F.F.); Mr E. G. Hills (Chiltern District Council); Mr A. V. Holden (M.A.F.F.); Mr A. Horne (Dacorum District Council); Mr D. Hunt; Mr H. G. Hurrell; the Earl of Iveagh; Dr M. Kennedy (Inland Fisheries Trust, Dublin); Mr W. Kerr; Mr B. King; Mrs J. C Laidlay; Mr D. F. Leney (Surrey Trout Farm and United Fisheries); Dr A. Leutscher; Dr R. S. J. Linfield (Anglian Water Authority); Dr I. J. Linn (Department of Zoology, University of Exeter); Sir Giles Loder, Bt; the Revd David Low (Conservation Officer, Isle of Wight Natural History Society); Mr V. P. W. Lowe and Dr P. S. Maitland (Institute of Terrestrial Ecology); Mr D. Marlborough (British Ichthyological Society); Mr B. McWilliams (Keeper, Natural History, City of Norwich Museum); Mr D. Miles (Managing Editor, *Nature in Wales*); Sir William Murray, Bt; Mr J. Newby (Leisure Sport Ltd); Mr F. J. Nixon (Editor *Surrey Life*); Mr M. A. Ogilvie (Wildfowl Trust); Major S. Ogilvie; Mr D. Owens (Educational Department, Edinburgh Zoo); Mr F. J. T. Page (Deer Group, Mammal Society); Mr E. J. Pavey (M.A.F.F.); Mr W. H. Payn; Major H. Peacock; Mr L. R. Peart (Berkshire Trout Farm); Major A. D. Peirse-Duncombe (Scottish Ornithologists' Club); Mr R. Penhallurick (Assistant Curator, County Museum, Truro); Mr T. V. Pickvance (University of Birmingham); Dr A. Powell (Department of Biology, City of London Polytechnic); Mr R. Prior (Forestry Commission); Mr A. T. Rawlinson (Chiltern District Council); Mr R. V. Redston (Director of Environmental Health, Bath); Mr P. J. Reynolds (Director, Butser Ancient Farm Project); Mr N. D. Riley (Department of Zoology, British Museum); Mr T. B. Rothwell; Mr C. D. W. Savage (World Pheasant Association); Mr M. J. Seago (Norfolk and Norwich Naturalists' Society); Dr J. T. R. Sharrock (Organizer, Ornithological Atlas, British Trust for Ornithology); Mr C. Sims (Keeper, Biology, Yorkshire Museum); Mr R. W. Sims (Senior Scientific Officer, Department of Zoology, British Museum); Mr K. Smith (Superintendent, Paignton Zoo); Dr D. Snow (Senior Principal Scientific Officer, Department of Zoology, British Museum); the Earl of Southesk; Dr I. F. Spellerberg (Department of Biology, University of Southampton); Mr N.

Stevens; Mr J. H. F. Stevenson (Peregrine Fur Farm); Dr J. P. Stevenson; Mr B. Stott and Mr I. Storey (M.A.F.F.); Mr K. Suiter (Leisure Sport Ltd); Miss E. J. Taylor (M.A.F.F.); Mr G. Taylor; Dr A. M. Tittensor; Lord Tollemache; Mr D. Tomlinson (*Country Life*); Dr F. A. Turk (Reader in Natural History and Senior Research Fellow, University of Exeter); Mr M. N. P. Utsi (Technical Adviser, Reindeer Council of the U.K.); Mr A. F. Walker, Mr B. G. S. Ward and Mr R. B. Williamson (M.A.F.F.); Mr P. Wayre (Pheasant Trust); Mr C. Wright (Shropshire Ornithological Society); Dr D. W. Yalden (The Department of Zoology, University of Manchester); Prof Lord Zuckerman and Mr V. J. A. Manton (Zoological Society of London).

Finally, I should like to thank Mrs Ann Thomson for her delightful drawings which greatly enhance the text; Miss Emma Hogan of Hutchinson & Co. Ltd for again so ably editing my manuscript; Mrs Helen Baz for providing the index; and last, but by no means least, Mrs Ann Trevor for her patience in once again bringing order to and typing my sometimes chaotic and well-nigh incomprehensible manuscript.

To Sir Peter Scott I extend my most grateful thanks for so kindly agreeing to provide a Foreword to this book.

Note

The individual distribution maps are intended to be an approximate guide only to where each species occurs in the British Isles, and are not based on the 10-kilometre-square Ordnance Survey National Grid. With a few exceptions they show only established communities, and do not take account of isolated records. Species for which no maps are given are either considered to be virtually ubiquitous, or are of doubtful viability.

Similarly, the endpaper maps of the world give only a general indication of the natural range of each species.

Metric conversions have, in most cases, been rounded up or down to the nearest whole unit.

Prologue

The indigenous wildlife of the British Isles is meagre indeed compared to that of continental Europe. In the final period of glaciation of the Ice Age the huge sheets of ice reached almost as far south as the valley of the Thames: only when they had receded could the fauna then living in warmer climes return from their havens in southern Europe. Colonization of Britain was simplified by the fact that at this time the gap now occupied by the English Channel was dry land. When, some 5000 to 10 000 years ago, the so-called 'Land Bridge' was invaded from the north and south by the sea and continental Europe split from the British Isles, those species already in Britain became divided by the Channel thus formed from those which had not yet reached so far north. Since that time, additions to the fauna of the British Isles have been forced – with or without the assistance of man – to face the barrier of the sea.

This book sets out to describe when, where, why, how and by whom the various alien vertebrate animals now living in a wild state in Britain were introduced, how they subsequently became naturalized,* and what effect they have had on the environment and on the native fauna and flora. The criteria for inclusion of a species are that it should have been imported from outside the British Isles deliberately or unwittingly by man, and that at the time of writing (1976) it should appear to be established in the wild in self-supporting and self-perpetuating numbers even if, as in the case of the porcupines (*Hystrix cristata* and *H. hodgsoni*), it may soon become eradicated, or, like the wood duck (*Aix sponsa*) and bobwhite quail (*Colinus*

---

*'Naturalized' is defined as 'established in the wild in self-maintaining and self-perpetuating populations unsupported by man': 'acclimatized' as 'able to survive in the wild only with the support of man': 'feral' as 'a previously domesticated or captive animal now naturalized'.

B

virginianus), it has not yet been formally admitted to the British and Irish List.*

One originally indigenous species, the capercaillie (*Tetrao urogallus*), which became extinct in Britain in about 1785, is included because the entire present British stock stems from a series of re-introductions made by the Marquess of Breadalbane at Taymouth Castle in Perthshire† in 1837–8. (The capercaillie is, incidentally, the only example of the successful re-introduction to Britain by man of a previously indigenous vertebrate.)

On the other hand, such domestic varieties as the Chinese goose, which is descended from the Siberian swan-goose (*Anser cygnoides*), and the Muscovy duck, whose ancestor is the Central and South American wild Muscovy (*Cairina moschata*), both of which sometimes breed ferally in Britain, are not discussed. The Barbary dove (*Streptopelia 'risoria'*) – a domesticated form, unknown in the wild, of the Eastern collared dove (*S. decaocto*)‡ or the African collared dove (*S. roseogrisea*) – which not infrequently escapes from captivity and has interbred in the wild with *decaocto* (which arrived in Britain naturally in 1954), is similarly excluded, since the dominant genes of *decaocto* soon obliterate *risoria* characteristics. The so-called 'wild ponies' (*Equus caballus*), which roam freely over the New Forest, Dartmoor, Exmoor and elsewhere, are also omitted, as they are subject to an annual round-up by their owners and are thus not truly 'wild'.§

The criteria for inclusion have resulted in the omission of such species as the gadwall, greylag goose, goshawk, and red-crested pochard. All are natives of continental Europe, occurring in the British Isles mainly as passage migrants and winter visitors, and all have to some extent artificial British breeding distributions.

*An official roll, maintained by the British Ornithologists' Union, to which there is no equivalent for other animals.

†Under the Local Government Act (1972) the names and boundaries of a number of counties in England and Wales were changed from 1 April 1974, and in Scotland from 1 May 1975. As many of the events described here antedate these changes, and as it would be confusing to use both old and new systems, the old names and boundaries have been adhered to throughout.

‡The Turks likened the call of the collared dove to *Allah-hu-Akbar* (Allah is the Greatest). The ancient Greeks invented a legend that a maidservant who received only eighteen *paras* (about 12p) a year in wages complained of her plight to Zeus, who created the collared dove to cry accusingly *decaocto* (eighteen) for the rest of time.

§The New Forest alone provides valuable grazing for some 4000 such ponies; these are owned by the Commoners (those living within the Forest boundaries) who, from Norman times, have enjoyed grazing and other rights in the Forest, which was at one time a royal hunting ground for deer, where private enclosure was forbidden.

Most of the gadwall (*Anas strepera*) now nesting in East Anglia (their principal British breeding area) are believed to be descended from a pair captured in a decoy at Dersingham in Norfolk in about 1850, which were subsequently turned down by the Revd. John Fountaine of Southacre, on Narford Lake in the Breckland district of south-west Norfolk. It is, however, probable that the gadwall of East Anglia are reinforced from time to time by continental migrants which stay on to breed. Those gadwall nesting in small numbers in various parts of Scotland (where the largest colony is based on Loch Leven in Kinross-shire) and Ireland probably represent a natural extension of range from their main breeding grounds in eastern Europe, southern Scandinavia, and Iceland; it may be significant that the recent increase in the Scottish and Irish populations has coincided with a marked expansion of the species in Iceland, from where large numbers of gadwall migrate south to winter in Ireland.

The greylag goose (*Anser anser*) is Britain's only native breeding goose, nesting, until fairly recently, in the northern and western Scottish Highlands and in the Outer Hebrides, but currently only in the latter in a truly wild state. It has been reinforced in mainland Scotland and introduced elsewhere in the British Isles (even to parts of southern England) in such numbers that there are now far more feral than wild British breeding greylag geese: it remains, nevertheless, an indigenous species.

The present British breeding population of the goshawk (*Accipiter gentilis*) is very small (in 1974 only three pairs are known to have nested successsfuly out of a total population of about thirteen pairs) and the majority are likely to be the offspring of birds which had been imported by falconers, from whom they had subsequently escaped or by whom they had been released: some degree of natural colonization from continental Europe is, however, possible.[*]

The red-crested pochard (*Netta rufina*) is an autumn and winter vagrant to Britain, where it has also long been kept in waterfowl collections. In April 1937 a pair bred in north-east Lincolnshire, which may either have been wanderers from Woburn Abbey in Bedfordshire or natural colonizers from Europe, where the species was at the time spreading north-westwards and was already well established as a breeding bird in northern Germany and possibly in the Netherlands. In 1958 an escaped pair bred ferally on a lake in Essex. From the mid 1960s to the early 1970s one or two pairs of

[*]P. A. D., Hollom, *British Birds*, 50, pp. 135–6, 1957.

red-crested pochards, which had probably escaped from the Wildfowl
Trust at Slimbridge, bred on Frampton Pools at Frampton-on-
Severn in Gloucestershire,* but never succeeded in establishing a
feral population. During the winter of 1974-5 seven red-crested
pochards were present at the Cotswold Water Park – an extensive
area of flooded gravel pits between Cirencester and Cricklade on the
Gloucestershire/Wiltshire border – and two pairs stayed on to breed
in 1975; ducklings were later seen on the Gloucestershire side of the
Park, but the pair which nested on the Wiltshire side were unsuccessful.† In the 1950s the red-crested pochard was a regular
autumn migrant to south-eastern England, but since 1962 it has
reverted to its earlier vagrant status. Red-crested pochards seen in
Britain today are much more likely to be escapees from waterfowl
collections than immigrant birds. Thus any future viable breeding
population in Britain – which does not at present exist – is more
likely to consist of feral birds than of natural immigrants which
remained to breed.

Exotic animals have been introduced to the British Isles for three
main reasons: economic, ornamental and sporting.

Species in the first category can be divided into those introduced
primarily to provide a source of food, and those imported principally
for their skins. The common pheasant (*Phasianus colchicus*), which was
acclimatized though not naturalized in Britain before the time of the
Norman Conquest, was introduced to augment the available supply
of food. In 1929 the coypu (*Myocastor coypus*) and the mink (*Mustela
vison*) were imported to Britain from, respectively, South and North
America to stock nutria and ranch-mink farms. The rabbit
(*Oryctolagus cuniculus*), which probably first appeared in Britain at
some time between the reigns of the Norman king, Stephen
(1135-54) and the Plantagenet, Richard I (1189-99), filled an un-
enviable dual role, being valued equally for its flesh as for its
fur.

The recently introduced, but so far fortunately only acclimatized,
Chinese grass carp or white amur (*Ctenopharyngodon idella*), falls into
a different category of animals introduced for economic purposes. In

*Bristol Bird Report, p. 233, 1972.
†Wiltshire Orn. Soc. News Bulletin, 3, August 1975.

the 1960s 1900 specimens from the Hungarian State Fish Farm at Dinnyés near Budapest were released by the Ministry of Agriculture in a number of enclosed waters in Britain, principally in the Fens. This herbivorous cyprinid fish, which superficially resembles the native chub (*Leuciscus cephalus*), has been used in several parts of the world to clear waters of unwanted macrophytic vegetation. According to D. G. Cross:* 'The grass carp rarely breeds outside its native rivers. Although the gonads develop to maturity in fish that have been introduced, breeding does not occur. Apparently the fish need a temperature over 20°C [68°F] and a current flow between two to five feet per second before they spawn. Such conditions do occur locally in Britain in the heated water effluent from power stations,'† and 'there is a remote possibility that the fish will breed in this country'. Should this happen, the effect on the ecology of those waters containing grass carp would prove serious. At present it is planned to rear the fish artificially under carefully controlled conditions at Hellesdon in a joint research programme conducted by the Ministry and the East Suffolk and Norfolk River Authority.

Animals introduced on account of their decorative qualities are mainly deer, waterfowl and 'ornamental' pheasants. The first of these include the Japanese sika (*Cervus nippon*) introduced to Powerscourt in Co. Wicklow by Lord Powerscourt in 1860, and the Indian and Chinese or Reeves's muntjac (*Muntiacus muntjak* and *M. reevesi*) and the Chinese water deer (*Hydropotes inermis*) introduced at the turn of the century to Woburn Abbey in Bedfordshire by the Duke of Bedford.

Since the seventeenth century attempts have been made to naturalize more species of waterfowl in Britain than any other single group of animals, but only four aliens have become successfully established. The Canada goose (*Branta canadensis*), which is known to have been contained in the Royal collection of Charles II in St James's Park in London, where it was seen by Willughby and Ray prior to 1672, is now widespread and common in many parts of Britain. The Egyptian goose (*Alopochen aegyptiacus*), which was admitted to the British and Irish List as recently as 1971, may also have been acclimatized in Britain by the late seventeenth century,

---

*'Aquatic Weed Control Using Grass Carp', *Jour. Fish. Biol.*, 1, pp. 27–30, 1969. See also B. Stott, 'Aquatic Weed Control by Grass Carp', *Proc. 3rd Brit. Coarse Fish Confr. L'Pool*, pp. 62–5, 1967: C. F. Hickling, 'On the feeding process in the White Amur', *Jour. Zool. Lond.*, 148, pp. 408–19, 1960.
†See pages 476–9; 497–9.

when it was figured by Willughby and Ray: its stronghold today is Holkham Park in north Norfolk, from where it is extending its range along the valley of the Bure and south into the Brecklands of Norfolk and Suffolk. The mandarin duck (*Aix galericulata*) was first introduced to Britain shortly before 1745, when a drake in the collection of Sir Matthew Decker, Bt; at Richmond Green in Surrey, was drawn by George Edwards for his *Natural History of Birds*: this delightful little duck, which gained admittance to the British and Irish List also in 1971, is now established in several stable but isolated communities in various parts of the country. The ruddy duck (*Oxyura jamaicensis*) has perhaps been the most successful of all, having colonized a considerable area of the west country and west Midlands since it first escaped from the Wildfowl Trust's collection at Slimbridge in Gloucestershire in 1952. It, too, was admitted to the British and Irish List in 1971.

Three alien 'ornamental' pheasants are at large in Britain today, the first two of which were also admitted to the British and Irish List only as recently as 1971; these are the golden pheasant (*Chrysolophus pictus*) and Lady Amherst's pheasant (*C. amherstiae*), both of which are natives of China; their strongholds in Britain today are respectively the Breckland area of west Norfolk and west Suffolk and Galloway on the border between Kirkcudbrightshire and Wigtownshire, and parts of the eastern Midlands – especially southern Bedfordshire and neighbouring parts of Buckinghamshire and Hertfordshire. The third 'ornamental' pheasant living ferally in Britain at the present time, which has not however as yet been admitted to full citizenship, is Reeves's pheasant (*Syrmaticus reevesi*), also a native of China, which is found in and around Woburn Park in Bedfordshire and in the neighbourhood of Kinveachy near Boat of Garten in Inverness-shire.

Apart from pheasants, the capercaillie – which has been briefly referred to above – and the red-legged partridge (*Alectoris rufa*) are the most successful alien British game birds. The red-legged partridge was first introduced to mainland Britain in 1673, when Charles II despatched his gamekeeper, Favennes de Mouchant, to the Château de Chambord in France to obtain birds with which to stock the Royal parks at Richmond and Windsor. Today the red-legged partridge is widespread in suitable areas of south, central and eastern England.

Just as alien animals have been introduced to Britain for three principal purposes, so there have been three main means by which they have become established here. They have either been introduced with the premeditated intention of naturalization; or they have escaped from captivity or domesticity; or they have used man as an unknowing means of transportation.

The little owl (*Athene noctua*) was first imported to Britain from Italy by the eccentric Charles Waterton, who, 'Thinking that [it] would be peculiarly useful to the British horticulturalist . . . in his kitchen-garden', unsuccessfully turned some out in his park at Walton Hall in Yorkshire in 1842–3. Lieutenant-Colonel E. G. B. Meade-Waldo released little owls in his park at Stonewall in Kent in about 1874, to 'rid belfries of sparrows and bats and fields of mice', and Lord Lilford turned some out in Northamptonshire from 1888. In its early days the little owl was stigmatized even by such eminent ornithologists as T. A. Coward and C. B. Ticehurst on account of its suspected diet, and was persecuted as *persona non grata* by ignorant landowners and gamekeepers alike. It was left to Miss Alice Hibbert-Ware, under the sponsorship of the British Trust for Ornithology, to show in 1937 that the feeding habits of the little owl have, on balance, a beneficial effect on the British countryside. Other animals which have been deliberately set free in Britain include the north American grey squirrel (*Sciurus carolinensis*) the earliest recorded introduction of which took place near Macclesfield in Yorkshire in 1876, and the edible dormouse (*Glis glis*) which Lord Rothschild liberated in Tring Park, Hertfordshire, in 1902.

Animals now living ferally in the British Isles which have escaped from captivity include, in addition to the deer already mentioned, the fallow deer (*Dama dama*), and Bennett's wallaby (*Macropus rufogriseus bennetti*): the latter was formerly in the collections of Captain Henry Brocklehurst at Roaches House near Leek in Staffordshire and Sir Edmund Loder, Bt; at Leonardslee Park near Horsham in Sussex, but now lives in complete freedom in the Peak District of Derbyshire and Staffordshire and in north-central Sussex. The wallabies escaped on the outbreak of war in 1939 when it became impossible to continue either to feed them or to maintain the fences of their enclosures.

Species which have escaped from domesticity and have become self-supporting ferally include the wild goat (*Capra hircus*), the Soay sheep (*Ovis aries*) of St Kilda, and those domestic cats (*Felis catus*) which now inhabit many rural as well as urban parts of Britain. The ultimate ancestor of the feral pigeon, which is found wherever man has constructed large enough urban communities for it to dwell in, is the native wild rock-dove (*Columba livia*). This bird was domesticated by man many years ago, and was allowed to escape from captivity at a time when meat became more readily available through improved methods of preservation and distribution.

In the autumn of 1957 six golden hamsters (*Mesocricetus auratus*) escaped into the unheated basement of a pet shop in Bath. One year later over fifty were caught, proving that the British climate is no bar to their successful breeding in the wild. Other escapes have been reported from Finchley in Middlesex (four in 1960): Barrow-in-Furness (1961); Bootle (1962); and Manchester (1964) – all in Lancashire – and from Bury St Edmunds in Suffolk, where no fewer than seventy were captured in 1964. In Holland and Germany the European hamster (*Cricetus cricetus*) has caused considerable damage to fruit crops and stored produce. It seems likely that hamsters could become a serious problem in Britain were they to become established in rural areas, as appears biologically possible.*

Two unmitigated pests have used man as the unwitting agent for their colonization of Britain. The black or ship rat (*Rattus rattus*), which is now confined entirely to one or two offshore islands and to some of the larger sea-ports, was by tradition introduced to Britain in the baggage of returning Crusaders in the twelfth century: it was certainly well established in England a hundred years later. The brown rat (*R. norvegicus*), which is undoubtedly the most troublesome pest with which this country has ever had to deal, first reached England in about 1728 or 1729, probably in ships sailing from Russian ports.

A number of factors exert influence on the success or failure of an animal to become naturalized in an alien environment. The most important of these are: climate; habitat; the availability of an adequate supply of correct foods; absence of predators; lack of competition from native species; fecundity; a sufficiently large initial stock; and, in the case of birds, the absence or abandonment

*F. P. Rowe, 'Golden Hamsters Living Free in an Urban Habitat', *Proc. Zool. Soc. Lond.*, 134, pp. 499–503, 1960: *Jour. Zool.*, 156(4), p. 529, 1968.

of the instinct to migrate. In many cases, when these conditions are fulfilled, an introduced animal will initially increase fairly rapidly to the maximum population that the colonized area will support. There may then be some contraction in numbers and range until a level is reached at which the distribution and population become stabilized.

A glance at the maps on the end papers will show that, not surprisingly, the majority of successfully naturalized alien British animals come from continental Europe, of which the British Isles are geographically a part and with which they share much the same kind of climate. An exception is provided by the coypu: one of the main reasons for its success in becoming firmly entrenched in the Broadland area of East Anglia – some 5000 miles or so from its nearest natural range – is the similarity of the Broads' climate and terrain to those of its native habitat in South America. The availability of adequate supplies of correct foodstuffs is one of the most important factors which determine the success of failure of exotic species: however adaptable an animal may be, the right kinds of food are essential for its survival. The absence of any real threat from predators – other than man – is one of the contributory factors towards the successful naturalization in Britain of the Canada goose which is a 'perfect example of a species introduced to a new country where there is ample suitable habitat, and no natural controls to ensure that its numbers do not reach a level at which it becomes either an agricultural pest, a destroyer of habitat for other birds, or a direct competitor with an indigenous species'.* The fact that at the time of its introduction, Britain lacked a mainly diurnal and principally insectivorous small bird of prey contributed in no small measure to the successful colonization of the little owl. The rabbit has become a byword for its fertility, which initially played a considerable part in its spread throughout Britain, and in more recent years helped it to make a remarkable recovery from the devastating effects of myxomatosis in the 1950s. The enormous numbers of grey squirrels liberated between 1876 and 1929 undoubtedly helped the species to gain its initial foothold in Britain, which unfortunately it has never lost and from which it is unlikely ever to be dislodged. One of the contributory reasons to the success of the Canada goose, the mandarin duck and the night heron (*Nycticorax nycticorax*) in establishing themselves in Britain has been their abandonment of the instinct

* *British Birds*, 66(4), p. 134, April 1973.

to migrate, although in 1964 it was discovered that the Canada goose
has developed a special moult migration pattern between central and
northern England and northern Scotland.

Much controversy exists among ecologists over the advantages and
disadvantages of introducing exotic animals outside their natural
range.* On the one hand the purists argue that it is wrong to attempt
to introduce, or even to re-introduce, any species to an alien country.
They claim that available financial resources are better channelled
into preserving existing native animals rather than in trying to naturalize new ones. They point out that the introduction of an alien animal
could result in ecological disturbance and habitat damage, such as
that caused by the Canada goose and coypu, and might also have a
detrimental effect on an endangered native species; they cite the case
of the grey squirrel which, although it did not, as is sometimes
believed, drive out the red squirrel (*Sciurus vulgaris*), did prevent the
latter from re-colonizing those areas which it had abandoned
following a virus epidemic in the early years of the present century.
The purists also draw attention to the risk of denuding one country
of its fauna in order to enrich that of another, and mention the
danger of transferring diseases which may prove harmful both to man
and to other animals. A seemingly less valid argument against the
introduction of exotic birds which subsequently escape from captivity, is the so-called 'incongruity', on zoogeographical grounds, of
having to admit certain aliens to the British and Irish List, and the
problem which escapees present to those compiling lists of rare
genuine vagrants. The difficulty for recorders certainly exists, but
could be lessened if all aviculturalists would adopt the current policy
of the Wildfowl Trust of placing identifying leg-rings on unpinioned
birds. It is surely not by itself a sound enough reason for limiting
such importations.

Clearly there must be more justification in attempting to re-
introduce to a country a comparatively harmless erstwhile native
species which has become extinct, such as the capercaillie, than in
bringing in an entirely foreign animal. It would be unreasonable to

*See C. Cottam, *The Effect of Uncontrolled Introductions of Plants and Animals*, 1950;
P. Duvigneaud, *The Introduction of Exotic Species*, 1949; G. A. Swanson, *The Need for
Research in the Introduction of Exotic Animals*, 1949; Int. Tech. Confr. Protect. Nat.
U.N.E.S.C.O.

criticize the attempts of the Great Bustard Trust, under the guidance of the Hon. Aylmer Tryon, to re-establish the great bustard (*Otis tarda*), which has not nested in the wild in England since the 1830s, on Salisbury Plain, or the so-far unsuccessful attempts by the Nature Conservancy Council to re-introduce the white-tailed or sea eagle (*Haliaetus albicilla*), which last nested in Britain on the Isle of Skye in 1916, to Fair Isle and Rhum. The efforts made since 1952 by the Reindeer Council of the United Kingdom to acclimatize in the Cairngorms the reindeer (*Rangifer tarandus*) – which probably became extinct in Scotland during the Ice Age – seem equally blameless.* On the other hand, it is hard to condone the deliberate liberation of such a harmful animal as the grey squirrel, and much blame must surely rest with those who allow to escape from captivity such potentially harmful creatures as the porcupine, mink and coypu which have proved themselves capable of wreaking havoc on our native fauna and flora. Similarly, few would surely regret the extinction of the house mouse, the rabbit, and the two rats. One potentially extremely dangerous reason for introducing an animal outside its natural range is the control of an already existing species; artificial attempts such as this to affect the balance of nature nearly always end in disaster. On the other hand, many ecologists think that to introduce a harmless alien animal for which a country has a clearly defined empty niche, such as the little owl, is not only permissible but praiseworthy. And surely no one could object to the introduction of harmless and beautiful animals which add to the attractions of the British countryside; such a one is the lovely mandarin duck, to which this book, with admiration and affection, is dedicated.

*Statements that the reindeer existed in Scotland until as late as the twelfth century result from the incorrect identification of the bones of the native red deer (*Cervus elaphus*).

The Acclimatisation Society

In 1860 a group, entitled with typical Victorian verbosity 'The Society for the Acclimatisation of Animals, Birds, Fishes, Insects and Vegetables within the United Kingdom' (The Acclimatisation Society), was formed, whose declared objects were:

1. The introduction, acclimatisation, and domestication of all innoxious animals, birds, fishes, insects, and vegetables, whether useful or ornamental.

2. The perfection, propagation, and hybridisation of races newly introduced or already domesticated.

3. The spread of indigenous animals, &c., from parts of the United Kingdom where they are already known, to other localities where they are not known.

4. The procuration, whether by purchase, gift, or exchange, of animals, &c., from British Colonies and foreign countries.

5. The transmission of animals, &c., from England to her Colonies and foreign parts, in exchange for others sent thence to the Society.

6. The holding of periodical meetings, and the publication of reports and transactions for the purpose of spreading knowledge of acclimatisation and inquiry into the causes of success or failure.

It will be the endeavour of the Society to attempt to acclimatise and cultivate those animals, birds, &c., which will be useful and suitable to the park, the moorland, the plain, the woodland, the farm, the poultry-yard, as well as those which will increase the resources of our sea shores, rivers, ponds, and gardens.

This enterprise was intended, in part at least, to increase the available supply of food. Meat at the time was an expensive luxury, and more intensive methods of cultivation resulted in the increasing spread of diseases and pests in crops: the potato blight of the 1840s, for example, continued to reappear. The Society was the brainchild of Francis Trevelyan Buckland (1826–80), son of Dean Buckland, a scholar of Winchester and Christ Church who, after studying medicine at St George's Hospital, received a commission as Surgeon in the Second Life Guards, later becoming Inspector of Fisheries.

Buckland described 'acclimatisation' as 'a term which may be said to comprise the art of discovering animals, beasts, birds, fishes, insects, plants and other natural products, and utilizing them in places where they were unknown before'. This idea was not new. One of the principal objectives of the Zoological Society of London, formed in 1826, had been 'to introduce new and useful animals to Britain'. In December 1847 the Society presented medals to Sir Roderick Impey Murchison (1792–1871), later to become a Patron of the Acclimatisation Society, and M. Dimitri de Dalmatoff for their joint efforts in introducing to this country the European bison (*Bison bonasus*). Murchison had been employed as a geologist by Tsar Nicholas I, who presented him with a pair of bison as a reward for his services, and a further pair were captured by de Dalmatoff while he held the post of Master of the Imperial Forests in the Government of Grodno, one of the Lithuanian governments of western Russia. Unfortunately, all four bison died shortly after their arrival from 'a murrain' which decimated the zoo's cattle. Among private individuals who attempted the acclimatization of exotic animals, Lord Stanley (1775–1851; from 1834 the 13th Earl of Derby), whose menagerie was described as 'the largest private collection of modern times', was pre-eminent. Stanley, who was President of the Zoological Society for more than twenty years until his death in 1851, despatched collectors to Central and North America, India, Singapore, Norway and Lapland to search for animals for his private zoo, and subsidized expeditions to West Africa over a period of some fifteen years.

Interest in the introduction of alien animals was quickened in Britain by the formation in Paris in 1854, largely at the instigation of the biologist Isidore Geoffroy Saint-Hilaire (1805–61), of the *Société Impériale d'Acclimatation*, which was able to claim among its members the Emperor Napoleon III and Pope Pius IX. By the time of Saint-Hilaire's death membership of the *Société* numbered over two thousand, including thirty-five members of Royal families 'from the Emperor of the French to the King of Siam, from the Sovereign Pontiff to the Emperor of Brazil'.

When Lord Derby, as he then was, died in 1851 he bequeathed in his will to the Zoological Society of London their choice from his menagerie of whichever species they believed would most successfully acclimatize in Britain. The Society's Secretary, Mr David W. Mitchell, chose the elands. The animals flourished and increased to such an extent in the Zoological Gardens that a number were presented to the Marquess of Breadalbane at Taymouth Castle in Perthshire, to Lord Egerton at Tatton, and to Viscount Hill – later to become a Patron of the Acclimatisation Society – at Hawkstone near Shrewsbury. One result of this successful acclimatization was the 'Eland Dinner', of which Frank Buckland wrote:

On January 21, 1859, I had the good fortune to be invited to a dinner, which will, I trust, hereafter form the date of an epoch in natural history; I mean the now celebrated eland dinner, when, for the first time, the freshly killed haunch of this African antelope was placed on the table of the London Tavern. The savoury smell of the roasted beast seemed to have pervaded the naturalist world, for a goodly company were assembled, all eager for the experiment. At the head of the table sat Professor Owen himself, his scalpel turned into a carving knife, and his gustatory apparatus in full working order. It was, indeed, a zoological dinner to which each of the four points of the compass had sent its contribution. We had a large pike from the East; American partridges shot but a few days ago in the dense woods of the Transatlantic West; a wild goose, probably a young bean goose, from the North; and an eland from the South. The assembled company ardent lovers of Nature and all her works: most of them distinguished in their individual departments. The gastronomic trial over, we next enjoyed an intellectual treat in hearing from the professor his satisfaction at having been present at a new epoch in natural history. He put forth the benefits which would accrue to us by naturalising animals from foreign parts, animals good for food as well as ornamental to the parks.

The glades of South Africa have been described by numerous travellers as reminding them forcibly of the scenery of many of our English parks, and here were the first-fruits of the experiments as to whether the indigenous animals of these distant climes would do well in our own latitudes. The experiment was entirely successful, and he hoped would lead to more, and that we might one day see troops of elands gracefully galloping over our green sward, and herds of koodoos and other representatives of the antelope family which are so numerous in Africa, enjoying their existence in English parks, and added to the list of food good for the inhabitants of not only England, but Europe in general.

The vice-chairman, Mr. Mitchell (Secretary of the Zoological Society), then instanced the case of the Indian pheasants, already in course of naturalisation at several places in England, and expressed his conviction that the American partridges we had just partaken of, as well as the European gelinotte, would thrive well in our woods and copses, particularly in Kent, and that there could not be any great difficulty in getting them over from America for this purpose. Elands, since the present experiment had become publicly known, had risen in the market, the demand much exceeding the supply, and there were numerous applications for them which he was sorry to say the Zoological Society could not meet. There were, however, plenty more elands in South Africa, to be had for the trouble of importing them; a fresh supply was much wanted, and he trusted that this subject might be taken up by those who had convenient pasture-ground for them in England, and would be patriotic enough to further the important cause of the acclimatisation of useful exotic animals in English parks and homesteads.

Professor Sir Richard Owen (1804–92), later to become another Patron of the Acclimatisation Society, and bitter critic of the theory of evolution by natural selection propounded by Charles Darwin (1809–82) in his magnum opus, *Origin of Species*, published in November of the same year (1859), a few days later wrote a letter to *The Times* in praise of eland meat, which he described as

The finest, closest and masticable of any meat . . . it is not too much to expect that in twenty years eland venison will be at least an attainable article of food: and seeing the rapidity with which it arrives at maturity, its weight and its capacity for feeding, it is quite possible that before the expiration of the century it may be removed from the category of animal of luxury to the more solid and useful list of the farm.

Owen's views, taken in conjunction with a concurring judgement given by David Mitchell in the *Edinburgh Review*, greatly impressed Buckland, 'for they showed us how science, even in her gravest moods, tends to utility, and that there was a grand uncultivated field open to those who would take up the subject in earnest'. As a result, he decided to form a British Acclimatisation Society, which held its inaugural meeting on 26 June 1860. The Duke of Newcastle, K.G., was elected President; Viscount Powerscourt, Lord Stanley, the Hon. Grantley F. Berkeley, Viscount Gage, Captain Dawson Damer, M.P., and Mr Edward Wilson were elected Vice-Presidents; Mr John Bush was appointed Treasurer and Frank Buckland and James Lowe became Secretaries.

In a lecture which he gave before the Society of Arts in November 1860, Buckland suggested American bison, eland, elk, Barbary, Virginian and Persian deer, reindeer, wapiti, yaks, beavers and kangaroos as among the animals most likely to become successfully acclimatized in Britain. The Society attempted the breeding of a number of hybrids – Lord Breadalbane, for example, mating a cow bison with a domestic bull* and unsuccessfully, a domestic cow with an eland bull. A number of branches of the Society were formed, especially in Australia, at Melbourne and Brisbane, and Glasgow and Guernsey, and the Admiralty issued instructions to the Royal Navy to assist the Society in the transportation to Britain of exotic flora and fauna.

On 12 July 1862, after receiving reports of a similar function held by the Australian Acclimatisation Society, over one hundred guests were invited to a dinner, organized by Lord Powerscourt, Berkeley, Buckland, Mr J. Crockford (a member of the Council) and Lowe, in Willis's Rooms in London. The menu included such exotic delicacies as Japanese sea-slug; kangaroo steamer; Chinese lamb; kangaroo and wild boar ham; Syrian pig; pintail duck; curassow leporine; Chinese yam; botargo; and tripang, a dried holothurian echinoderm popular in south-east Asia, portions of which were described as 'shrivelled objects resembling pieces of horse's hoof' which, having been marinated all day and then slowly simmered all night, 'looked like large black slugs'.

At a meeting of the British Association in Bath in 1864 Dr J. E. Gray was of the opinion that

Some of the schemes of the would-be acclimatizers are incapable of being carried out, and would never have been suggested if their promoters had been better acquainted with the habits and manners of the animals on which the experiments are proposed to be made.

Burgess refers to an extract from a hoax annual report of the Society, which the editor of the *Gardeners' Chronicle* was, somewhat gullibly, deceived into printing:

In Birds a great success has been obtained by the Hon. Grantley Berkeley who has succeeded in producing a hybrid between his celebrated Pintail Drake and a Thames Rat: the Council considers that this great success alone entitles them to the everlasting gratitude of their country-

---

*In 1975 North American breeders announced that they had successfully crossed a domestic cow with a North American buffalo bull: it is claimed that the progeny – termed a 'beefalo' – feeds and thrives on poor land; that the meat is of a better quality, is quicker to cook and cheaper to produce than beef; that it contains less fat and 10–12 per cent higher protein value than beef, and that it reduces the cholesterol level in consumers.

C

men, as this hybrid, both from peculiarity of form and delicacy of flavour (which partakes strongly of the maternal parent) is entirely unique.

Perhaps it was a combination of views such as those held by Dr Gray and ridicule like that printed by the *Gardeners' Chronicle*, which helped to hasten the end of the Acclimatisation Society, whose final annual report appeared in 1865.

It would be tedious to list in detail all the different species of animals which the Society attempted to acclimatize, but some examples will serve to show the eclecticism of the members' choice.

In April 1860 – two months before the Society was officially formed – sixteen bobwhite quail from Canada were received by Lord Malmesbury, a Patron of the Society, at Heron Court in Hampshire: these were subsequently reinforced by a further nine sent by Mr F. J. Stevenson. In March 1861 Grantley Berkeley was sent eight prairie grouse from the United States, and 'a pair of diminutive sheep from Brittany' were despatched to another Patron, Miss Burdett Coutts.

In January 1862 the Society received 'from Shanghae (per favour of Messers Matheson & Co.) 5 ewes, 2 rams and in March following 6 ewes, 2 rams, 6 lambs, total 21' Chinese sheep, which were distributed among various members.

Between 1860 and 1863 Lord Powerscourt introduced sambur, Japanese sika, Sardinian moufflon (wild sheep), roe and wapiti to his park in Co. Wicklow, and Grantley Berkeley turned out prairie grouse; bobwhite quail; Brazilian geese; pintail, wood and Bahama duck; and scaup.

In 1863 Mr John Bush received, on behalf of the Society, six Honduras turkeys from Mr Robert Marshall in South America; two Japanese fowl from Mr A. D. Bartlett; and a pair of agamia or trumpeters (*Psopheo crepitans*) from Central America. Other birds received in 1863 included a pair of jungle fowl in December by Colonel Denison; a fireback pheasant in May by Mr William Wienholt; in August an Indian partridge by Dr Miller; seven dusky ducks by Mr Bush; and, on 23 June, from the Acclimatisation Society of Queensland, a pair of bronze-winged pigeons and a cock brush turkey (*Talegalla lathami*). In the same year Lady Dorothy Nevill received some silkworms which she introduced to Dangstein Plantation near Petersfield in Hampshire, and some Murray cod or cod-perch were presented on 21 May by the Victoria branch of the Society.

In 1864 the same branch sent three wombats to Mr Bush, and Mr Marshall despatched a pair of curassows from South America, which were handed over to Mr H. D. Carré, Lieutenant-Bailiff of Guernsey. In 1863 and 1864 the Victoria branch sent '3 lots' of wonga-wonga pigeons (*Leucosarcia picata*), totalling twenty-three birds, which were distributed between Dr Bull of Hereford, the Revd. William Collings, Seigneur of Sark, Mr H. J. B. Hancock, and Lady Dorothy Nevill. In March the Queensland Society sent three Australian speckled doves, two green-winged pigeons, and one 'Emeu'. Seven prairie grouse were received from Mr Simms of Montreal on 20 February; a pair of talegallas arrived from Brisbane on 2 August, and a pair of demoiselle cranes from Hamburg on 12 December. In September, Sir Stephen Lakeman brought fourteen European wels (*Silurus glanis*) – the survivors of a consignment of thirty-six – from Bucharest; 'the ownership of these fish', the Society later sorrowfully reported, 'we regret to say, is now disputed'.

On 2 April 1865 the Society received what may well have been its last consigmnents of animals; four Romanoff sheep (one ram and three ewes); and a pair of Kalmuck sheep, from the Imperial Society of Moscow; a pair of Spanish asses from Catalonia, and three 'Bennett's kangaroos' (wallabies) from Australia.

The wide variety, and in many cases total unsuitability for purposes of acclimatization, of many of the animals imported by the Society, lends some substance to Dr Gray's disparaging claim. The efforts of the Society as a group were largely unsuccessful, as is shown by its early demise. It may, however, have helped to encourage a number of private individual landowners to make their own experimental introductions of animals. Such men as the Earl of Fife, Lord Powerscourt, Colonel Edward H. Cooper, Lieutenant-Colonel E. G. B. Meade-Waldo, Lord Lilford, Lord Tweedmouth, Lord Breadalbane, the Duke of Bedford, Mr Alfred Ezra, Mr J. H. Gurney and Lord Iveagh, among others – some of whom were never members of the Society – introduced many exotic species to Britain with varying degrees of success. Their efforts, and those of their congeners, are described in the following chapters.

I Mammals

1. Red-Necked Wallaby

(Macropus rufogriseus)

The red-necked or scrub wallaby – variously known as *Macropus rufogriseus, Wallabia rufogrisea, Protemnodon rufogrisea and Macropus ruficollis* – represents a group of large wallabies (a number exist which are no bigger than hares) or 'brush' kangaroos, which includes species smaller and more brightly coloured than the true kangaroos. The red-necked wallaby, which is a native of New South Wales and Victoria, is one of the largest species and is capable of leaping nearly as far as the true kangaroo.

The red-necked wallaby was one of the first marsupials to be discovered in Australia, being observed and described in the Port Jackson (Sydney) region of New South Wales by Governor (Captain Arthur) Phillip in about 1788. From the country in which it lived it was known to the early colonists as the 'brusher'.

The usual habitat of the red-necked wallaby is woodland, forest edge, inland brush and coastal scrub. It is greyish-fawn or rusty/reddish-grey in colour, with an indistinct whitish face stripe but no black dorsal stripe. The neck and rump have a reddish patch – that on the neck being particularly noticeable. The tip of the tail and the toes are black. An adult red-necked wallaby measures about 3 ft 6 in. (106 cm) from the nose to the root of the tail, which is a further 2 ft 6in. (76 cm) in length. The species is both a browser and a grazer, living mainly on grass and leaves, and in common with other members of the kangaroo tribe – *Macropodidae* ('large-footed') – it is a pseudoruminant. Only one young is carried in the pouch at a time.

Red-necked wallabies depend for shelter on dense thickets of undergrowth, from which they normally only emerge to feed, and are therefore particularly susceptible in Australia to the clearances which

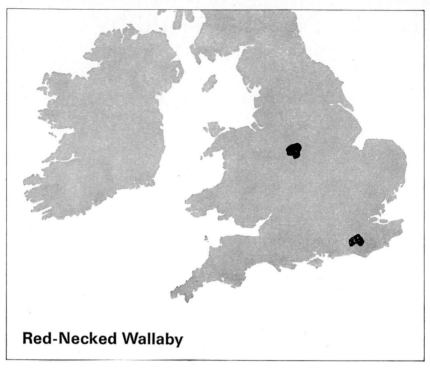

Red-Necked Wallaby

precede the introduction of domestic stock. If, however, enough cover is left to provide the necessary shelter, incomplete clearance can prove beneficial to the wallabies, whose populations are therefore difficult to control where they are causing damage to crops and young trees. In 1870 one male and two female red-necked wallabies were introduced to Canterbury, New Zealand; between 1948 and 1965 government control measures alone accounted for some 70000 animals, and many more were killed by private individuals, who continue to hunt them today.

The sub-species of the red-necked wallaby living ferally in England – sometimes known as Bennett's wallaby (*M. r. bennetti*) after a distinguished former Secretary of the Zoological Society of London – is a native of Tasmania. It is rather smaller and stockier than mainland *M. rufogriseus*, and is duller in colour, with a longer and more uniformly dark greyish-brown coat, indicative of its damper and cooler island home: this contrasts sharply with the silvery-grey tail which has a conspicuous black tip. A pale stripe runs along the

upper lip, and there is a spot of the same shade over each eye. The nape of the neck has the characteristic reddish-brown patch.

According to Gould, the haunches and loins of Bennett's wallaby were much esteemed as food by the early colonists of Van Diemen's Land, who also exported considerable quantitities of skins to England for making into boots and shoes. Notices were published from time to time in Hobart newspapers advertising skins at between 4d. and 6d. each. In the early 1860s the then Governor of Tasmania reported to the newly-formed Acclimatisation Society:

The only Tasmanian quadruped of much economic value is the brush kangaroo, very superior leather being made from the skins. Like most of our indigenous marsupial quadrupeds, this animal is fast disappearing before the creatures introduced from Europe.

On 2 April 1865 the Society received a consignment of 'three Bennett's kangaroos from Australia'; in its 5th Annual Report, published later in the same year, the Society stated that

There can be no doubt whatever that kangaroos will form very highly ornamental animals in closed spaces, etc; if the trees be guarded from their gnawing. They are exceedingly tame and very amusing in their habits. They will breed freely in England, as has been proved not only in the case of our own animals, but by numerous young ones born at the Zoological Gardens.

The kangaroo or wallaby would surely rank high on most people's list of the least likely species of exotic animal to find at large in the British countryside. Yet in the 1850s several wallabies escaped into the woods surrounding Northrepps Hall near Cromer in Norfolk, where they were kept by Mr J. H. Gurney, and there were rumours of feral wallabies in the Pennines in the early years of the present century, when they were also being kept in complete freedom by Lord Rothschild at Tring in Hertfordshire. Fitter records that several 'kangaroos', which were probably wallabies, were turned down by the 4th Marquess of Bute at Mount Stuart near Rothesay on the Isle of Bute in about 1912; one of these became self-supporting for about three years, and used to lead the line of beaters at pheasant drives. In the late 1920s Mr Martin Harman introduced a number of wallabies which were later accidentally drowned, to the grounds of Millcombe House, on Lundy Island in the Bristol Channel.

In the mid-1930s the late Lieutenant-Colonel (then Captain) Henry Courtney Brocklehurst established a small private zoo in the grounds of Roaches House near Leek in Staffordshire.* On 11 June 1936 he obtained a pair of Bennett's wallabies for his collection from Whipsnade; a single young was born in March 1937, and two more – one of which was drowned in the following October – in March 1938; a fourth wallaby was born in May 1939. These five animals escaped from their enclosure in 1939 or 1940† – when because of the war it proved no longer possible to continue to feed them or to maintain the fences of their enclosures – and were the ancestors of the stock of feral wallabies which have been established since then in the Peak District of Derbyshire and Staffordshire.

The wallabies of the Peak District exhibit a certain amount of individual colour variation; one animal has a pale fawn body – paler even than its grey tail – and lacks the reddish nape patch; its eartips, muzzle, feet and the tip of its tail, are a darker brown. There are

---

*Not at Swythamley Park near Macclesfield (the home of his brother, Sir Philip Brocklehurst, Bt;) as stated by Fitter. For information on this herd and its descendants I am indebted to the paper by D. W. Yalden and G. R. Hosey (see Bibliography).

†A single yak (*Bos grunniens*) escaped at around the same time, and survived until about 1951.

unconfirmed rumours of a whitish or silvery-grey animal, which could well be true, as elsewhere albinos are not uncommon. Very little is known about their breeding habits: at the Zoological Gardens in Regent's Park, Bennett's wallabies have bred almost throughout the year – the peak months being May and June. At Whipsnade, where a colony of between 200 and 270 wallabies roams freely over the 500 acres (200 hectares) of the park, the principal breeding season is in late August or early September. At Chester Zoo, however, the majority of births have occurred in February, March or early April, which corresponds more closely with the records at Roaches in the 1930s: the gestation period is normally between thirty and thirty-one days.

The Peak District wallabies live on heather moorland and in thick scrub in the area of the National Park north-west of Leek in Staffordshire, and in woodlands near Hoo Moor in Derbyshire, some 10 miles (16 km) to the north. They eat mainly heather, grass, young bracken-shoots, pine- and birch-scrub and bilberries. Some are extremely wary of man, while others allow a cautious approach to within a few yards.

The earliest published reference to feral wallabies in the Peak District appeared in a letter to the *Buxton Advertiser* on 16 March 1951: it reported that a single animal had been seen near Errwood Hall – 8 miles (13 km) north of the source at Roaches House – in 1940, and that on 11 March 1951 another had been observed at Hindlow Quarry near Buxton, which died two days later after being caught, exhausted in the snow, at Harpur Hill Quarry, 6 miles (10 km) north-east of Roaches. In 1944 a lone wallaby was seen at Barmoor Clough near Chapel-en-le-Frith, 12 miles (19 km) north-north-east of the original source. A letter in the *Buxton Advertiser* on 16 July 1954 reported that a wallaby had been seen earlier in the year in Burbage near Buxton, 7 miles (11 km) north-north-east of Roaches. In the following year Mr F. Heardman saw one near the Yorkshire Bridge (A.57) by the Ladybower Dam, 18 miles (29 km) north-east of Roaches. Miss S. Evans reported that a wallaby spent most of the summer of 1956 at Chapel-en-le-Frith, where it grazed on the lawn in front of Ferodo Ltd; it was also referred to in the 16 November issue of the *Buxton Advertiser*. In the winter of 1963 a wallaby was captured near Sheen – 6 miles (10 km) east of the source at Roaches – but again died, exhausted, in the snow. The *Daily Telegraph* of 25 July 1970 reported that a wallaby had been seen earlier in the month near

Werrington, 9 miles (14 km) south-south-west of source. (A wallaby reported in the *Derbyshire Times* to have been killed in 1963 or 1964 on the railway line near Ashover – some 20 miles (32 km) west of Roaches – probably escaped from a small local collection). It is interesting to note that, with one exception, all these records are from the north and east of the original source at Roaches House.

The hard winter of 1947 probably checked the increase in population of the Peak District wallabies. In the early 1960s, however, their numbers were estimated, by two independent sources, to have reached between forty and fifty, and one observer reported seeing seven animals bolted by terriers from cover, one after the other. The exceptionally severe winter of 1962–3 caused serious casualties among the wallabies, as it did among most other wild animals in Britain. A number of dead wallabies were discovered in the following spring – six or seven being reported from one small area: one animal died half-way over a wall, having scrambled up one side and failed to get down the other. A few, however, managed to survive, and Mr R. F. Billing observed four or five wallabies together in the spring of 1965.

Between August 1967 and September 1970 Dr D. W. Yalden and G. R. Hosey of the Department of Zoology at Manchester University paid a total of thirty-five visits – each of about five hours' duration – to the area frequented by the wallabies; they recorded thirty-four separate sightings of the animals on only fourteen occasions. The wallabies are particularly difficult to find in summer, when the trees and shrubs are in full leaf; in winter, partly through lack of cover and partly because they can at times be tracked in the snow, the animals are easier to observe. They appear to lead a mainly solitary existence, although on one occasion in February 1970 a party of four was observed feeding on a south-facing slope, where the sun had melted much of the snow-cover. In that month at least eight individual wallabies were observed, and it was estimated that the total population was approximately a dozen.

A second, but so far less well-documented, colony of feral Bennett's wallabies has been living for some thirty-five years or so in north-central Sussex. The source of this colony is presumed to be Leonardslee Park, Lower Beeding, near Horsham in Sussex, where a number were introduced around 1908 by Sir Edmund Loder, Bt.

In the early 1940s numerous reports were made of a small but apparently fully naturalized and breeding colony of wallabies in the Ashdown Forest and St Leonard's Forest district. On 24 September 1965 a feeding wallaby was disturbed in a cornfield at Laybrook Farm Thakeham, near Storrington, some 8 miles (13 km) south-west of Leonardslee: it was followed for some distance in a Land Rover at speeds of up to 40 m.p.h. (64 k.p.h.). No wallabies were seen between 1965 and 1968, when Sir Giles Loder, F.L.S., Sir Edmund's grandson, was quoted in *The Sussex Mammal Report* as believing that 'conditions now make it impossible for abnormal animals to survive in a feral state even if the natural habitat is suitable'. Nevertheless, in the following year, five reports of wallabies were received from the north and north-east of Leonardslee: from St Leonard's Forest, Horsham, 4 miles (6 km) from source; Worth Forest, Crawley (6 miles; 10 km); Turner's Hill and Crawley Down, Crawley (10 miles; 16 km), and from Felbridge, East Grinstead (13 miles; 21 km). In 1970 a wallaby was found dead near the electric railway line at Christ's Hospital, Horsham, 5 miles (8 km) north-west of Leonardslee, and a live specimen was observed in late July at Lower Beeding, 1 mile (2 km) north-west of source. No reports of wallabies were received in 1971, but they are undoubtedly still present in small but self-perpetuating numbers in the St Leonard's Forest and Worth Forest area south of Crawley New Town, which provides an effective barrier to their further expansion north, as does Horsham to the west.

The winter of 1962–3 was an exceptional one: as both colonies of wallabies were able to come through that with a breeding nucleus intact, it is not unreasonable to suppose that they are capable of surviving any British winter. Yalden and Hosey point out that while it is popularly supposed that wallabies – coming as they do from the antipodes – need a mild climate in order to survive, the subspecies *M. r. bennetti* is found at considerable altitudes in the Tasmanian mountains. Professor G. B. Sharman reports that he was told by local farmers that in the severe Tasmanian winter of 1946 'nearly every bush at 4000 ft [1215 m] altitude had a dead wallaby lying beneath it'. Many too, have been killed in Tasmania by poison in regenerating forests of *Eucalyptus regnans,* which the wallabies severely damage while the trees are young. Since the near-extinction in Tasmania of the great grey kangaroo (*M. giganteus*) by hunting and through tree-felling for the wood-chip industry, Bennett's wallaby is often referred to on the island as a kangaroo.

Population biologists have attempted to formulate theories concerning the most efficient reproductive habits of animals in different environments. Kangaroos and wallabies are closely related species which interbreed but tend to live in differing habitats. Dr B. J. Richardson of the Australian National University has recently compared their breeding characteristics, basing his research on a theory propounded by Drs R. H. McArthur and E. O. Wilson of Princeton University, who have suggested that a dense population favours what they term 'K' selection, in which the careful husbanding of available resources is the primary consideration, while relatively thinly populated areas produce 'r' selection, in which a high rate of reproduction is the most important factor. Dr Richardson has shown that the theories of Drs McArthur and Wilson are largely confirmed by the known breeding habits of red and grey kangaroos.

The red kangaroo (*M. rufus*) lives mainly in the dry 'bush' of central Australia, whereas the great grey kangaroo – like the red-necked and Bennett's wallabies – lives in the more predictable climate of the densely populated forests of south-eastern Australia and Tasmania. In theory, therefore, red kangaroos should adhere to 'r' selection and grey kangaroos and red-necked and Bennett's wallabies to 'K' selection.

Drs McArthur and Wilson suggest that 'r' selection is dominant in those areas in which there is little competition for food and water – either between animals of the same or different species – but where availability is subject to unforeseeable circumstances. In this instance, mortality among the young is the result not of fluctuations in the population, but of extraneous conditions. This suggestion is confirmed in practice by the red kangaroo, the young of which have a high survival rate in normal conditions but a low survival rate in times of drought and famine. The theory of 'r' selection also suggests a certain amount of migration in times of hardship, and a steady reproductive cycle, to both of which suppositions red kangaroos appear to conform. They reach sexual maturity at an exceptionally early age, and breed continually throughout the year. The young develop rapidly, and although they leave the pouch after between thirty-five and forty-two days they are not normally weaned until the age of about fourteen months.

Great grey kangaroos and red-necked and Bennett's wallabies, on the other hand, living as they do in the more stable climate and more heavily populated conditions of woodland and forest, are comparatively sedentary animals. The adults are less prolific and frequently do not breed until a year or more after reaching sexual maturity: they have a distinct breeding season, and carry and suckle their young in the pouch for a longer period than does the red kangaroo. The young are subject to a steady mortality rate of at least one in two.

Red and grey kangaroos and the red-necked wallaby are increasing in numbers in Australia. As the grey kangaroo and the wallaby are conditioned to overcrowding, which the red kangaroo is not, they should theoretically possess a definite ecological advantage. In practice, however, it is probable that culling by hunters will limit the populations of all of these species, before they come into competition for the natural resources available.*

Although wallabies have been living ferally in the Peak District and in north-central Sussex for some thirty-five years, and are now clearly fully established and naturalized as breeding colonies, they have not so far spread far from their original homes at Roaches House and Leonardslee Park. The principal reasons for this are probably threefold: firstly, persecution by mindless people with rifles and shot-guns; secondly, the suitability of the habitats where they live which – combined with their small populations and sedentary nature – makes further colonization unnecessary; and thirdly, their vulnerability when attempting to cross railway lines and roads; in the autumn of 1973 two were run over by cars in Derbyshire and killed. Dr Yalden continued to pay irregular visits to the Peak District colony between 1971 and 1975, and counted up to four or five wallabies on most occasions. In the autumn of 1973 he saw a single immature animal, and it seems possible that the colony is breeding more freely than might be expected. The wallabies of the Peak District are living on private ground previously owned by a sympathetic landowner, whose recent death has inevitably cast some doubts on the future of the estate. The area is moreover, increasingly

*'r' and 'K' Selection in Kangaroos, *Nature*, 255, p. 323–4, 22 May 1975.

frequented by sightseers and tourists. There is thus a considerable danger that the wallabies will be driven into less suitable habitats, or into the surrounding farmland where they are likely to be killed.

Providing they can be kept away from vegetable crops and young timber plantations, these endearing creatures cause no harm whatever to the countryside; as our only wild marsupial, they are an exceptionally interesting addition to the British fauna, and it is therefore to be hoped that they will be allowed to continue to live unmolested by man.

2. Greater White-Toothed Shrew
(Crocidura russula)

Lesser White-Toothed Shrew
(C. suaveolens)

Three species of white-toothed shrews inhabit continental Europe, but none occurs on the mainland of Britain. The pygmy white-toothed shrew (*Suncus etruscus*) of the Mediterranean region is the smallest known mammal, measuring barely $1\frac{1}{2}$ in. (3–4 cm) in length with a tail of about a centimetre less.* The greater or common white-toothed or musk shrew has become known in recent years for its 'caravanning', which may, however, only take place in captivity, and probably then only when its nest is disturbed. When a litter is nearly weaned they follow in a line behind their mother, with the first in line holding on to the base of the mother's tail with its teeth. The following young each take hold of the tail of the next in front in the same manner, each gripping so firmly that the mother can be lifted from the ground with her family dangling beneath her. Shrews are extremely fierce little animals, fully deserving the description of a sixteenth-century author who referred to them as 'ravening beasts, feared by all'. They will kill and devour animals considerably larger than themselves – e.g. voles (*Microtinae*) – though they mainly eat insects, worms, slugs, larvae, snails, carrion and – not infrequently – each other. They probably have several litters of between five and eight young a year and have a maximum life-expectancy of only some eighteen months.

Shrews have been the subject of much prejudice and supersitition

*In the entrance hall of the British Museum (Natural History) in London is a small transparent plastic dome containing a pair of pygmy white-toothed shrews. Standing beside them is the world's largest terrestrial animal – the African elephant (*Elephas africanus*).

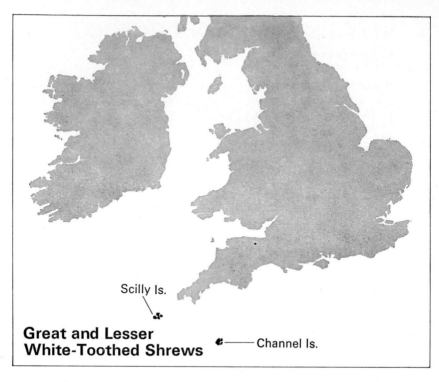

Scilly Is.

**Great and Lesser
White-Toothed Shrews** ⟡————— Channel Is.

concerning the evils and misfortunes which they are capable of inflicting upon man and his domestic animals. In a letter of 8 January 1776 to the Hon. Daines Barrington, Gilbert White (1720–93) described how it was believed by country people that these troubles could be cured by the use of a 'Shrew-Ash':

Now a shrew-ash is an ash whose twigs or branches, when gently applied to the limbs of cattle, will immediately relieve the pains which a beast suffers from the running of shrew-mouse over the part affected: for it is supposed that a shrew-mouse is of so baneful and deleterious a nature, that where ever it creeps over a beast, be it horse, cow, or sheep, the suffering animal is afflicted with cruel anguish and threatened with the loss of the use of the limb. Against this accident, to which they were continually liable, our provident forefathers always kept a shrew-ash at hand, which when once medicated, would maintain its virtue for ever. A shrew-ash was made thus: – Into the body of the tree a deep hole was bored with an augur, and a poor devoted shrew-mouse was thrust in alive, and plugged in, no doubt with several quaint incantations long since forgotten.

A subspecies of the greater white-toothed shrew known as

C. r. peta occurs on Alderney (where it has been known since 1898) and on Guernsey and Herm (since 1908) in the Channel Islands. A subspecies of the lesser white-toothed shrew *C. suaveolens,* was first discovered on Jersey and Sark in 1925. In the previous year a number of unidentified shrews from the Scilly Isles were sent for classification to M. A. C. Hinton. They were described as a new sub-species which was named the Scilly Shrew (*C. s. cassiteridum*): this animal differs slightly from the continental species, although they are clearly very closely related. It is about the same size – about 3 in. (7–8 cm) as the common shrew (*Sorex araneus*), but has larger and more prominent ears and long silky hairs on the tail: the fur is greyish or silver-brown and the teeth are white – there being only three unicuspids. It is now known that this variety is the only one found in the Scilly Isles.

White-toothed shrews – so named because their teeth lack the red pigmentation of the common shrew, the lesser or pygmy shrew (*S. minutus*) and the water shrew (*Neomys fodiens*) – are found throughout south and central Europe from northern Spain to Russia, and probably as far east as Asia: they also occur on a number of islands off the north-western coast of France. They are unknown as fossils in the British Isles, where today they are found only in the Channel and Scilly Isles. It may possibly be that they are the descendants of a population of *Crocidura* which became isolated somewhere to the south-west of the British Isles during the last glacial period, and was wiped out, except on the Channel Islands and the Scillies, by the rising sea-level and/or by competition for land, food and shelter. On the other hand, most authorities now consider that the present population reached these islands by chance introduction. As Dr G. B. Corbet points out, the usual habitat of the lesser white-toothed (Scilly) shrew 'among the kelp on the storm beaches, coupled with the extensive use of kelp for manure and reduction to ash, suggest clearly how it could have been distributed in the islands. . . . Considering the overall distribution and ecology of these [white-toothed] species, the impoverished fauna of these islands and their recent geological history, it seems very probable that these populations have all arisen from introduction by man.'

3. Orkney and Guernsey Voles

(Microtus arvalis)

Voles are distinguished from other small rodents by their blunter nose and shorter furry ears and tail. The commonest species on the British mainland and on a number of off-shore islands is the short-tailed or field vole (*M. agrestis*) which measures about 4 in. (10–12 cm) in length. The breeding season extends from March to September: between three and eight young are born, in from three to four litters a year, after a gestation period of around three weeks, in a grass-lined nest either on or just under the ground. The principal food of the field vole consists of grass-leaves and stems and bulbs.

Among those off-shore islands from which the field vole is absent are the Orkneys and Guernsey. Here it is replaced by forms (respectively *orcadensis** and *sarnius*) of the closely allied but larger and darker *M. arvalis* which, although unknown elsewhere in Britain, is one of the commonest species of vole in continental Europe. It has been known in the Orkneys since at least 1805.

In 1952 Dr Harrison Matthews suggested that

the ancestral form [*M. corneri*, which has been extinct since pleistocene times] came into Britain from the south-east and penetrated to all parts of the country, but was later driven out by the arrival of another species. [The Scottish short-tailed vole, *neglectus*, which arrived in Britain in late pleistocene times, and was itself pushed north by the common race, *hirtus*, during the post-pleistocene period]. It was only in those islands that became separated from the mainland after the ancestor had colonized them, but were cut off before the arrival of the competing species, that its descendants have survived; and these have become differentiated into distinct species and subspecies.

Dr G. B. Corbet, however, points out that no evidence exists to

*This form has been divided into five separate sub-species, which inhabit respectively: mainland Orkney, South Ronaldshay, Rousay, Westray, and Sanday.

prove the presence of *M. arvalis* in the British Isles – other than in the Orkneys and on Guernsey – in post-glacial times. Orkney is separated from the Scottish mainland by a 30–40 fathom (180–240 ft: 55–73 m) channel which provides an effective barrier to ground predators, and its fauna suggests a very early isolation from the mainland. Bones recently excavated on Orkney, which were associated with other material dated by the radio-carbon method to around 2000 B.C., were identified in 1975 by Dr Corbet as belonging to this species. 'There is no historical evidence,' writes Dr Corbet, 'for the introduction of these island populations, but there are strong reasons for suspecting such introduction.' This could have taken place, no doubt accidentally, from continental Europe, either by Neolithic or early Bronze Age man.

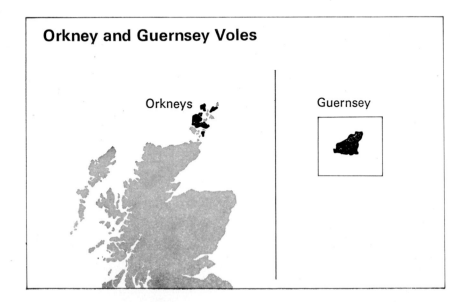

Orkney and Guernsey Voles

Orkneys

Guernsey

4. Yellow-Necked Mouse

(Apodemus flavicollis)

St Kilda Field Mouse

(A. sylvaticus hirtensis)

The genus *Apodemus* is typified in the British Isles by our most numerous mammal, the long-tailed field mouse or wood mouse (*A. sylvaticus*). The brown, yellow and reddish hair on the upper parts of the wood mouse is distinctly separated from that of the undersides, which is a pale shade of silvery-grey. A yellowish-orange spot, which may extend along the abdomen, is usually present on the chest between the forepaws. The wood mouse is slightly larger than the house mouse, measuring about 4 in. (10–12 cm) long, with a tail of about the same length. From four to seven young are born in up to five litters between April and October, in a grass-lined nest which may be either above or below ground. Wood mice live in woods, fields and gardens where they may cause some damage to horticultural and agricultural crops. They frequently store fruit, nuts and berries in subterranean stores for winter eating, and also consume bulbs, roots, insects, grubs and carrion.

The yellow-necked field mouse is larger and more brightly coloured than the wood mouse, with redder upper-parts, whiter under-parts, a larger chest patch and a longer tail. In Europe it ranges north to Finland and Sweden, east to the Russian steppes, south to the hilly regions of the Mediterranean, and west to eastern France and the Netherlands. The subspecies found in Britain, *A. f. wintoni,* was only clearly recognized as distinct from the wood mouse as recently as 1894.

The yellow-necked mouse is found locally throughout the south and south-east of England, in the Severn Valley, and on the Welsh marches, where it does not replace but overlaps with the wood mouse.*

*The British Museum collection contains a specimen taken near Sunderland prior to 1911. Most northerly records are from Cheshire (1957), Derbyshire (*c.* 1950), Leicestershire (1950) and Lincolnshire (1956).

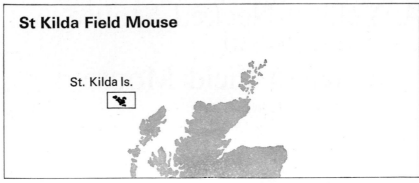

St Kilda Field Mouse

St. Kilda Is.

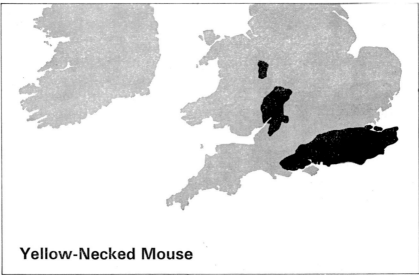

Yellow-Necked Mouse

In 1952 Dr Harrison Matthews wrote:

The inter-relations of the British forms of *Apodemus* can be regarded in two quite different ways, as pointed out by Hinton. Fossil forms believed to be closely allied to *A. flavicollis* because of their large size are known from British deposits, but they do not occur until the late Pleistocene, and consequently this species may be regarded as a comparatively late invader of the territory of *A. sylvaticus*. On the other hand, although fossil forms resembling *A. sylvaticus* are known from as far back as the late Pliocene, none are known from deposits between the early and late Pleistocene; consequently the earlier forms may have died out and been replaced much later by new immigrants entering at the same period as did *A. flavicollis*.

As Dr G. B. Corbet states, however, 'the peculiar distribution [of the yellow-necked mouse in England and Wales] could perhaps be

interpreted as the result of slow and incomplete recolonization after deforestation had reached its peak, but the possibility of accidental introduction by man should not be ruled out'.

The St Kilda field mouse (*A. s. hirtensis*) is a large sub-species of the wood mouse, closely related to the Hebridean mouse (*A. s. hebridensis*), which inhabits all the large islands of the Hebrides with the exception of Skye. The St Kilda field mouse, which lives on the islands of Soay, Hirta and Dun in the St Kilda group, is the largest of all the field mice, with an exceptionally large head. When Dr W. Eagle Clarke visited the islands in the autumns of 1910 and 1911 he remarked that it was 'most abundant where coarse grass prevails, but is to be found almost everywhere –in the crofted area, in the neighbourhood of the houses, on the faces of the cliffs, and on the sides and hilltops; finding congenial retreats in the rough stone-built "cleits" . . . and in the walls surrounding the crofts.' This mouse is known to be a form of the species *A. sylvaticus,* and R. J. Berry has shown that it more closely resembles the field mice of Norway than it does those of mainland Britain: similarly, the now extinct St Kilda house mouse (*M. m. muralis*) was more like the species of the Shetland and Faeroe islands than the typical British *M. m. domesticus.* It thus appears probable that both the St Kildan mice were introduced accidentally by the Norsemen from Scandinavia over one thousand years ago, possibly in the hay which they imported to feed their domestic stock.

5. Mongolian Gerbil
(*Meriones unguiculatus*)*

The home of the Mongolian gerbil or clawed jird is, as its name suggests, the People's Republic of Mongolia in Central Asia – especially the Gobi Desert region – whence its range extends into adjacent parts of the U.S.S.R. and China.

The hairs on the sides and back of the Mongolian gerbil are slate-grey at the base, pale brown or ochreous-buff in the centre, and are tipped with black: the hairs on the underside are generally white or tan, merging to grey at the base. The terminal half of the long and hairy tail bears a crest of long dorsal hairs and ends in a pronounced black paint-brush-like tuft. The eyes, which are large, black and protruding, are encircled by an indistinct buffish-white ring which extends as far back as the base of the ears which are pricked and thickly-haired. The nails are black, and the soles of the small feet are fully haired apart from a bare patch near the heel: the young tend to leap or hop on their slightly elongated hind legs, although the adults usually prefer to proceed on all fours. When at rest the Mongolian gerbil frequently sits up on its hind legs using its tail as a prop, very much in the manner of a kangaroo. In size it is a little smaller than a hamster, the head and body measuring about 4½ in. (11–12 cm) long, with a tail of about the same length.

The Mongolian gerbil – which is both diurnal and nocturnal throughout the year – is a gregarious animal living in colonies in elaborately constructed underground burrows in dry sand or clay, usually on the steppes or savannahs or in desert regions or sandy-grasslands. It builds its circular nest – which is lined with the leaves of buckwheat, millet, grasses or sedges – near the centre of its labyrinthine tunnel system. Food-stores are constructed in the surrounding passages, which from September to March contain seeds of these plants for winter feeding; in spring and summer the

*'Clawed' or 'nailed'.

Mongolian gerbil's diet consists of fresh leaves of the same plants. It is capable of surviving lengthy droughts by extracting the liquids from roots, tubers, bulbs and insects. Mongolian gerbils breed throughout the year – principally between April and September – an average of four to six young being produced after a gestation period of twenty-four to twenty-six days. During the breeding season – though not exclusively so – a noise described by M. P. Anderson, who collected some wild Mongolian gerbils for the British Museum, as 'very much like the distant galloping of a horse on a hard road' is produced by the males who rapidly stamp and drum their feet on the ground within the burrows.

The first specimens of the Mongolian gerbil to reach Europe arrived in 1867, when Henri Milne-Edwards in the Museum of Natural History in Paris received a number from the eminent French missionary and naturalist, Père (Jean-Pierre) David,* who had captured

*Two years previously David had succeeded in discovering, near the Non-Hai-tzu (Southern Lake) in the Imperial Hunting Park south of Peking, the only surviving herd of what came to be known as Père David's deer (*Elaphurus davidianus*). This species, which had once been widespread in north-eastern China from north-east of Peking south to Hangchow and east to Lo-yang in the Province of Honan, became extinct in the wild during the Shang Dynasty (1766–1122 B.C.) when the swamplands of the Chihli Plains where it lived came under cultivation. From that time the animal survived solely in captivity in parks. In 1866 David despatched two skins to Paris, where they were described by Milne-Edwards. These were subsequently followed by a number of living specimens, and they and their offspring were distributed among several zoological gardens. In 1894 the Hun Ho River burst its banks and breached the wall of the Imperial Hunting Park; the deer escaped into the surrounding countryside where most were eaten by the starving peasants, the majority of the few survivors being killed six years later during the Boxer Rebellion. The small number which survived the Rebellion were transported to Peking, where by 1921 all had died. The existence of the species today is due almost entirely to the efforts of the 11th Duke of Bedford who by 1922 had succeeded in establishing a herd of sixty-four at Woburn, some of which after the war were distributed among other collections. In 1964 – forty-three years after the species became extinct in China – the Zoological Society of London presented four Père David's deer to the zoo in Peking.

Mongolian Gerbil

some 'yellow rats' in the previous year while travelling in the Province of north-western Shansi – 'la Mongolie Chinoise'. In his diary, David described his discovery as follows:

April 14th: Thermometer 25 degrees F. Sky serene with cumulous clouds, calm. This morning I acquired . . . three yellow rats I do not yet know the name of, having long hairy tails . . . According to [the Chinese] the yellow rat . . . is abundant in the desert as in cultivated areas: it lives in companies of several individuals in shallow holes: its habits are diurnal: it amasses stores of grain for the winter and during that season emerges from its retreat from time to time. April 23rd: Today the weather is excellent. We have left the high plateau and it is very warm. Here the plain of Kweisui begins, crossed by a little river which we have followed this morning. The plain is fertile and well cultivated and is scarred by some sandy places inhabited by a great number of yellow rats with black claws, which gambol in front of their holes.*

In 1954 the Japanese Central Laboratory for Experimental Animals sent eleven pairs of Mongolian gerbils – which it had somehow managed to collect from the remote and politically closed savannahs of north-eastern China – to Dr Victor Schwentker at the West Foundation in the United States, for laboratory experiments. Nine pairs bred successfully, and the animals were reported to adapt readily to a wide range of environmental conditions and to show a

*David's Journals were published in the *Bulletin des Nouvelles Archives du Muséum d'Histoire Naturelle* (vols III and IV: 1867–8), where the 'yellow rat' is described as '*Meriones unguiculatus*, found in the sandy plains of Mongolia'. See also: H. Fox, *Abbé David's Diary*, Harvard Univ. Press, 1949.

marked ability for temperature regulation, as well as great curiosity, gentleness, and tameness. (M. P. Anderson wrote of some wild Mongolian gerbils: 'I frequently succeeded in approaching within about 8 feet (2–3 m) of a sitting individual, during which manoeuvre the animal would eye me steadily and finally with one rapid movement plunge into his hole but reappear after a few moments if I remained perfectly still'). Since then, Mongolian gerbils have become increasingly popular in the United States and in Britain as pets and for laboratory experiments, for both of which purposes they are exceptionally well suited.

In 1973, at the conclusion of the filming of *Tales of the River Bank* on the Isle of Wight by an independent company for B.B.C. Television, some white rats, hamsters, guinea-pigs, and several Mongolian gerbils were permitted to escape or deliberately released. The rats and guinea-pigs soon disappeared, but the gerbils evidently found the area – at Coathy Butts near Fishbourne, $2\frac{1}{2}$ miles (4 km) west of Ryde – greatly to their liking; here they became established in and around five houses, a barn, some waste-ground and a wood-yard, owned by Mr Clifford Matthews; by February 1976 the population had increased to about one hundred.

These attractive and endearing but destructive little creatures are capable of consuming vast quantities of seeds and are known also to be possible carriers of bubonic plague, bilharziasis, and rabies; as Mrs Gulotta writes: 'The use as pets of species such as the Mongolian gerbil, which probably could adapt to some areas in North America [and, of course, Britain] if released, carries a grave risk of damage to the environment and to human agriculture and health.' Their escape or irresponsible release is greatly to be deplored.*

*My thanks are due to Mr Henry R. Arnold of the Biological Records Centre at Monks Wood for drawing my attention to this colony of gerbils.

6. Rabbit

(Oryctolagus cuniculus)

Before the arrival in England in 1953 of the myxomatosis virus which decimated its population, the ubiquitous rabbit was probably the most familiar larger mammal in the British Isles. Yet it is not, as many people imagine, a native species, but rather a naturalized alien which was imported to this country from northern and central Europe.

Even today the rabbit is so familiar a sight in the countryside that it needs only the briefest description. The colour of the fur (from which felt is made) is greyish-brown with a russet undertone: the underparts and the lower surface of the tail ('scut') are white. Black, fawn, white, silvery-grey and parti-coloured specimens are not uncommon, and are usually the result of inter-breeding with escaped domestic stock. The male (buck) and female (doe) are similar in size and appearance: each measures about 17 in. (43 cm) in length and weighs approximately 3 lb (1·5 kg).

Perhaps the best-known characteristic of the rabbit is its fecundity. In the British Isles wild rabbits breed throughout the year, although the principal breeding season extends from January to June. Normally some half a dozen litters are born in a nest underground annually, each averaging the same number of young; these are born after a gestation period of about four weeks, and the doe is ready to mate again some twelve hours after giving birth. The eggs of the doe are not released from the ovary automatically at oestrus, but appear some twelve hours after mating, thus ensuring a high percentage of fertilization. The resulting potentially explosive rate of increase is fortunately counterbalanced by a phenomenon which occurs regularly in only one other mammal, the hare. According to Dr Harrison Matthews, 'at least 60 per cent of all litters conceived are never born, the developing embryos dying before they reach full growth. They are . . . broken down within the uterus . . . the material forming them being taken back into the body of the mother'.

Rabbits cause immense damage to farm and garden crops: they consume almost all vegetable matter, and in winter especially gnaw the bark of trees. They live gregariously in extensive warrens, appearing above ground during the daytime mainly in hot weather, but otherwise emerging only at dusk and dawn to feed. Their enemies – among whom are badgers, foxes, stoats, raptors and, not least, man – are so numerous that notwithstanding their prolificacy, it is surprising that they can achieve such a high-density population.

Fitter advances an interesting hypothesis concerning the ancestry of our wild rabbits: the animal was originally a western Mediterranean species, being especially common in Spain and Morocco: it was unknown in Italy until the third century B.C., and did not make its first appearance in northern and central Europe until the Middle Ages.

Why did an animal so eminently suited to life in the British Isles and other parts of northern and western Europe not spread there by natural means long before? . . . The answer to this conundrum must be that the wild rabbit with which European man has plagued much of the earth's surface is not a true wild rabbit at all, but a feral domesticated rabbit. . . . It seems . . . likely that the Romans, who are known to have domesticated rabbits, turned the Spanish stock into a hardier and more adaptable strain. It is this strain, it would seem, that was imported into northern Europe in the Middle Ages. . . .

Until the eighteenth century the term 'rabbit' applied solely to the young of the species, the adult being known as a 'coney'. In *Promptorium Parvulorum* (1440) a 'rabet' is defined as being a 'yonge conye' (*cunicellus*). The *Boke of St. Albans* (1486) refers to the 'bery' (burrow) of 'conyis' (adults) and to the 'nest of rabettis' (young). In *The Noble Arte of Venerie or Hunting* (1575), we learn that 'the conie . . . beareth hyr rabettes xxx dayes, and then kindeleth'. Rabbit-burrows were referred to as coneygarths, conygers or conigrees. A warren was an area of land used for breeding rabbits for domestic needs: in about 1540 the antiquary, John Leland (*c.* 1506–52), wrote in his *Itinerary*, that Stonor in Oxfordshire was encompassed by '. . . a fayre parke, and a waren of connes, and fayre woods'.

For the same reason that they may well have brought the first edible dormice – *glis* – and the first edible frogs – *esculenta* – to Britain, the Romans quite possibly may have been the earliest importers of rabbits to these shores. In his *Rerum Rusticarum,* compiled in 54 B.C., Marcus Terrentius Varro (116–27 B.C.) wrote that they brought rabbits from Spain to Britain, where they were reared in *leporaria*. Rabbit embryos, known as *laurices*,[*] were a highly esteemed delicacy by Roman gourmets, especially during times of fasting, and Gregory of Tours (540–94 A.D.) records that his contemporary, Roccolenus, ate unborn rabbits while he was at Poitiers. John Whitaker[†] appears to have been the earliest writer to suggest that the Romans first introduced rabbits to Britain, but no evidence exists to show that our present population is descended from Roman stock. As Fitter points out, there is no Anglo-Saxon or Celtic word for the rabbit, and there is no mention of rabbit-warrens in Domesday Book (1086), which suggests that the animal was not established here in the eleventh century.

Miss Elspeth M. Veale is responsible for having discovered what are to date the two earliest references to rabbits in the British Isles:[‡]

---

[*]*Laurex* was the Balearic Islands' name for a rabbit foetus.
[†]*The History of Manchester*, 1771.
[‡]In his *British Mammals* (1920–1) Archibald Thorburn quotes G. F. Browne: 'Dr

In 1176 there were rabbits in the Scilly Isles, where Richard de Wyka granted to the abbey of Tavistock his title *de cuniculis*, 'which for some time I had unlawfully withheld, believing that tithes were not payable on things of this sort'.* At some time between 1183 and 1219 the tenant of Lundy Island was entitled to take fifty rabbits a year from certain *chovis* (coves?) on the island.†

Evidence also survives as to the existence of rabbits in the early thirteenth century on the Isle of Wight, where in 1225 there was a *custod' cuniculorum* in the manor of Bowcombe, Carisbrook. . . .‡

When, however, in May 1199 King John (reigned 1199–1216), as Earl of Moreton (Mortain in Normandy) gave permission for his tenants outside the forest of Dartmoor in Devon, to take hares and other animals, no mention was made of rabbits.§ It thus appears that even as late as the end of the twelfth century rabbits were still very scarce on the English mainland.

These dates tend to suggest that the rabbit was not introduced to Britain, as has usually been suggested, by the Normans (1066–1154), but between the reigns of the Plantagenet kings Henry II (reigned 1154–89) or Richard I (reigned 1189–99), who may possibly have brought them back to England on their return through Europe from the Crusades. The evidence, however, is not conclusive, and it remains possible that rabbits may have made their first appearance in Britain towards the end of the Norman period, perhaps during the reign of Stephen (1135–54).

The earliest rabbit remains so far discovered on the mainland of England appear to date from the late twelfth to early thirteenth cen-

Browne, writing in his *Life of Bede*, describes one of the robes – presumably made between 1085 and 1104 – and used when the body of St Cuthbert was removed from Lindisfarne to Durham, as having pictured on the border of the garment a horseman "with hawk in hand [Browne actually wrote "with hawk and hound"] and a row of rabbits below" ' – the implication clearly being that rabbits were known in the British Isles by this date. Thorburn, however, omits to add that Browne continues ' . . . these magnificent robes have usually been held to be the work of the Arab weavers of Sicily . . . though recent opinion appears to be settling upon Syria and the Mesopotamia region . . .'. See G. F. Browne, *The Venerable Bede*, p. 167, plate XII (S.P.C.K., 1879).

*H. P. R. Finberg, 'Some Early Tavistock Charters'. *Eng. Hist. Review*, LXII, 1947, p. 365.
†Exeter City Archives, Misc. Deeds D.614.
‡P.R.O. Exchequer, Foreign Rolls 8 Henry III. E.364.
§*Rowe's Perambulation of Dartmoor*, 1848.

E

turies They were found in a midden at Rayleigh Castle in Essex*
'which was', writes Miss Veale, 'in royal hands from 1163–1215. . . . It
seems probable that the castle itself fell into disrepair some time during
the first quarter of the thirteenth century, and it was no longer occupied after about 1220. . . . Possibly the rabbits once eaten there had
come from the islands just off the Essex coast, such as Foulness or . . .
Wallasey, which were manors in the Honour of Raleigh.' Sheail
refers to rabbit-bones discovered in the Buttermarket in Ipswich
which are also thought to date from about the twelfth century.

In 1221, 6000 rabbit skins – which may have been English but
could also possibly have been imported – were referred to in a Devon
plea. The first mention of certainly native rabbits on the English
mainland occurs in 1235, when Henry III made a gift of *decem
cuninos vivos* from the royal park at Guildford: in 1241 he ordered
hay to be taken from his *cuningera* at Guildford – the first mention
we have of a coneygarth on the English mainland. In 1242 the King
ordered the collection of thirty to forty rabbits *secundum quod
invenerint prefatam cuneram fertilem* from the same coneygarth.

During the next decade or so the number of Royal and other
coneygarths appears to have multiplied, and rabbits began to
feature increasingly on the menus at feasts and banquets, and to be
transported alive from one part of the country to another. Miss
Veale's researches have shown that for the Royal Christmas feast
of 1240

100 [rabbits] were to be supplied from the lands of the bishopric of Winchester, 200 from those of the earl of Warenne, and 200 by the King's
escheator. In 1241 the sheriffs of Hampshire, Sussex, Surrey and Kent
were to produce 100, 50, 100 and 500 respectively. In 1243 180 rabbits
were required from the estates of the bishop of Winchester, 100 coming
from the Isle of Wight, and 300 from those of the archbishop of Canterbury: 300 were to come from the lands of the bishop of Chichester in
1244, and 200 in 1245. Similar orders were going to the sheriffs of Essex,
Hertfordshire and Middlesex in 1248, and to those of Buckinghamshire
and Bedfordshire in 1249. The coneygarth belonging to the manor of
Kempston, Bedfordshire, held by the earl of Chester, is referred to as early
as 1254.

In 1241, the keepers of the bishopric of Winchester were ordered to take

*E. B. Francis, 'Raleigh Castle', *Trans. Essex Arch. Soc.*, xii p. 184, 1913; and
M. A. C. Hinton, 'On the Remains of Vertebrate Animals found in the Middens of
Raleigh Castle', *Essex Naturalist*, xvii, pp. 16–21, 1912–13.

100 rabbits within the bishopric . . . alive to Sugwas, the manor of the bishop of Hereford. In the same year the keepers of the lands of the bishopric of London supplied the King's uncle, Peter of Savoy, with eighty live rabbits from Clacton, Essex, for his warren at Cheshunt, and by 1244 the King himself had begun to stock his park at Windsor. The sheriff of Surrey sent some rabbits from Guildford; the keepers of the bishopric of Chichester and the earl of Derby produced others . . . the earl of Aumale sent some to the royal park at Nottingham at the same time . . . from Lincolnshire . . . or the Holderness estates.

During these, and doubtless many other movements of rabbits about the countryside, it seems not improbable that considerable numbers escaped from captivity, and became established in the wild. Even at this stage, their status as a pest was being recognized, for as early as 1254–7 the burgesses of Dunster in Somerset were making complaints about their destructiveness.

Sheail mentions a tax-return for Sussex of 1340 in which there are complaints from the inhabitants of West Wittering that rabbits belonging to the Bishop of Chichester had destroyed their wheat. At Ovingdean in the same county one hundred acres of arable land were ravaged by rabbits owned by the aptly named Earl of Warenne. Severe damage to crops in the Breckland area of Suffolk was caused by rabbits throughout the medieval period, and the manor of Methwold was said to have fallen into disuse by 1522 because the farmland was in close proximity to a rabbit warren. In Scotland rabbits were proving to be a menace in at least one district by 1511, when 'Schir Robert Egew, Chaiplan to My Lord Sinclair' complained that 'There will be too many cunnings within two years – they have riddled all the banks of the Links right well'. There are more complaints about the destructiveness of rabbits in the uprising of 1549 led by Robert Ket. On the other hand, there remained some districts where rabbits were still scarce enough to be welcome arrivals; Reyce in his *Breviary of Suffolk* (1618) refers to 'the harmless Conies, which do delight naturally to make their abode here with great increase and rich profitt to all good housekeepers. . . .'

The medieval historian Giraldus Cambrensis (alias Gerald de Barri; *c.* 1146–1220) when discussing the Irish hare in his *Topographia Hibernica* compiled between 1183 and 1186, wrote: 'There are hares

. . . closely resembling rabbits [*cuniculi*]', although, as Fitter points out, he may have seen rabbits when he visited Paris in 1167.

In a Pipe Roll of 1245 in the Hampshire Record Office* a rabbit (or possibly a hare) is depicted being attacked by a species of raptor. The first documented case of trespass involving rabbits occurred in 1268, when Richard, Earl of Cornwall and King of Almain, complained that his coneygarth at Isleworth in Middlesex had been plundered. A severe case of trespassing took place during the reign of Richard II (1377–1400), when an unspecified number of poachers were excommunicated for stealing some 10000 rabbits from the manor of North Curry in Somerset. In 1290 rabbit-warrens (*cunicularia*) are mentioned in *Fleta,* a book compiled by an anonymous author. In the roughly contemporary *Britton* – the earliest summary of the law of England written in Norman French probably by either John le Breton, Bishop of Hereford (d. 1275) or by Judge Henry de Bracton (d. 1268) – mention is made of '*De veneysoun et de pessoun et de conyis*'.

Miss Veale cites what appears to be the earliest reference to the rabbit as an article of export when she refers to two hundred skins shipped from the port of Hull in 1305. In 1389 'conynges', together with warrens, 'connigeries' and ferrets, first appear on the Statute Book (13 Richard II. I. c. 13). Sheail refers to another early representation of the rabbit in English art – a pair depicted in a fourteenth-century painting at Longthorpe Tower, Peterborough.† Rabbits appear regularly on the menu at feasts and banquets during the late fourteenth and early fifteenth centuries, such as those at the coronation of Henry IV in 1399; the enthronement of the Archbishop of Canterbury in 1443; and that given for 'George Nevill, Archbishop of York and Chancelour of Englande' in 1465, when no fewer than 4000 'conyes' were consumed. In his *Book of Nurture* (1460) John Russell (d. 1494), English bishop and chancellor, gives an early example of instructions for cooking rabbits – 'the cony, ley hym on the bak in the disch, if he have grece' – and for carving them; 'unlace that cony'.‡

From the time of their introduction until towards the end of the fourteenth century rabbit skin and flesh must, from the prices

---

*MS. Eccl. 2, 159287, cited by Sheail.
†E. Clive-Rouse and Audrey Baker, 'The Wall Paintings at Longthorpe Tower, near Peterborough', *Archaeologia,* 96, pp. 1–58, 1955.
‡See also *Boke of Kervinge,* published by Wynkyn de Worde in the sixteenth century.

obtained for them, have been regarded as luxuries. These high prices also suggest that rabbits were still of only local distribution.

At Farnham in Surrey, for example, rabbits were sold for an average of 2½d. each between 1253 and 1376. In 1270 in Cambridge-shire rabbits were sold at 5d. each, and even by 1395 rabbits for a Determination Feast at Merton College, Oxford, were procured at between 6d. and 8d. a couple plus the transportation cost of ½d. each from Bushey in Hertfordshire; this suggests that they were unknown – or at least uncommon – in Oxfordshire at the time. An inquisition on Lundy Island in 1274 during the reign of Edward I shows that some 2000 rabbits were killed annually; they fetched £5 10s. 0d. or 5s. 6d. 'each 100 skins because the flesh is not sold'.* In 1272 the capture of rabbits with hawks, dogs and ferrets† at Waleton (where they were sold for 2½d. each) is mentioned by Rogers in his *History of Agriculture and Prices in England*, which also refers to prices of between ten rabbits for 1s. to twenty-six for 1s. 11d. obtained in Oxford market from 1310 to 1313. For the induction banquet for Ralph de Borne, Abbot of St Austin's Abbey, Canterbury, in 1309, rabbits were purchased at a cost of 6d. each. By the early sixteenth century one dozen 'rabbet ronners' were being sold for 2s., and by 1530 they were fetching 5d. a couple in Yorkshire. (In Scotland at this time a rabbit was being sold for 'two shillings unto the Feast of Fastens Eve [Shrove Tuesday] next to come, and from thenceforth at twelve pennies'). Twenty years later the price in England had risen again to 4d. each, and on 9 November 1610 Thomas Cocks of Canterbury noted in his diary that he paid 'for a rabett 9d.'. By the end of the eighteenth century it is calculated that the value of a rabbit skin in proportion to that of the carcase was higher than that of an ox or a sheep.

The history of the rabbit in the rest of the British Isles is likely to have been much the same as in England. During the reign (1214–49) of Alexander II of Scotland, the Royal Warrens were protected from poachers by statute, and in 1264 in the succeeding reign (1249–86) of Alexander III, the keeper of the king's warren at Crail in Fife received a salary of 16s. 8d. for his services. In the first year of the reign (1329–71) of David II, the King's Chamberlain paid 8s. to four men for catching rabbits on the Isle of May in the mouth of the Firth

*Steinman; *Collectanea Topographica*, 1837.
†See pages 129–32.

of Forth: the same king granted one William Herwart, keeper of his warrens in Fife, charter in life-rent of the office of 'Keeper of the King's Muir' in Crail and of its 'cuningare' at a salary of 40s. per annum. As early as 1377 rabbits are mentioned as an article of commerce in Berwickshire*, and a warren is referred to on the links-land of Aberdeen in 1424, when an Exchequer Roll records that a duty of 12d. per hundred was levied on exported 'cunning' skins. In 1457 the killing of rabbits when snow was on the ground was made the subject of 'dittay',† with a fine on conviction of £10.

The rent book of 1474 for Cupar Abbey shows that the monks employed a 'warander of kunynyare', and in the following year one Gilbert Ra or Rae promised to maintain the Abbey's 'conygar from all harm . . . and put it to all profit within his power'.

The Earl of Orkney's rent book for 1497–1503 records that rabbits ('114 cunnings' and '1274 cunningis skinnis') formed part of his annual rental received in kind. In 1551, because of 'the great and exorbitant dearth', young rabbits were protected by statute for a period of three years, except from hawking by the nobility. A charter of 1583 during the reign of James VI, granted to the Provost and Baillies of Aberdeen, refers to a 'cunicularium de Abirdene' near the 'Gallowhills' of that city.

On the Scottish off-shore islands rabbits are known to have been present on Little Cumbrae between Bute and Renfrewshire as early as 1453, and on Orkney, Lambholme and Sanday by 1529. In his *Description of the Western Isles of Scotland called Hybrides* (1530) Donald Monro refers to rabbits on Mull: on 'Inche Kenzie' (Inchkenneth) which was said to be 'full of conyngis about the shoiris'; and on the unidentified 'Sigrain-moir-Magoinein, that is to say the Cuninges Isle, wherein there are many cuninges'. A number of islands in the Firth of Forth were 'verie full of conies', and the animals were also reported 'on a little isle, with a chapel on it, called Cavay' in the Orkneys. The German writer and traveller, Von Wedel, saw rabbits on the Bass Rock when he visited North Berwick in 1584. Great Cumbrae and Ailsa Craig were colonized by 1612, and by 1677 rabbits had spread as far west as the Outer Hebrides. Seven years later warrens were established on Burray Island in the Orkneys, and on Sanda Island off the Mull of Kintyre.

Proc. Berwicks. Nat. Club, 1863–8.

†The matter of the charge or the grounds for an indictment against someone for a criminal offence.

Rabbits were introduced to several districts on the Highland mainland during the eighteenth century by a number of misguided lairds. In 1750 they made their earliest appearance in Morayshire, when they were to be found on the golf-links at Lossiemouth; they were believed locally to have escaped from a ship moored in Branderburgh harbour. Rabbits were introduced to the island of Gigha in 1763, and to Caithness at some time between 1743 (when they were unknown) and 1793 (when they were said to be abundant). They first appeared in Dumfriesshire in 1815, in Kintyre in 1845 and in Wester Ross in 1850. Between 1840 and 1890 a number of introductions were made to Raasay and to North and South Uist. In 1865 rabbits became established on an islet in Loch Seaforth on the Isle of Lewis, and two years later they were brought by lobster-fishermen to Lunga in the little Tresnish Isles of the west coast of Mull. In the 1870s some pet rabbits were released on Foula in the Shetland Islands, where they soon became established. One of the most recent Scottish introductions was that made to Hascosay in the Shetland Islands in 1900.

Dr Colin Matheson has traced the early history of the rabbit in Wales. The first written evidence dates from 1282, when Richard le Forester was paid 3s. 6d. for catching rabbits and keeping ferrets* for the King at Rhuddlan Castle in Flintshire.† In 1284 the commote of Estimaner in Merionethshire paid 8s. for the maintenance of a warren (*haracium*).‡ In a Pipe Roll of 1325-6 rabbits are referred to on the islands of 'Schalmey, Schokolm [?Skokholm] and Middleholm', and between 1386 and 1388 a total of 6120 was obtained from these three islands. Rabbits also doubtless occurred on a number of other islands in the Bristol Channel, e.g. Caldy, and on Skomer off the coast of Pembrokeshire. In 1376 there was a rabbit-warren at Castle Kerdyf (Cardiff), and in 1492 there was one on Flat Holm, a small island opposite Lavernock Point in the Bristol Channel. Dr Matheson quotes a list of rabbit-warrens in Glamorgan in 1578, given by the Elizabethan historian Rice Merrick; Llandaf; Barry Yland; Mynidd Glew; Wyke; Wenny; Morgan; and Britton Fery – all on or close to the coast of the Bristol Channel.§

*See pages 129–32.
†*The Antiquary*, August 1911, cited by Barret-Hamilton.
‡*Archaeologia Cambrensis*, 1884.
§*A Booke of Glamorganshires Antiquities*, ed. J. A. Corbett.

In 1517 the Prior of Pill on the mainland in Pembrokeshire granted a forty-year lease of property including a rabbit-warren to Morris Butler, reserving for himself and the monastery the right to hunt within the warren thrice yearly.* Writing in about 1540, John Leland refers to numerous 'conies' on St Tudwall's Island, St Dwynwen (Llanddwyn) on the Isle of Anglesey, and on Hibre Island near Hoylake at the mouth of the river Dee. During his tour in Wales in about 1773 Pennant observed rabbits on Priestholm (Puffin Island) and on the Skerries, both off the coast of Anglesey. He also remarked on the 'vast and profitable warren' noted 'for the delicacy of the rabbits, by reason of their feeding on the maritime plants', owned by Sir Pyers Mostyn at Talacre on the coast of Flint.†

Miss Veale has discovered the earliest reference in the British Isles to rights *in warennis cunigariis*, which occurs in a charter granting lands in Connaught to Hugh de Lacy in 1204. An early reference to rabbits in Ireland is contained in a poem dating from the thirteenth or fourteenth century: '*Da choinin a Dhúmha duinn*' ('Two conies from Dumho Duinn'). As the first documentary evidence of rabbits in Ireland dates from about the same time as that in England, it seems probable that they were introduced either direct from the continent by the later Normans, of from England in Plantagenet times. In 1282 twenty rabbit-skins from Balisax (Ballysax) in Co. Kildare were sold for 1s. 4d.: five years later one hundred 'great coneys' from the same source fetched 13s. 4d. In 1324 the profits made by hunting in the *cunicularium* at Rosslare, Co. Wexford, formed a part of the tax return made on his land by Aymer de Valence. During the fourteenth and fifteenth centuries a not inconsiderable export trade of rabbit-skins was built up, as described in *The Libel of English Policy* (*c.* 1430). Fynes Moryson, Secretary to Lord Mountjoy when he was Lord Deputy of Ireland (1599–1603) refers to 'Felles of kydde and conies grete plente' and again to 'great plenty . . . of conies'.‡

Sir William Brereton saw rabbits on the banks of the river Slaney when he visited the manor of Ollort (Oulart) at Ferns, Co. Wexford, on 16 July 1635, and at the park near Wexford there was an 'abundance of rabbits, whereof here there are too many, so as they pester

*Emily M. Pritchard, *History of St. Dogmael's Abbey.*
† *Tours in Wales*, 1810.
‡ *The Description of Ireland*, 1735.

the ground'. In the late 1770s Arthur Young reported that rabbits were then common in many parts of Ireland.* As recently as 1906 rabbits were deliberately introduced to Clare Island in Clew Bay, Co. Mayo and in 1911 to Rathlin Island, off the north coast of Co. Antrim.

Although the history of the rabbit in the British Isles is now fairly well documented, it still remains uncertain as to when the animals first began to escape from their warrens to form feral populations in the countryside. In their early days, rabbits were largely confined – especially in Scotland and Wales – to the sandy parts of the links-land between the sea and hinterland, and to a number of off-shore islands. The reasons for this are probably two-fold: first, that rabbits took more naturally to burrowing in soft sandy soil; and second, that the larger carnivores do not normally flourish on small islands which usually lack the necessary supply of food and shelter.

It seems likely that rabbits first began to escape from domesticity in any numbers in the mid-thirteenth century, when livestock were being transferred from one part of the country to another. Fitter suggests that the Black Death of 1349, with the resulting reduction in the country's labour force, may have been a contributory factor. He mentions the Bohemian traveller, Schaschek, who in 1465 saw 'rabbits and hares without number' in Clarendon Park, Wiltshire, although this may still have been an enclosed warren. Rabbits were clearly not regarded at this period as animals worthy of the chase, for the fifteenth-century *Master of the Game* scathingly records that 'of conies I do not speak, for no man hunteth them unless it be fur hunters, and they hunt them with ferrets and with long small nets'. By 1551 the Swiss naturalist, Conrad Gesner (1516–65) was able to write of '*copia ingens cuniculorum*' in lowland England, and that 'there are few countries wherein coneys do not breed, but the most plenty of all is in England'. Even here, however, there were many districts where the rabbit was extremely scarce or unknown until as late as the early 1800s.

The second half of the nineteenth century witnessed a phenomenal explosion of the rabbit population in Britain. Professor Ritchie has put forward two main reasons for this: first, the considerable advances made in agriculture which provided additional food supplies,

*Tour in Ireland, 1776–79.

especially in winter when they were most needed; and second, the fashion for game-preservation which led to the wholesale slaughter of avian and mammalian predators. To these reasons may be added the many deliberate introductions of rabbits made by misguided landowners, especially in parts of Wales and in the Highlands of Scotland.

By the late 1930s it has been estimated that the wild rabbit population in Britain amounted to almost 30 million; by the early 1950s, doubtless aided by the lack of control during the war years, the figure had risen to a staggering 60–100 million. Then, in the late summer of 1953, one of the most devastating scourges ever to attack an animal population struck the rabbits of Europe.

In June 1952, a retired French physician, Dr Armand Delille, obtained some myxoma virus – *myxomatosis cuniculus* – from a former Swiss colleague, with which he inoculated and then released a pair of rabbits in the Département of Eure-et-Loir, west of Paris. It is to be hoped that he had little conception of the misery which his action was to cause. The disease spread throughout Europe with amazing rapidity; the first recorded cases in England occurred in August or September 1953, and were reported to the Ministry of Agriculture by Mr F. H. Feeke, a gamekeeper at Bough Beech near Edenbridge in Kent. The virus was officially recognized as myxomatosis on 13 October 1953. It did not by itself advance very rapidly, but especially in 1954 was extensively spread by human agency, and within a few years it is estimated to have destroyed an almost unbelievable 99 per cent of the wild rabbit population, without, however, apparently reducing the animal's range.

Myxomatosis is a virus disease, the most obvious symptoms of which are a general swelling of the eyelids and base of the ears, which eventually renders the animal unable to see or hear: it has been shown that it is transmitted from one animal to another by the European rabbit-flea (*Spilopsyllus cuniculi*), which is the main vector in Britain, although on the continent and in Australia mosquitoes are notorious carriers. It was first discovered in Uruguay in 1896, since when a number of attempts have been made to introduce it deliberately throughout the world. Because it is such a disfiguring and distressing disease, there was an outcry in Britain – led by the R.S.P.C.A. – against suggestions to attempt to transmit the virus artificially. This resulted in an amendment to the Pests Act 1954, which made the deliberate spreading of the disease illegal.

By the late 1960s and early 1970s the rabbit population was beginning to show definite signs of staging a recovery. The disease must, however, now be regarded as enzootic in Britain, although its virulence varies from year to year and from one part of the country to another. It may be that the animal's self-preservation mechanism has come into play, giving it a newly-acquired propensity to live and, occasionally, to breed above ground, thus to a limited extent escaping the attentions of the disease-spreading flea, although fleas are known to be exchanged between rabbits on the surface as well as underground. The evidence for this, however, is inconclusive; all that can definitely be said is that when the animal becomes too numerous in a given area, myxomatosis usually breaks out to decimate the population once more. The behaviour of the virus is still not fully understood, but in some districts it appears capable of killing what would otherwise be the entire annual increase in the wild rabbit population. As a result, we are unlikely ever to see again the vast hordes of rabbits which infested the countryside in the first half of the present century.

7. House Mouse

(Mus musculus)

As the rabbit was at one time the most familiar larger wild mammal at large in the British Isles, so the house mouse is perhaps our best-known smaller quadruped. Yet it too, like the rabbit, is not indigenous to these islands, but is an alien species, derived from *M. m. wagneri*, the wild subspecies found on the dry steppes of the Russian Turkestan/Persian border. The subspecies which inhabits Europe, including the British Isles, west of a line running approximately 10° East, is known as *M. m. domesticus*.

Both sexes of the house mouse are similar in appearance: each measures about 3 in. (7–8 cm) long, with a tail of about the same length. The general colour of the fur is greyish brown above, shading to greyish-buff beneath. There is no particular breeding season but, depending on each individual's environment, between five and ten litters, each averaging about six young, are produced every year, after a three-week gestation period. The species is virtually omnivorous, favouring particularly insects, grain and human food-scraps.

The house mouse is a largely commensal animal, depending for its food and shelter to a considerable extent on man. In Britain there are two distinct populations – urban and rural – which often overlap, although normally house mice are very restricted in their individual ranges. Although the urban house mouse is highly commensal, the rural populations are markedly less so; these mice live and breed, unsupported by man, in banks, fields and hedgerows, sometimes taking to hay-ricks and grain-stores in winter: insects appear to form a considerable proportion of their food, and they tend to breed more in summer and less in winter. The house mouse is an extremely adaptable animal, flourishing in such apparently unlikely environments as coal mines and frozen-food stores; this helps considerably to ensure the survival of the species.

Dr Corbet writes:

The house mouse was probably the first mammal to be introduced to Britain by human agency. None of the very early historical references to mice allows clear identification of species and information on its date of introduction can therefore only come from subfossil finds. Many of these are difficult to interpret because of the possibility of intrusion through burrows into older strata, but one recent record, hitherto unpublished, seems to confirm its presence in the pre-Roman Iron Age. This is based on the rostrum of two animals and one mandible, identified by the author and now in the British Museum (Natural History), from a site at Gussage All Saints, Dorset. They were in a sealed layer with other small mammals and with no sign of subsequent disturbance.*

E. and H. K. Schwarz, in their study of wild and commensal house mice throughout the world, found that there were originally four wild subspecies, three of which are the ancestors of present commensal forms which, given suitable conditions, may revert again to the wild type.

The tendency of isolated offshore islands to allow the development

*The Distribution of Mammals in Historic Times, Systematics Association special Volume No. 6. The Changing Flora and Fauna of Britain, edited by D. L. Hawksworth, pp. 179–202, Academic Press, 1974.

of different races of one species displaying new genetic characteristics, is typified by the now extinct St Kilda house mouse (*Mus
musculus muralis*). Although this animal, which had a slightly lighter-
coloured underside and rather larger feet than the mainland form,
must have inhabited St Kilda for many hundreds of years, it was not
known to science until the late nineteenth century. In June 1894
J. Steele Elliott found two kinds of mice on St Kilda; one was a field
mouse and the other a species of house mouse which was 'fairly
numerous among the dwellings': these two mice were described and
named by G. E. H. Barrett-Hamilton in 1899.*

Exactly seventy years later, R J. Berry showed that the St. Kilda
house mouse more closely resembles the mice of the Faeroe and
Shetland Islands than those of mainland Britain. The likelihood,
therefore, is that the Norsemen accidentally brought the species
from Scandinavia, in cargoes of hay which they imported to feed
their domestic stock, over a thousand years ago.

In 1905 James Waterston found the house mouse infesting all the
houses and the *cleitans* or 'cletts'† on St Kilda. It had a considerable
range of size and coloration of the underside, which varied from a
smoky-grey to 'a lovely creamy yellow'. In 1910 and 1911 W. Eagle
Clarke spent almost three months on St Kilda, during which he
prepared some forty skins of the house mouse 'which was very
abundant in the houses': it continued so until the human population
of the island, numbering but thirty-six people, was evacuated on 28
August 1930. The combined Oxford and Cambridge Universities
Expedition of the following year caught a total of twelve St Kilda
mice, and estimated the total population at not more than twice that
figure. In their reports of 1932 and 1933, T. H. Harrison and J. A.
Moy-Thomas concluded that the species must inevitably become
extinct, partly because of the resident population of feral cats, but
also because the house mouse could, in conjunction with the field
mouse, only survive as a commensal of man. Without the competition
provided by the field mouse, the house mouse's adaptability might
well have assured its survival even after the departure of its hosts.
Further expeditions (Robert Atkinson in 1938, James Fisher in
1939 and John Naish with Atkinson and Fisher again in June 1947)
found no further trace of the St Kilda house mouse.

*'On the species of the genus *Mus* inhabiting St. Kilda', *Proc. Zool. Soc. Lond.*,
pp. 77–88, 1899.
†Stone cells between 12 and 18 feet long and some 6 to 7 feet high, used by the St
Kildans as drying chambers for a wide variety of produce.

Between 1966 and 1968 the Ministry of Agriculture reported a growth in house-mouse infestation in both urban and rural areas throughout the country; this was thought possibly to be due to an increasing resistance to anticoagulant rodenticides such as warfarin. More notifications of house-mouse infestation were received by the Ministry than in the previous few years, probably from those who sought help when their own efforts proved ineffective. Even if this probable acquired resistance to warfarin is not solely responsible for the rise in house-mouse infestation, it is thought to be a material contributory factor. A number of authorities reported some success against wafarin-resistant mice with other poisons such as alpha-chloralose and zinc phosphide, but to what extent, if any, these will restore mouse infestation to its previous level remains to be seen.

8. Black Rat
(*Rattus rattus*)

When the first brown rat (*Rattus norvegicus*) arrived in England in the late 1720s, the black or ship rat was sometimes referred to as the 'Old English' rat to differentiate it from the newcomer. The black rat is not, however, a native species, but was introduced to this country from its home in south-east Asia, possibly via Asia Minor.

The black rat is a smaller and lighter animal than the brown, measuring from 7 to 9 in. (18–23 cm) in length, with a relatively longer tail which is equal to or longer than the head and body. The head of the black rat is narrower than that of the brown, and the ears and eyes are larger. The colour of the true form (*R. r. rattus*) is black above, shading to greyish-black beneath. In Britain there are two other races of the black rat, the tree rat (*R. r. frugivorus*) which is darkish brown above and white beneath, and the Alexandrine rat (*R. r. alexandrinus*) which is lightish brown with a darkish belly. According to Hinton and Harrison Matthews, *R. r. frugivorus* was the original wild form, *R. r. alexandrinus* having evolved for commensal living with man. 'These races on reaching temperate and northern Europe are believed to have given rise to the black rat, *R. r. rattus* Linnaeus'. Like the brown rat and the house mouse, the black rat is almost omnivorous and breeds throughout the year.

Exactly when the black rat first arrived in Europe is lost in the mists of time: it was certainly unknown to the ancient Greeks and Romans. By tradition the animal first reached western Europe – including of course Britain – in the baggage of returning Crusaders in the twelfth century: its spread was also doubtless aided by natural migration west through the steppes corridors, as in the case of the brown rat some 600 years later. By the fourteenth century it was widely distri-

buted enough in Europe to have been the cause, via the flea it carried, of the dreaded Black Death.

If the story of the Pied Piper of Hameln is more than just a legend, rats were firmly established in Germany by 1284. In that year, we are told, the town was infested by a terrible plague of rats. One day there appeared a piper clad in a suit of fantastic clothes, who offered for a reward to charm the vermin with his music into the river Weser. After completing his task, the citizens reneged on their promise of payment. The piper returned to the town and, playing another tune on his pipe, lured all the children – save one who was a cripple and was thus too slow to follow him – into the interior of the nearby Koppelberg Hill.

Abbot Aelfric 'the Grammarian' (*c.* 955–1020) in his *Vocabulary* compiled around 1000 A.D. defines *raturus* as meaning 'raet'; *rata*, however, was the Provençal name for the house mouse, to which he was almost certainly referring.

The earliest probable reference to rats in the British Isles is that made by Giraldus Cambrensis in his *Topographia Hibernica* where he writes: 'There is another thing remarkable in this island [possibly Aran, but according to Millais more probably Inishglora or Caher

F

Black Rat

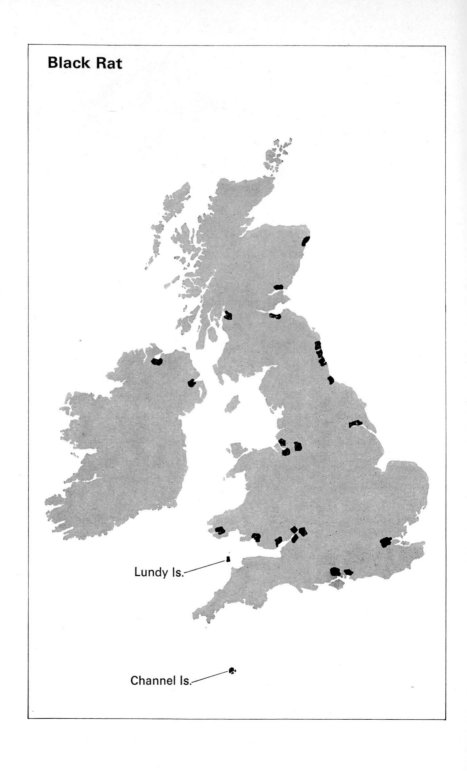

Lundy Is.

Channel Is.

in Co. Mayo] – although *mures* swarm in vast numbers in other parts of Ireland, here not a single one is found. No *mus* is bred here nor does it live if it be introduced.' *Mus* and *mures* have in many translations been assumed to refer to the house mouse, although the probable meaning was 'rat'. Ranulf Higden in his *Polychronicon*, written in the first half of the fourteenth century, added the pejorative *nocentissimos* ('most harmful'). Millais mentions that Giraldus Cambrensis also describes how Bishop St Yvorus* cursed *mures* and expelled them from the province of Leinster because they had gnawed his books. Millais and a number of more recent authors have claimed that *mures* here refers to rats rather than to mice, ignoring the fact that the black rat did not reach Europe before the twelfth century. In this instance, therefore, *mures* clearly refers to mice, as it also undoubtedly does to the animals reputedly expelled from a number of Irish islands, *e.g.* Tory Island, Co. Donegal and Inishmurray, Co. Sligo, by St Colombcille (521–97 A.D.). Some years later in Wales, Giraldus Cambrensis speaks of *murium majorum, qui vulgariter Rati dicuntur* ('the larger species of mice, commonly called Rats'), which were said to have attacked a man, and to have eaten a woman and her child at Merrion, south-west of Pembroke.

By the late thirteenth century rats were clearly becoming a nuisance in many parts of England: in 1297 a number were caught at a village named Weston, where they were sold for ¼d. each, and more were trapped in Oxford in 1335 and again in 1363. The contemporaneous Black Death (bubonic plague), which was carried by the rat-flea and which in some places killed up to 90 per cent of the population, was prevalent over a large area of England, which suggests that the animal was extending its range rapidly during the fourteenth century. In the *Vision of Piers Plowman*, written between about 1362 and 1392, occurs the following disparaging couplet:

> Had ye rattes youre wille
> Ye couthe nouyt reule youreseule.

This work goes on to speak of a 'ratoner', and of a 'route of ratones and smale mys mo than a thousande with hem'. Geoffrey Chaucer (1340–1400) in the *Pardonere's Tale* tells of a citizen who

---

*Also known as St Ibhar, Iberius or Ivory, Bishop of Bergery or Bergerin: born early in the fifth century, he was probably a pupil of St Patrick: he lived first in the Aran Islands in Galway Bay, then in Geshille Plain, King's County (now Co. Offaly), and finally on the island of Bergerin in Wexford Haven, where he died in about 500 A.D.

> Forth he goeth, no lenger wold he tary,
> Into the toun unto a Pothecary,
> And praied him that he him wolde sell
> Som poison, that he might his ratouns quell.

In about 1377 damage caused by rats or mice is mentioned in the Register of Archbishop Sweteman of Co. Armagh, and a rat-trap is listed as an item of expenditure in the accounts for 1469 of the churchwardens of St Michael's, Cornhill, in the city of London.

In 1578 came the first reference, albeit a negative one, to rats in Scotland, when Bishop Leslie wrote '. . . a wonder, the rattoun lyves not in Buquhane [Buchan] . . . in this cuntrey no Rattoune is bred, or brought in from any other place, there may lyve'. Even by 1630 it was said 'There is not a ratt in Sutherland, and if they doe come thither in shipps from other parts (which often happeneth) they die presentlie, as soon as they doe smell the aire of that cuntrey. But they are in Catteynes [Caithness], the next adjacent province, divyded onlie be a little strype or brook from Sutherland'. The same tradition appears to have been maintained in Ross-shire, for when in 1656 an ex-Cromwellian trooper, Richard Franck, found himself in that county, he wrote 'The inhabitants will flatter you with an absurd opinion that the earth in Ross hath an antipathy against rats . . . to the best of my observation I never saw a rat; nor do I remember of any one that was with me ever did . . .' The parishes of Dunnet in Caithness, Annan in Dumfriesshire and Liddesdale in the Lowlands were also by reputation free of rats, and their soil was much in demand for flooring granaries in the belief that they would thus be made rat-proof.

One of the earliest positive references to rats in Scotland appears to be that made by Martin Martin in his *Description of the Western Islands of Scotland* (1703), where he saw 'a great many Rats in the Village Rowdil [on Harris], which have become very troublesome to the Natives, and destroyed all their Corn, Milk, Butter, Cheese, etc. They could not extirpate these Vermin for some time by all their endeavours.'

The earliest mention of rats in Ireland appears to be a doggerel quoted by Fynes Moryson in his *Description of Ireland:*

> For four vile beasts Ireland hath no fence,
> Their bodies lice, their houses rats possess.
> Most wicked priests govern their conscience,
> And ravening wolves do waste their fields no less.

The fortunes of the black rat appear to have declined rapidly
following the introduction to England of the brown rat in 1728 or
1729. Robert Smith, rat-catcher to Princess Amelia, describes in *The
Universal Directory for Taking Alive and Destroying Rats and All
Other Kinds of Four-footed and Winged Vermin in a Method Hitherto
Unattempted: Calculated for the Use of the Gentleman, the Farmer
and the Warrener* (1768), how

the black ones do not burrow and run into shores and sewers as the others
do but chiefly lie in the ceilings and wainscots in houses, and in outhouses
they lie under the ridge tiles, and behind the rafters, and run about the
side-plates: but their numbers are greatly diminished to what they were
formerly, not many of them being now left, for the Norway rats always
drive them out and kill them wherever they can come at them; as a proof
of which I was once exercising my employment at a gentleman's house,
and when the night came that I appointed to catch, I set all my traps
going as usual, and in the lower part of the house in the cellars I caught
the Norway rats, but in the upper part of the house I took nothing but
black rats. I then put them together in a great cage to keep them alive
until the morning, that the gentleman might see them, when the Norway
rats killed the black rats immediately and devoured them in my presence.

In his *British Zoology* (1776) Thomas Pennant was of the opinion
that 'the Norway-rat has also greatly lessened their [i.e. the black rat]
numbers, and in many places almost extirpated them'. He goes on
to say that 'among other officers his *British* majesty has a *rat-catcher,*
distinguished by a particular dress, scarlet embroidered with yellow
worsted in which are the figures of mice destroying wheat-sheaves'.
Thomas Swaine, a rat-catcher who worked in fifteen counties,
mentions in *The Vermin Catcher* (1783) that he found black rats in
two counties only – at High Wycombe in Buckinghamshire, and
in Middlesex.

By 1845, however, the black rat was uncommon enough in London
for a rat-catcher of St Giles – described condescendingly in *Notes and
Queries* for 1854 as 'an intelligent man and not a bad naturalist for
his station in life' – to be able to charge collectors three guineas each
for specimens from his cage outside the National Gallery in Trafalgar
Square. In 1850 the animal was still to be found in a number of
ancient granaries in London, and in 1875 a house in Cornhill was
found to be heavily infested.

In 1813 in Scotland the black rat was the only rat known in
the town of Forfar and it was described as 'not rare' throughout

the county. In 1830 it was said to be still common in rural Aberdeen-shire (it lingered on at Cairnton of Kemnay until 1855) and in Moray-shire where, however, it had become extinct according to Charles St John, by about the middle of the 1840s. In 1838 it was said that in the bordering county of Banffshire 'in Keith, which is at a greater distance from the coast, it is not very uncommon'. Black rats were still to be found infesting 'the garrets of the high houses in the old city' of Edinburgh until 1834. In 1879 they were rare enough in Dunkeld for Millais to mention the fact that he had personally seen two in that town.

On the Scottish offshore islands black rats were still colonizing the Orkneys in some numbers in 1808, where on South Ronaldshay they were comparatively abundant in 1813: the species continued to infest Benbecula in the Hebrides until the 1880s and the Shetlands as late as 1904. In the Channel Islands black rats were said to be still 'common on Alderney and Hern' in 1862.

In Ireland a single specimen was found in Co. Cork in 1842, and unconfirmed sightings were received from Co. Kerry, Co. Armagh, Co. Dublin and Co. Antrim. In 1876 a litter was discovered in Leviston, Co. Kildare, and as late as 1911 a colony was found in a warehouse in Dungarvon in Co. Waterford. In 1935 a single specimen was observed on Lambay Island north of Dublin.

In the nineteenth and early twentieth centuries the black rat popu-lation was reinforced on a number of occasions by new blood from wrecked ships from foreign ports: in 1866 a fruit-carrying vessel, from which a number of rats escaped, was driven ashore at Seascale in Cumberland; after the Italian grain-ship *Espagnol* was wrecked in Acton Cove, near Marazion in Cornwall in 1875, 'the whole of the surrounding district was swarming with these little rats': more rats escaped from an Austrian barque wrecked neat Port St Mary on the Isle of Man in 1883, and from a Greek ship stranded on Lundy Island in the Bristol Channel in 1928. Although the black rat in Britain is still re-inforced from time to time by fresh blood from abroad, the precautions taken to prevent this from happening make it clear that the resident population is able to maintain a self-perpetuating existence without alien assistance.

By 1890 – although it was still not uncommon in parts of Cheshire and north Wales – the black rat was thought to be more or less extinct in the rest of mainland Britain; so much so that the discovery by Mr Arthur Patterson of a colony of black rats in the docks at Yarmouth in Suffolk in 1895 was looked upon with some surprise.* In fact, the black rat was probably never even near extinction; almost certainly its two brown forms *frugivorus* and *alexandrinus* were frequently mistaken for brown rats, although the corollary is also probably true – that the black form of *norvegicus* was often identified as *rattus*. Between 1925 and 1938 Matheson collected records of black rats killed in the ports of Bristol, Swansea, Weymouth, Southampton, Plymouth, Cardiff, Grimsby, Middlesbrough and Liverpool, and lesser numbers elsewhere. Fisher records that during the war years the black rat was common not only, as would be expected, in the docks and quays of the port of London, but also in clubs, restaurants, theatres and flats in the West End: at the outbreak of war some 90 per cent of rats in the City were black ones.

After his survey of the black rat in the United Kingdom between 1951 and 1956 when it was 'considered to be always present somewhere' in forty-eight localities, E. W. Bentley was able to write: 'There is little likelihood of completely eliminating the species from Britain. It can be expected to persist for a considerable time yet in the commercial quarters adjacent to the docks in London, Bristol and Liverpool (and perhaps Edinburgh and Belfast) where conditions for its continued existence seem most favourable.'

In 1961 Bentley conducted a second survey of the status of *R. rattus* in Britain. He found that it had existed on Lundy Island for several years, and that it was still present on Sark and probably on Alderney. Between 1956 and 1961 the number of permanently infested localities in Britain had declined from forty-eight to twenty-eight. Outside London the black rat was reported to be permanently present in 1961 only in the Channel Islands, on Lundy Island, at Bristol and Avonmouth, and at Sharpness, Gloucester, Salford, Wallasey, Liverpool, Bootle, Aberdeen, Dundee, Edinburgh, Leith, Glasgow, Faslane (Dunbartonshire), Belfast and Londonderry: a decrease in population – though not in distribution – was reported for Sharpness, Aberdeen, Dundee (60 per cent), Leith, Glasgow and Faslane. The ports of Hull, Portsmouth, Southampton, Cardiff and Lerwick in the Shetland Islands each reported five or less infestations, and a single black rat was caught in Middlesbrough. Sunderland, South Shields and Eston (near Middlesbrough) each reported single

*'A Plague of Black Rats': *Notes of an East-Coast Naturalist*, pp. 262–6, 1904.

infestations. Several were killed in Ellesmere Port, but all were from the same warehouse.

Black rats occurred in 1961 in the following places from which they had been absent five years earlier: on the Clyde; at Walthamstow (a single specimen); at Milford Haven; in a lock-keeper's house on the Severn at Martley in Worcestershire; and a single rat at Congleton which is thought to have arrived by road in a consigment of grain. Reports of possible black rats were also received from Hardington Mandeville near Yeovil; Higham Ferrers (Northamptonshire) and Cranbrook in Kent.

In London in 1961 black rats were permanently resident in Battersea, Southwark, Stepney, Bermondsey, Finsbury, Holborn, the City, Poplar, and West Ham. In all but the first three of these districts a decrease in numbers was reported, and in Poplar the infestation was confined to a single wharf. In Southwark, however, a small increase in distribution further away from the river was reported. In 1961 black rats were also said to be permanently present in the vicinity of a railway goods yard partially within the borders of Bethnal Green. A single infestation was reported from Erith in 1961, where in 1956 there had been none. More than five infestations were recorded from the City of Westminster in 1961, and five or less from Greenwich and Shoreditch. Where smaller numbers or no black rats were reported in 1961 the decline was thought to be due to intensified control measures; more rigorous inspection of ships from foreign ports; increased cleanliness of warehouses and wharfs, resulting in better protection of edible merchandise; improved trawler design; and the continued requirement of a Rodent Control Certificate for coastal shipping. In the conclusion to his 1961 report, Bentley wrote; 'It is clear . . . the status of the ship rat in Britain has further diminished and the species is on the way to losing its permanent foothold in the United Kingdom. . . . A further reduction in status can be expected over the next few years.'

His view was prophetic, for by 1975 – apart from occasional sightings as e.g. at Burnham-on-Crouch – the species was only present in any numbers on some offshore islands such as Lundy; in the Channel Islands; and in the docks and warehouses of some major ports, especially those of London and Liverpool: how long will it be before the black or ship rat finally renounces for ever its eight-hundred-year-old British citizenship?

9. Brown Rat

(Rattus norvegicus)

The brown or common rat – sometimes also known as the Norway rat from the mistaken belief that it was in Norwegian timber ships that it first reached the British Isles – is a native of south-east Asia south of the Himalayas.

Both sexes of the brown rat are similar in appearance: each measures an average of between 8 and 9 in. (20 to 23 cm) in length, with a slightly shorter tail. The typical colour is medium-brown above and greyish-white beneath; a so-called 'black' variety – which was at one time given the name *hibernicus* and is dark-grey above shading to dusky-grey beneath – was first noticed in Ireland in 1837 where – and in the islands of the Outer Hebrides – it was confined until making its appearance in England in the 1890s. It now seems to be spread thinly throughout the country; in London, for example, it forms perhaps 2–3 per cent of the total brown rat population. Brown rats breed throughout the year, an average of eight or nine young being born after a gestation period of some three weeks. This species lives partly in commensal communities with man, when they principally infest commercial and industrial premises, drains and sewers, and partly in wild communities in the surrounding countryside.

Brown rats are without doubt the worst pest in Britain, and everyone's hand is turned against them: they will consume almost any form of edible matter, and foul much food which they do not eat; they are extremely vicious – to man as well as to domestic and other wild animals – and transmit a number of dangerous diseases.

The brown rat appears to have reached Europe via Russia early in the eighteenth century. The German-born naturalist, Peter Simon Pallas (1741–1811) records that vast hordes migrated to the west from Asia following a severe earthquake in 1727: they infested houses in the town of Astrakhan, swam the river Volga and soon spread further west to the remainder of European Russia and the Baltic. They arrived in Copenhagen, probably in Russian ships from Asia, in 1716, and by 1750 had reached Paris and parts of eastern Prussia. They were first recorded on the Norwegian mainland (where by 1776 they were reported as being commoner than the black rat) in 1762, on the Faeroe Islands – which they are said to have reached in the wreckage of the *King of Prussia* which foundered off the Isle of Lewis – in 1768, in Brunswick and in Greenland around 1780, and in Sweden in 1790. They arrived in Spain and Italy in the mid to late eighteenth century and in Switzerland in about 1809. Brown rats first appeared in the United States – having crossed the Atlantic presumably in English ships – in about 1775, a sound enough reason, had no others existed, for wishing to secede from the mother country. According to John James Audubon (1780–1851) brown rats had not penetrated as far as the Pacific coast by the time of his death, although they must have arrived there shortly afterwards, having presumably travelled either by ship or overland in the baggage of the early settlers.

In western Canada brown rats were reported to be well established in Vancouver, New Westminster and Victoria by 1887, and seven years later they had penetrated as far east as Chilliwack, some 60 miles (96 km) up the Fraser river. Further east still brown rats crossed the border from North Dakota into Manitoba around the turn of the century, where by the outbreak of the First World War they were established as far north as the Assiniboine river and had crossed over into Saskatchewan. During the two decades between the wars almost all the urban areas of Manitoba and Saskatchewan became infested.

An old legend records that the first brown rats to arrive in England came over from the continent in the ship that landed William of Orange at Torbay in 1688. Another story – propagated by opponents of the House of Hanover such as Jacobites, Tories and Roman Catholics – tells that the animal first reached England in the ship which brought George I from Germany in 1714, when it accordingly received the name 'Hanoverian rat'. Charles Waterton (1782–1865) – the eccentric Roman Catholic 'Squire' of Walton Hall in Yorkshire, and an inquisitive naturalist – fostered this idea which he inherited

from his father. He describes the brown rat as 'a little grey-coloured short-legged animal . . . known to naturalists as the Hanoverian rat'. His father, he continues, was 'always positive that it actually came over in the same ship which conveyed the new dynasty to our shores'. Waterton believed that the newcomer would soon exterminate 'the original rat [*Rattus rattus*] of Great Britain', and only ever saw 'one solitary specimen' of the latter species which he apostrophized as 'Poor injured Briton! hard, indeed, has been the fate of thy family! In another generation at fatherest it will probably sink down to the dust for ever!'*

In fact, according to Pennant, the first brown rats reached England around 1728 or 1729, probably in ships from Russian ports but certainly not in Norwegian bottoms since, as mentioned above, the animal did not reach Norway until 1762. Thereafter it spread throughout the country with amazing rapidity. In his *Mammalia Scotica* (published in 1808 but compiled between 1764–74) John Walker wrote: '*Mus fossor* . . . the Norway Rat . . . First brought, as they say, to Scotland in ships from Norway . . . Wheresoever it pitches its abode, it pitches out the Black Ratten utterly . . . the Black Ratten . . . and the Norway Rat were previously entirely unknown in Annan . . . however about twenty years agone [i.e. in

*Charles Waterton; *Essays* (1838, 1841, 1858) and *Autobiography*, 1858.

about 1744–54], the Norway Rat was cast on shore from ships driven to the mouth of the River Annan, and now is scattered through almost the whole region of Annan.'

The New Statistical Account of Peebles for the parish of Newlands relates that

The brown or Russian or Norwegian rat . . . a good many years ago invaded Tweedale, to the total extirmination [sic] of the former black rat inhabitants. Their first appearance was at Selkirk, about the year 1776 or 1777 [where the townspeople feared that their burrows would undermine the buildings], and passing from Selkirk, they were next heard of in the mill at Traquair; from thence . . . they appeared in the mills of Peebles; then entering by Lyne Water, they arrived at Flemington Mill, in this parish . . . about the year 1791 or 1792.

Brown rats are first recorded in the Highlands in 1814, when the Revd. G. Gordon saw some in Morayshire.

According to John Rutty in his *Natural History of Dublin* brown rats – reputedly the size of cats and rabbits – 'first began to infest these parts about the year 1722'. On the Isle of Anglesey in Wales brown rats were so numerous in August 1762 that they are said to have eaten ears of standing corn while it was being reaped.

A number of offshore islands have at one time or another been over-run by brown rats – almost invariably to the detriment of the existing animals. On Priestholm (Puffin Island) in the Menai Straits between Anglesey and Caernarvonshire, rats which came ashore from a wrecked Prussian ship in about 1816 almost wiped out the resident population of rabbits and puffins. The mammalian and avian inhabitants of Ailsa Craig and Little Cumbrae in the Firth of Clyde also suffered severe depredations from invading brown rats, and Fitter believes that they may have been responsible for the failure of the manx shearwater to breed successfully on Lundy Island in the Bristol Channel.

The success of the brown rat in becoming established in such a comparatively short period of time has been quite remarkable. Being larger and fiercer than the black rat, it has had little difficulty in usurping the position in the country held by the latter in all districts away from docks and quaysides. It has been helped in its success by its prolificacy, its omnivorousness, and its adaptability to a variety of habitats. As a subterranean-dwelling animal it has proved less vulnerable to its enemies – including man – than the surface-living black rat, in spite of increasingly sophisticated methods of bringing about its downfall.

According to the Ministry of Agriculture brown rat infestation tended to decrease in the north and west of Britain and in urban areas between 1966 and 1968; there was, however, during the same period a corresponding increase of infestation in rural districts. A number of authorities reported some success in the use of fluoroace-tamide against sewer-living rats, and emphasis was placed on the importance of preventing new infestations on building sites. Brown rats today occur in almost every type of habitat throughout mainland Britain and Ireland, as well as on the Isle of Wight, the Isle of Man and many other offshore islands. They are perhaps most numerous in southern and eastern England. Unless they are attacked by some virulent disease – such as the myxomatosis which decimated the rabbit population in the early 1950s – it is unlikely that their numbers in the British Isles will ever be materially reduced.

10. Edible Dormouse
(Glis glis)

The squirrel-tailed, fat or edible dormouse is found from France and Spain eastwards to the Caucasus and Turkestan; as far north as central Germany and southern Russia; south through Switzerland to Italy, Sicily and Sardinia, and south-east to Yugoslavia, Moldavia, Asia Minor, and Iran.

The long hairs on the upper parts are generally silvery-grey in colour with brown tips, and are darkest on the spine. The under-fur is dark ashy-grey, while the underparts, inner sides of the limbs, cheeks, throat, chest and belly are light yellowish-grey. The dark hairs on the back continue as stripes down the outside of the legs, which are short and possess sharply clawed and mobile toes. The ears are broad, hairy and projecting. The eyes, which are large in size, are surrounded by a ring of black hairs, and have a dark and horizontal pupil, which is unique among rodents. The slightly flattened and bushy tail is thickly covered with long hairs of much the same colour as the body, and is lighter below than above: it is rather brittle, and the tip easily breaks off. The head and body of the edible dormouse reach a length of from 6 to 7 inches (15 to 18 cm) and the tail measures about an inch (2–3 cm) less. An adult male weighs from 5 to 7 oz (140 to 200 gm).

The edible dormouse was a favourite food of the Romans, by whom it was much esteemed as a delicacy. Marcus Terentius Varro (116–27 B.C.) gives details of how to husband the animals *in villatica pastio* ('within the precincts of the villa') in his *Rerum Rusticarum*, compiled in 54 B.C. Their natural history is discussed by Pliny the Elder (*c.* 23–79 A.D.) in his *Naturalis Historia* published in 77 A.D., which is largely derived from Varro. Lucius Junius Moderatus Columella wrote of edible dormice in his first-century-A.D. *De Re*

Rustica, which formed the basis of Rutilius Taurus Aemilianus Palladius's similarly named book of about 350 A.D. The dormice were reared in oak and beech groves, and were fed on currants and chestnuts. For final fattening they were placed in earthenware vessels called *gliraria* (from *glis*, a dormouse) where they sometimes grew so fat that it became necessary to break the *gliraria* in order to remove them. They were then cooked and served with a sweet sauce, sometimes made from honey seasoned with poppy seeds. Barbara Flower and Elizabeth Rosenbaum write: 'stuffed dormice [were] either . . . placed on a tile and cooked in the oven (*furnus*) or else put in the *clibanus* – a small portable oven of earthenware, iron, bronze or occasionally of more precious metals'.* They quote and translate a Roman recipe of the first century A.D.:

Glires isicio porcino, item pulpis ex omni membro glirium, trito cum pipere, nucleis, lasere, liquamine farcies glires, et sutos in tegula positos mittes in furnum aut farsos in clibano coques.	*Dormice*: stuff the dormice with minced pork, the minced meat of whole dormice, pounded with pepper, pine-kernels, *asafoetida*† and *liquamen*,‡ sew up, place on a tile, put in oven, or cook, stuffed, in a small oven (*clibanus*).

**The Roman Cookery Book – a critical translation of the Art of Cooking by Apicius*, pp. 31–2, 205, Harrap, 1958.

†The Persian variety of the herbal plant *silphium, laserpitium* or *laser*, also known as 'Devil's Dung', which retains its importance in the Middle East to this day.

‡A sauce made from dried fish and salt.

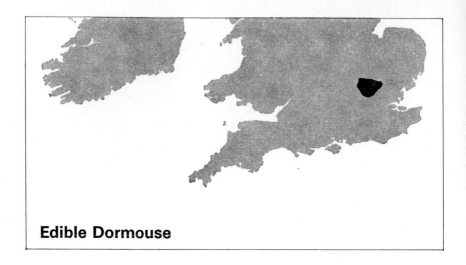

Edible Dormouse

Among the Romans it was fashionable to be an authority on dormice, in much the same way as it is today to be knowledgeable on wine, and scales were often brought to the table to weigh the animals before eating them.

It thus seems quite possible that edible dormice were first imported to Britain by the Romans. If so, they would not of course be the ancestors of our present stock, which was introduced to England in 1902; B. Lloyd (1947) quotes Major Stanley Flower, O.B.E., F.Z.S.; 'On or about February 4th, 1902, some individuals of *Glis glis* from the Continent of Europe (probably Germany or Switzerland) were turned loose at Tring Park by [the Hon.] Walter (later Lord) Rothschild.'

The dormice multiplied very fast and caused considerable damage to corn and other crops, and also to thatch. As a result a campaign was conducted which was thought nearly to have exterminated them. Between 1910 and 1924 six specimens were sent to the Zoological Gardens in Regent's Park, but few others were seen. Then from 1925–7 a number were noticed in Tring Park – where they used bird-boxes in which to make their nests – and also in the surrounding countryside, In 1926 Pendley Manor near Tring was over-run and thirty-nine edible dormice were caught. Cisterns had to be covered to prevent them from drowning themselves, and a short circuit was caused by dormice gnawing through an electric cable.

Since 1926 they have only been recorded occasionally at Pendley: one or two specimens were sent to the county museums of Hertfordshire and Buckinghamshire.

On 11 November 1929 Major Flower exhibited the skin and skull of a female edible dormouse, found by Mr Frank McIver Riching at Hastoe in Hertfordshire on 13 October, before the Zoological Society of London. He said that he had heard that the species, described locally as the 'chinchilla' and 'Spanish rat', had occurred during the previous twenty years in the parishes of Aldbury, Tring and Wigginton in Hertfordshire, and Drayton Beauchamp and Aston Clinton in Buckinghamshire.

By the autumn of 1931 edible dormice had spread to Whipsnade in Bedfordshire, where some were caught in the lofts of a bungalow, and they have been captured there intermittently ever since. In 1933 one was found in the hollow walls of a house at Wendover. In 1936 seven were caught in scrub on an escarpment between Lion Pit and Holly Findle, and on the nearby downs. They were found in some numbers in a house at Aldbury in 1935–6, and in the latter year, when three nests were found in the roof, no less than seventy-five were caught. By 1938 they were recorded from as far afield as Worcestershire and Wiltshire, and unconfirmed reports were received from Berkshire, Gloucestershire, Hampshire, Northamptonshire, Oxfordshire and Surrey: they were also reported from Ludlow, Shropshire in 1941–2, and from Coventry in 1945. It is, however, unlikely that they travelled so far from their main haunts naturally, and they were probably assisted, unwittingly or deliberately, by man.

Edible dormice have been observed sporadically at Great Pednor, Buckinghamshire, since 1941, when a number were found drowned in a cistern. Since 1945 they have been reported in small numbers in houses in Amersham and since the following year there have been a few at Berkhamsted, where they were first seen in Incents House at the school.

Early in 1951 a census was made of edible dormice sites in England, which resulted in the discovery of not less than twenty-four new colonies, some of which were in entirely new districts such as Ashley Green, Cholesbury, Hyde Heath, Great Missenden and Pitstone in Buckinghamshire, and Ashridge Park, Little Gaddesden, Ringshall and Rossway in Hertfordshire.

The following table shows the total number of edible dormice killed in England between 1902 and 1962:

G

Date	Number of dormice killed
1902–20	?
1921–30	43
1931–40	152
1941–50	321
1951–60	327
	843

Between 1945 and 1951 the Amersham Rural District Council (now the Chiltern District Council) accounted for 215 edible dormice in the area under their control, as follows:

Date	Number of dormice killed
1945	47
1946	21
1947	24
1948	36
1949	24
1950	49
1951	14
	215

The edible dormouse is a mainly nocturnal animal, although it may also occasionally be seen by day. It haunts woods and gardens, climbing about in low trees and bushes with great agility, sometimes making amazing jumps. Dormice seldom descend to the ground except to enter houses, where they can be very noisy in attics and lofts; they eat stored apples, rolling them along the floor and bouncing, thumping and pattering after them, all the while making strange squeaking chirps and twitters.

In autumn the edible dormouse becomes very fat before going into hibernation, usually between mid-September and mid-November. The males hibernate first, followed by the females and then the young. The nest, composed of vegetable fibres, moss and dead leaves, is normally built in deep burrows or holes in trees, and less often in thatch, lofts, attics and cellars. Hibernation may be either solitary or communal, with up to eight dormice in a single nest, and quietness is essential, for during hibernation the temperature of the body is lowered considerably, and if disturbed the animal may easily die. It does, however, wake up periodically throughout the winter, but invariably behaves in a thoroughly lethargic manner, in captivity sometimes nibbling rather half-heartedly at an apple. In April 1949 two hibernating edible dormice were found under a bedroom floor at Great Missenden: in 1950 a keeper at Ashridge Park found two in a nest of leaves and paper on a shelf in a cottage: a pair were once discovered four feet down a fox's earth. They may also hibernate in rabbit burrows, as twelve were ferreted near Wendover in 1936, and about fifty were trapped in holes near Amersham between October and December 1950. Hibernation lasts for about seven months (the species is known in Germany as '*Der Siebenschläfer*' – the 'Seven Sleeper') and seldom ends before late April or early May. Breeding takes place in June, when between five and nine young are born in a nest usually built in the crown of a mature tree. Three nests with an unrecorded number of young were found in a loft at Aldbury in 1936, and one in Ashridge Park in 1950 contained nine. Only a single parasite has so far been discovered, an American squirrel flea, *Orchopeas wickhami*, brought here by the grey squirrel, and first found on dormice at Tring in 1949 and at Chesham in 1951.

Edible dormice eat fruit (mainly apples and cherries), berries, seeds and nuts, as well as insects, especially cockchafers. In England, though not on the continent, they appear to eat mainly stored fruit: apart from some grapes and peaches taken from greenhouses at Tring from 1925 to 1927, there are comparatively few records of dormice eating growing fruit in England. The bark of willow and wild plum trees is also sometimes eaten, and at times possibly small birds. Beech-mast is reported to have been eaten at Ashridge, and milk has been taken from bottles at Aldbury. In captivity edible dormice will eat dog-biscuits but seldom, if ever, green vegetables.

As can be seen from the tables on page 98, edible dormice colonized only a small number of places in the first three decades of this century, and the spread to their present limits took place almost entirely during the fourth decade. At present they are confined to a triangle with an area of approximately 100 square miles in the Chilterns formed by Beaconsfield, Aylesbury and Luton.* Edible dormice are often confused with grey squirrels and brown rats, and are sometimes known locally as 'little chinchillas'. This confusion has led to occasional reports of feral colonies of chinchillas (*Chinchilla laniger*), although this species is very susceptible to damp and is much too delicate to survive in the wild in the British climate.

In June 1963 it was first noticed that edible dormice had caused a considerable amount of damage by stripping the bark from spruce, larch, Scots-pine and birch trees in Wendover Forest; an area of some 10 acres (4 hectares) was attacked over a period of five weeks, and approximately 160 Norway spruce valued at upwards of £200 were destroyed; as a result eighty dormice were killed there in that year and a further twenty in the following year. The Chiltern District Council caught between thirty and forty dormice a year from 1970 to 1975, and it appears possible that the range and population of the species may be decreasing slightly. A curious and so far unexplained phenomenon is that if one particular house in a street is occupied by and then cleared of dormice, a year or so later the same house may be re-invaded, while the neighbouring houses remain immune.

Edible dormice cause considerable damage to crops and timber abroad, so a careful watch must be kept on their activities in England. They are preyed upon by rats, cats, birds and stoats, and are easily poisoned and trapped, so control of their numbers should not prove difficult, especially as they have so far shown no propensity to extend their present range. It will be interesting to see to what extent the species is able to colonize the country further in the future.

*Isolated records from outside this area include Ibstone and Bledlow Ridge on the Buckinghamshire/Oxfordshire border, and Potters Bar, Middlesex.

11. Grey Squirrel
(Sciurus carolinensis)

The grey squirrel is a native of the north-eastern United States and south-eastern Canada, where it is found from Quebec, Ontario, New Jersey and Pennsylvania westwards to Ohio.

The general greyish ground colour of the grey squirrel is the result of the blending of various shades of colour on the hairs, which are mainly black at the base, then rufous or russet, and finally black tipped with white. There is often considerable colour variation between individuals, and even on the same specimen at different seasons of the year. In winter, the upper surfaces of the back, head and feet, frequently assume a distinctly brownish tone, which has occasionally led to confusion with the native red squirrel (*Sciurus vulgaris*), or to the suggestion that the two species may interbreed, for which, however, there is no evidence. The flanks and tail are generally grey, while the underparts are white; there are no tufts of hair on the ears, but a conspicuous narrow line of bright rust-red is sometimes developed on either side of the body where the flanks meet the chest and belly. Melanism (which in England occurs mainly in Hertford-shire, probably as a result of twelve melanics introduced to Woburn) and albinism (found principally in Kent, Surrey and Sussex), although occasionally reported, is uncommon. The sexes are similar in both size and coloration; each weighs about 21 oz (590 g) and measures approximately a foot (30 cm) in length with a tail some 9 in. (23 cm) long.

The most characteristic sign of the grey squirrel is its drey, an untidy-looking nest of twigs lined with grass, leaves or moss normally built between 20 and 50 ft (6 to 15 m) from the ground in the fork of a tree: here the young, usually numbering between three and five per litter, are born between January and April and from June to August after a gestation period of between thirty and forty days;

Grey Squirrel
(Principal areas only shown)

they are able to forage for themselves in about seven or eight weeks. Grey squirrels do not hibernate, and may be seen searching for food at all seasons of the year; their diet is very varied, consisting mainly of nuts, especially acorns, beechmast, chestnuts and walnuts; all kinds of fruit and cereal crops; buds and young shoots of many trees, especially broadleaved hardwoods, as well as inner bark and sap; bulbs, fungi, birds' eggs, nestling birds, young rabbits, fish, carrion and honey.

The first recorded occurrence of the grey squirrel in Britain was at Llandisilio Hall, Denbighshire, in October 1828. In 1830, in a letter to the editor of the *Cambrian Quarterly Magazine*, a correspondent stated that for some time prior to 1830 grey squirrels had been seen at Llanfair Caereinion, Llan Eurvyl and Cwm Llwynog (Fox's Dingle) in Montgomeryshire. The origin of these specimens is unknown, but they were in all probability of the American variety *S. c. leucotus*.

The earliest recorded introduction of the grey squirrel to Britain was in 1876, when Mr T. V. Brocklehurst liberated a pair, imported from America, at Henbury Park near Macclesfield, Cheshire. It is not known how these fared, but a pair shot in 1884 at Highfields, Nottinghamshire, may have come from Henbury, as grey squirrels are great travellers. An apparently abortive attempt at naturalization was made by Mr J. H. Gurney at Northrepps Hall, near Cromer, Norfolk. In 1889 Mr G. S. Page of New Jersey imported a small stock of grey squirrels from the United States, of which five were released in Bushy Park, Middlesex, where they apparently failed to obtain a footing as no more were seen there until the middle of the present century. Ten grey squirrels imported by Mr Page were set free at Woburn Abbey by the 9th Duke of Bedford in 1890, where they rapidly increased. Two years later, a pair were liberated at Finnart on Loch Long, on the borders of Dunbarton and Argyll: from here they spread northwards to Arrochar and Tarbert by 1903: east to Luss (1904) and Inverbeg (1906); south-west to Garelochead (1907) and Rosneath (1915), and south as far as Helensburgh, Alexandria and Culdross by 1912. Three years later they had penetrated to the eastern side of Loch Lomond and had become established at Drymen in Stirlingshire. Thus, within a period of twenty-five years they succeeded in colonizing an area of over three hundred square miles. An introduction of grey squirrels to the Edinburgh Zoo in 1913 resulted in the establishment of a number of escapees and their progeny by the end of the decade outside the zoo grounds at Corstorphine.

In 1902 an introduction of a hundred grey squirrels was made at Kingston Hill, Surrey, by an American who released them in Richmond Park, and at about this time others were liberated in Rougemont Gardents, Exeter. From 1902 to 1929 a veritable wave of introductions took place in all parts of the country; about a hundred of the specimens set free came from America, while some one hundred and fifty were taken from the stock at Woburn. Unfortunately, in several instances no accurate records were kept of which new points of colonization received squirrels drawn from the already existing Woburn stock, and which received entirely new blood from America.

From 1876 to 1929 (the year of the last recorded introduction), the most important areas of liberation were:

Date	Place	Origin
1876	Henbury Park, Cheshire	T. V. Brocklehurst; from America
1889	Bushy Park, Middlesex	G. S. Page; from America

Date	Place	Origin
1890	Woburn Abbey, Bedfordshire	Duke of Bedford (through G. S. Page)
1892	Finnart, Loch Long, Dunbarton, and Argyll	G. S. Page
1902	Kingston Hill, Richmond Park, Surrey	America
1903	Rossett, Wrexham, Denbighshire North Wales	—
1905–1907	Zoological Gardens, Regent's Park	Woburn Abbey
1906	Scampston Hall, Yorkshire	—
1908	Kew Botanical Gardens	—
1908	Cliveden, Buckinghamshire	Viscount Astor
1909	Farnham Royal, Buckinghamshire	—
1910	Chiddingstone, Kent	Lieutenant-Colonel Meade Waldo
1911–1912	Bramhall, Cheshire	—
1913	Castle Forbes, Co. Longford	Earl of Granard
1913	Corstorphine, Edinburgh	Edinburgh Zoo
1919	Pittencrieff Park, Dunfermline, Fife	—
1919	Bournemouth, Hampshire	—
1928	Ballymahon, Co. Longford	—
1929	Bestwood and Hartsholme, Nottinghamshire	—
1929	Needwood Forest, Staffordshire	—

In 1938 it was declared illegal in future to import grey squirrels to Britain or to keep them in captivity.

The single female grey squirrel released at Farnham Royal, Buckinghamshire in 1909 came from South Africa; she was probably of the American species, a number of which were introduced to that country at Groote Schuur in about 1880 by Cecil Rhodes.

Up to 1929 grey squirrels were also set free in small numbers in Ayrshire; at North Queensferry in Fife; and in Berkshire, Northamptonshire, Oxfordshire, Staffordshire, Devon and Warwickshire from where they soon spread into Gloucestershire and eastern Wiltshire.

It is difficult to estimate the relative importance of each fresh introduction during this period because of the considerable numbers involved, and because extensions of range from different original centres overlapped. In some places new colonies merely served to augment the already existing local population. By 1930 grey squirrels were firmly established throughout south-east England and as far north as Northamptonshire and Warwickshire.

Between 1930 and 1945 the species became entrenched in the Midlands and in most of the south of England apart from Cornwall: no grey squirrels, however, were reported from East Anglia, probably owing to the unsuitability of the fens and marshlands, and to the fact that most of the forests there are composed of coniferous trees which are favoured by the red squirrel, whereas the grey prefers deciduous varieties. By the outbreak of war the grey squirrel covered a large area roughly extending from Sussex and Kent in the south and east past the outskirts of London, as far as Cheshire and Wrexham in north Wales in the north and west; by the end of the war the species had further extended its range and had become firmly established in the North and East Ridings of Yorkshire, from Middlesbrough to Hull. There were, however, still few grey squirrels to be found in East Anglia and in north-west England: isolated records came from Alnwick, Northumberland in 1930; Loch Shiel, Inverness-shire (1939); Galloway (1937); and Selkirkshire (1944).

From 1945 to 1956 grey squirrels increased their numbers in Cheshire and Yorkshire, and moved further east into southern Essex and west into south Wales, south-west Dorset, Devon and Somerset. They appeared in Westmorland for the first time in 1944–5, in Suffolk (1947–52), Cardiganshire (1950), southern Lancashire (1950–2), Merionethshire (1951), and Cornwall (1955), and spread further into Brecknockshire, Derbyshire, Durham, Montgomeryshire, Radnorshire, Shropshire, Staffordshire, Worcestershire, Lincolnshire, and eastern Cambridgeshire. In Scotland, grey squirrels spread east through Fife and northwards into Perthshire.

Between 1945 and 1955 the county agricultural executive committees issued free cartridges to squirrel-shooting clubs, and paid a bounty of 1s. for each squirrel killed between 1953 and 1955. Although the amount of the bounty was subsequently doubled, and by 1958 a total of £100000 had been paid out, the grey squirrel population showed no sign of decreasing.*

*'Englishmen, who have been menaced for half a century and more by the American grey squirrel, are astonished to find that, in its native country, it has a long close season

By the early 1960s it was easier to say where grey squirrels were not found than where they were: they colonized hitherto vacant areas in Devon, Wiltshire, eastern Cornwall, the south-West Riding of Yorkshire, and parts of Norfolk, Essex, eastern Suffolk, Westmorland, Cumberland and Northumberland. In Wales they have now spread to the Isle of Anglesey, and in England to the far west of Cornwall. There are still, however, northern and western regions comprising parts of Cumberland, Westmorland, and northern Lancashire where the grey squirrel is infrequent or unknown. In the east, parts of Norfolk, Suffolk and northern Essex have so far escaped invasion. Elsewhere today distribution is general throughout the country, although the most heavily infested districts remain the south and south-east, the south Midlands and the southern regions of the Welsh marches.

In Ireland by 1938 the descendants of the stock liberated at Castle Forbes, Co. Longford in 1913 had spread into Leitrim, Roscommon and Westmeath: by 1956 they had reached Armagh, Down, Fermanagh and Tyrone, and are continuing to extend their range. In Scotland the species is now fairly common in parts of Argyllshire, Clackmannanshire, Dunbartonshire, Falkirk, Fife, Kinross-shire, East and Mid-Lothian and Perthshire.

Wherever the grey squirrel is not to be found, its absence is probably due to a combination of such factors as the configuration of the terrain, where rivers and mountains form impassable barriers; unsuitable sites, such as fen-, marsh-, heath-, and moorland; the presence of coniferous rather than deciduous forests, and the growth of urban areas and industrial belts.

In some counties there is an apparent correlation between the disappearance of the red squirrel and the presence of the grey: there is, however, no definite information on this point, although the general impression is that the grey squirrel is exerting some unfavourable influence on the red: this, however, was not the case in the early years of the life of the grey in Britain. Red squirrels were decreasing in numbers in many parts of the country, almost cer-

and that there is a daily limit imposed on its hunting. Grey squirrels, indeed, are regarded as high-class sporting animals, and "either fried or in a stew, are a tasty treat".' J. N. P. Watson, 'Where Hunters Still Come First . . .', *Country Life*, pp. 1098–1102, 23 October 1975.

tainly due to disease, between 1904 and 1914, before the grey had become properly established. Both red and grey squirrels are highly susceptible to virus diseases, which from time to time cause wide fluctuations in their numbers. By the time that the reds had recovered from their epidemic in the 1920s the greys had become sufficiently well established to be able to resist their return, and now may even be driving the reds out of their remaining strongholds. Today there are few red squirrels in the Midlands and southern England, where the grey squirrel is so common, whereas in the west country, East Anglia, and those parts of northern England, Scotland and Ireland where the grey is seldom found, the red still maintains a somewhat precarious foothold. In parts of Scotland the grey squirrel now appears to be firmly established.

The grey squirrel, as well as being the cause of much damage to orchard, garden and farm crops, is a serious woodland pest. As grey squirrels feed largely on tree-shoots, bark and seeds, and build their dreys with young shoots and green leaves, it follows that wherever they are found they will injure tree growth to some extent The really serious damage, however, results from the creature's habit of stripping bark for a limited period of the year, usually around mid-summer, but also in severe winter weather. The outer bark is discarded, the squirrel's objective being the inner bast and cambium layers. The reasons for the attacks are various, and may be a combination of drought, seasonal famine (the attacks usually take place between the time of spring shoots and autumn fruits), and the desire for the vitamins which the sap contains. It has also been suggested that there is an element of social behaviour in bark-stripping, which may act as a display to other squirrels of territorial rights.

Grey squirrels show a marked preference for hardwood trees from twenty to forty years old, and have a characteristic mode of attack which enables the cause of the damage to be easily identified: it is thought to result from the animal's habit of sitting on the ground or on a side branch to attack the main stem. The tree is either barked for about a foot above ground level, or at several points above the main branches, but seldom between the near-ground zone and the main crown. The barked patches at the foot of the tree do not always completely encircle the trunk, and the tree may eventually recover. Those higher up almost invariably run right round the trunk, and the whole crown above that point is killed. The species most often attacked are, in order of preference, sycamore, beech, oak, ash, birch, larch and Scots pine. It was in order to attempt to control these attacks that the Grey Squirrels (Warfarin) Order, 1973, was enacted, which allows the use of this poison between April and July, when the growing trees are most at risk. It will be interesting to see whether bark-stripping squirrels become as resistant to warfarin as have grass-eating rats, as the result of an increased intake of vitamin K.

The introduction of the grey squirrel to the British Isles provides a classic example of the damage which can be done by the unthinking and wanton liberation outside its native environment of a harmful alien species with no redeeming characteristics. Apart from man, squirrels have few enemies, although some, especially the young, probably fall victims to stoats, weasels and foxes. Weather is an important controlling factor, and a cold, wet spring causes a heavy mortality among the first brood. During the winter months squirrels can best be killed by poking out their dreys, and by shooting, trapping and snaring, but to be fully effective these operations must be carried out over a wide area. Over the years numerous campaigns aimed at the extermination of the grey squirrel have been waged, notably by the Forestry Commission, but have met with little success. The only appreciable reduction in the population occurred in the winter of 1930–1, when some virulent disease (possibly coccidiosis) killed large numbers. Unless some further contagion attacks grey squirrels, as myxomatosis did the rabbit population in the early 1950s, it would seem that this trans-Atlantic immigrant must be accepted as permanently established in Britain.

12. Coypu

(Myocastor coypus)

The coypu is a native of South America, where it is found in the rivers and lakes of Argentina, Uruguay, Paraguay, southern Brazil, south-eastern Bolivia, Peru and Chile: it is also common in the Chonos Archipelago and on Chilöé island, south to the Straits of Magellan.

The coypu is one of the largest rodents, attaining a length of some 2 ft (61 cm) or more excluding the tail, and weighs on average between 10 and 20 lb (4·5 to 9 kg). It belongs to a group of several genera which are distinguished from other American species by the exceptionally harsh nature of the fur. The coypu itself is further characterized by its very large orange-coloured incisor teeth, and by the fact that the teats of the female are situated high up on the sides of the body, thus enabling the young to be suckled in the water, although normally suckling takes place on land. The ears of the coypu are small and round; the tail, which is scaly with a thin coating of short hairs, is about two-thirds the length of the head and body: the feet have five toes each, which in the hind limbs are connected by webs; the outer fur is long, and beneath is a dense and soft under-fur which is known in the trade as nutria.* The colour of the upperparts is a mixture of dusky and brownish-yellow, the sides and underparts being pure brownish-yellow, the tip of the muzzle and chin white with a yellowish patch beneath each ear, and the feet dusky-brown.

In its habits the coypu is not unlike a beaver, being thoroughly aquatic and making its burrow in the banks of the rivers and lakes it frequents; when, however, the banks are not high enough to allow this, a platform-like nest is constructed among the reeds. The burrow,

---

*Nutria was the early Spanish-American colonists' name for the coypu (from the Spanish *nutria*, meaning an otter).

which is about 8 or 9 in. (20 to 23 cm) in diameter, may extend to a depth of 20 ft (6 m) or more and expands at the end into a chamber some 2 ft (61 cm) in diameter; here the young, usually numbering between two and nine per litter, are born after a gestation period of between 127 and 138 days. The females sometimes produce two litters a year, which may be born at any season. Although it is an excellent swimmer and diver, the movements of the coypu on land are awkward and ungainly: it usually selects for its haunts the stillest parts of rivers, lakes or ponds, where it subsists mainly on the foliage, seeds and roots of the waterplants growing nearby, although molluscs are also occasionally eaten.

The coypu was introduced to the continent many years before it came to Britain – to France probably in about 1880, and to southern Russia around 1926. Here coypus were deliberately liberated in suitable areas in the hope that they would breed rapidly in the wild: many were freed near the rivers Kuban and Kura in the Transcaucasus and in west Georgia, where they have apparently thrived and multiplied ever since. They have also bred extensively in Germany, Austria, Norway and Belgium.

In North America feral coypus are widely distributed throughout the Pacific north-west having escaped from fur-farms in Washington, Oregon and Vancouver, British Columbia.

It was not, however, until 1929 that the first coypus were introduced

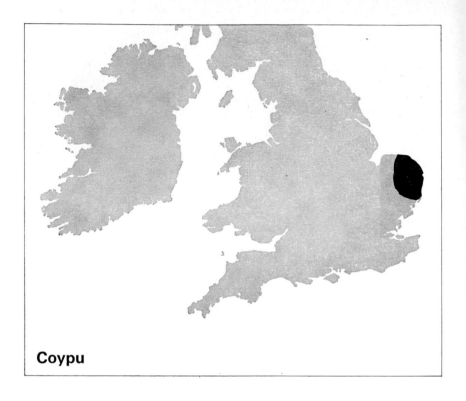

Coypu

to England to form nutria-farms, mainly in Surrey, Sussex, Hampshire
Norfolk and Devon, and by 1932 the first escapes were recorded
from these areas. The farms, totalling about fifty in number, were
mostly situated in low-lying districts, where rivers and streams are slow
running with muddy banks. Several small nutria-farms were established by owners of large estates through whose grounds ran suitable
rivers or streams, but none were conducted on any large scale. Some
bigger farms were, however, to be found where coypus were kept
on a commercial basis, notably in Norfolk, Sussex and Hampshire.
By the 1940s coypus were known to be established in any numbers
only in Norfolk and Suffolk, and on a sewage farm at Dorney, near
Slough in Buckinghamshire, where they remained until about 1954.

When nutria-farming was first started in Britain in 1929 the farms
of Surrey, Sussex and Hampshire outnumbered those of the whole of
East Anglia, and the escapes of coypus between 1932 and 1939 were
correspondingly more numerous in these southern counties. The first
records of escaped coypus in Britain were from Horsham, in Sussex,
in 1932, and from Tiverton, Devon, and Horsham again in 1933.

These were soon followed by reports of escapes from farms at Ring-wood, Hampshire and Walresley, Huntingdonshire in 1934, and from Whiligh, Sussex and Horsham again in the following year. Isolated instances of escaped coypus were also reported in Bedford-shire, Buckinghamshire, Cheshire, Essex, Gloucestershire and Staffordshire between 1936 and 1944, and on Sedgemoor in Somer-set in 1942–3. In 1950 two were killed at Barcombe Mills on the Ouse in Sussex by the Crowhurst otterhounds, and two more were seen there in October of the following year. From about 1940, however, feral coypu colonies largely disappeared in the south – those in East Anglia alone surviving. From two nutria-farms in Yorkshire and Cumberland, which were started between 1929 and 1939, there were, strangely enough, no recorded escapes. Coypus have also been bred in Scotland at Auchinroath near Rothes on the Spey, Moray-shire, from which there were some escapes in 1934, and at Perth: in Wales a farm was situated near the confluence of the rivers Vyrnwy and Severn, Montgomeryshire, from which escapes were reported in 1936.

In Norfolk the first coypus escaped from fur-farms in the valley of the river Yare in 1937, and during the war years that followed, when nutria-farming was abandoned, they spread up and down that river and its tributaries and along the rivers Wensum and Tas. These early escapes were probably from farms at Weston near the Wensum, at East Carleton between the Yare and Tas, and at Brundall near the lower reaches of the Yare. By 1943 coypus had spread up the Yare to Keswick and Cringleford, and in 1944 had reached Bawburgh and Marlingford, and had colonized some 40 miles (64 km) of the Yare, Wensum and Tas. At about this time a feral coypu colony appeared on 9 miles (14 km) of the Wensum between Drayton and Lenwade, near the site of the original Weston farm, and another was discovered near Tasburgh on the Tas. In the same year it was clear that coypus were becoming increasingly numerous in the marshland around Surlingham, Wheatfen and Rockland Broads, where they found an ideal habitat. In October 1944 a single coypu was seen at Gorleston, Great Yarmouth in the estuary of the Yare; no others were reported from this area, but it is interesting to remember that in South America coypus are as much at home in brackish water as in fresh. By 1945 coypus had spread east to Cantley, Reedham and the Langley marshes, and south-west to Northwold on the river Wissey, and were also to be found in small numbers at Wroxham.

H

In 1946 the discovery was made, on a young coypu caught near Norwich, of two lice, one of each sex, which were subsequently identified as specimens of *Pitrufquenia coypus*: this species belongs to the family *Gyropidae*, and was first recorded on a nutria-farm at Pitrufquen near Temuco in Chile. It is now known that virtually all British coypus are infested with this parasite.

Coypus were first seen at Hickling in 1948, and at Horsey in the following year. Soon afterwards, having followed little becks from the upper reaches of the Waveney, Yare, Wensum and Bure, they appeared in many inland localities away from the main rivers, and in some instances even settled down in isolated ponds and gravel-pits. By 1955 coypus had reached the river Glaven in north Norfolk and had followed the Wensum beyond Guist, while by 1956 colonization of Suffolk was well under way. In the late 1950s coypus were seen on Oulton Broad and the rivers Ant and Thurne; in east and west Suffolk and Cambridgeshire as far south and east as Holbrook near Ipswich; as far south and west as Mildenhall (on the river Lark), Wisbech and King's Lynn, and north to Melton Constable. Isolated coypus were observed at Thrapston in Northamptonshire and at Mountnessing in Essex.

It was at one time thought that coypus were a beneficent influence on the environment of East Anglia, and that by eating reeds they helped to keep the waterways of the Broads clear and prevented silting; but whatever good the animals may do is far outweighed by the considerable amount of damage they cause. They raid gardens for green vegetables, and farmland for cereal crops and roots such as swedes, mangolds, potatoes, and especially sugar-beet. It has been estimated that damage to crops is only caused when the density of the coypu population has reached four per acre (0·4 hectare). When they first escaped it was anticipated that coypus, like muskrats (*Ondatra zibethicus*), would undermine the banks of rivers and dykes with their burrows; this indeed happened, and was the primary reason for the control campaigns waged against them. Coypus also add to the danger of flooding where their burrows occur in the banks of rivers and dykes which are above the level of the surrounding land. They trample down and crush much marsh vegetation, and eat the young shoots of reeds (*Phragmites communis*), cutting up the rhizomes and often completely clearing entire beds, thus converting them into open water. They eat reed-mace (*Typha latifolia*) and lesser reed-mace (*T. angustifolia*), pulling them up by the roots and eating the submerged parts – mainly the rhizomes – leaving the remainder floating: this is also done to various species of rushes. The reed meadow grass (*Glyceria maxima*) is also eaten extensively, but this plant appears to be able to withstand the ravages of coypus better than does *P. communis*, and often replaces the latter in those areas where the animals abound. Coypus also consume the water-parsnip (*Sium erectum*); the cowbane or water hemlock (*Cicuta virosa*); the saw-sedge (*Cladium mariscus*); the tufted sedge (*Carex elata*); the great pond sedge (*C. riparia*); the burr-reed (*Sparganium erectum*); the yellow water-lily (*Nuphar lutea*); the reed-grass (*Phalaris arundinacea*), and the great water dock (*Rumex hydrolapathum*). Damage to ozier beds and larch saplings has also been recorded. The destruction of large areas of reed-beds between about 1957 and 1962 – when the coypu population was at or near its peak – threatened the habitat of many species of marshbirds – some already extremely rare – and also reduced the supply of Norfolk reed for thatching.

In Britain, coypus have few enemies other than man, although occasionally young ones may be taken by foxes, stoats and herons. The coypu's size and weight make it conspicuous and clumsy on land, and in spite of the rather alarming appearance presented by its incisors it is not a fierce animal, and is easily taken in live-traps. Between 1943 and 1945 a trapping campaign organized by the War Agricultural Executive Committee resulted in the capture of just under 200 coypus. From 1945 to 1962, apart from a severe set-back received in the hard winter of 1946–7, coypus steadily increased in range and numbers in the east of England, reaching a peak of about 200000 in 1960. In July of that year the 50 per cent grant made available by the Government to rabbit clearance societies was extended to cover the control of coypus by these societies and by the East Suffolk and Norfolk River Board. The campaign did not fully get under way until 1961, but by August of the following year some 97000 coypus had been accounted for, without, however, any significant reduction in either the total population or its range.

In its report for 1960–1 The Select Committee on Estimates recommended that 'the Ministry should undertake an immediate full-scale campaign against coypu, with the object of total extermination'. In fact, as in the case of the mink, 'total extermination' was considered impossible, and the main objectives were defined as (a) to stop the further spread of coypus; (b) to eliminate coypus from outside the Norfolk Broads; and (c) to reduce as much as possible the population within the Broads, and to contain it within Broadland.

This third campaign against coypus began in August 1962; until that time the animals continued to increase their range westward and began to be found increasingly outside Norfolk and Suffolk, occurring in Lincolnshire (Holland), Cambridgeshire, the Isle of Ely, Huntingdonshire, north-east Bedfordshire, south-west Hertfordshire, and in various parts of Essex and Sussex. Isolated reports of coypus were received from Lincolnshire (Lindsey), where an 18 lb (8 kg) male was killed at Thurlby Fen in April 1962 and an 11 lb (5 kg) male at Bardney Lock in March 1963; Gayton, Northamptonshire (March 1963); Sheffield and Leicestershire (both 1963); and Derbyshire (1961 and 1964). During this campaign, which ended on 31 December 1965 and covered all Norfolk (except the extreme west) and the northern half of Suffolk, a total of 40461 coypus were killed, including nearly 1500 from outside the campaign area, at a cost of over £70500. From a comparison of trapping records before and after the severe weather of January and February 1963 it is estimated that between 80 and 90 per cent of the coypu population died in those two months from starvation and exposure. Up to the end of 1962 an average of some 3700 coypus were being trapped each month; by April this figure had dropped to about 350. The harsh weather caused coypus to migrate to new areas where they sought higher ground for food and shelter among farm-buildings, hay-stacks, and even in rabbit burrows. Thus, although the weather helped to reduce the coypu population considerably, the resulting dispersal of the surviving animals made the work of the trappers more difficult still: approximately 18000 out of the campaign total of 40461 were caught after the end of that winter, i.e. between March 1963 and December 1965.

This third campaign against the coypu proved to be fairly successful, and few have been reported since 1965 from outside the campaign area; three were killed in disused clay-pits at Welney in the Isle of Ely in March 1965; one on the Thame near Aylesbury (1965); a pregnant female near Latimer on the Chess (1965); and one at Spalding, Lincolnshire in October 1965. Possible sightings were reported from the moat at Michelham Priory, near Wartling, and from Brede and Staple Cross – all in Sussex. In Cornwall a female was killed on 6 February 1962 at Reskajeage Farm, Gwithian, and another, which had escaped from a farm at Troon near Camborne, was caught at Penryn. A number of large holes discovered in March 1965 in the banks of a river at St Erth were probably made by coypus.* On 3 January 1961 an adult male weighing over 12 lb (6 kg) was captured at Gibbet Moss, Hawkshead, in north Lancashire, and on 29 March 1966 an adult female weighing 10 lb (4·5 kg) was caught at Branthwaite Mill on the river Marron near Cockermouth in Cumberland: both these specimens, which are in the Carlisle Museum, are believed to have escaped from a short-lived nutria-farm opened in the area in the early 1960s.

At the conclusion of the 1962 to 1965 campaign an association was formed, now known as 'Coypu Control', which is still operating today. From 1966 to 1969 the coypu population steadily declined. In 1970, however, approximately the same number were trapped as in the previous year, and the figures for 1971, 1972, and 1973 showed dramatic increases in the number caught. (The average monthly totals of coypus trapped were: 1966—150; 1967—125; 1968—125; 1969—90; 1970—90; 1971—175; 1972—260.) In spite of this considerable increase in the coypu population there has fortunately so far been little evidence of expansion of their present range, and the situation is being carefully monitored by the Coypu Research Laboratory in Norwich. Writing in July 1973 Dr L. M. Gosling of the Laboratory estimated the coypu population in late 1972 at between 8000 and 11 000 and forecast that by the end of 1973 there would be an increase of some 7000 unless trapping efforts could be intensified: he projected the hypothesis 'that the population increase of 1970 to 1972 was caused by improved reproductive success and juvenile survivorship and that these factors are closely linked with climatic variation and the resultant variation in food quality'. It was recommended that the number of coypu-trappers should be increased; this was done, the total being raised to fifteen in 1973 and to eighteen by 1976. As a result, the coypu population was stabilized at around 10 000 in the former year, and was reduced to about 7000 by the middle of 1976, a total which in the future is likely to be reduced still further.

*F. A. Turk, 'Notes on Cornish Mammals', *Rep. Roy. Poly. Soc. Corn.*, 1962–6.

Today coypus are mainly confined to the marshy tracts of country and Broads of eastern Norfolk and Suffolk, whence it is to be hoped that they will spread no further but from where it seems unlikely that they will ever finally be dislodged. The recent eutrophication in the Broads – due to excessive amounts of phosphorus, nitrogen and other nutrients in the water probably arising from the seepage of agricultural fertilizers and/or sewage effluent – has resulted in a serious decline of many species of aquatic flora. The effect on the coypu population, however, is likely to be only marginal owing to the abundance of alternative wetland food supplies. Together with the mink and grey squirrel, the coypu provides a classic example of the folly of introducing to this country an animal so destructive of our environment and its native flora and fauna.

13. Himalayan or Hodgson's Porcupine

(Hystrix hodgsoni)

Crested Porcupine

(H. cristata)

The Himalayan or Hodgson's porcupine* is found in the central and eastern Himalayas in north-eastern India, in southern China, and in parts of Malaysia, at altitudes up to 5000 ft (1520 m). The crested porcupine† ranges from east and west Africa as far north as parts of southern Europe.

Apart from the beaver (*Castor fiber*), porcupines are the largest old-world rodents, the European species measuring between 26 and 28 in. (66 and 71 cm) in length. Their heads, which are small in size, end in a blunt nose surrounded by exceptionally long and sensitive whiskers. Most species of porcupine are generally a brownish-black in colour and have stout bodies and shortish legs. The quills covering the body – which are the porcupine's chief distinguishing characteristic – are mostly marked with broad black and white rings – both ends being white. The rump-quills are chiefly black while the open quills on the end of the tail are mostly white. As its scientific name implies, the species *cristata* has a crest of long brown and white bristles on the neck, which is shorter and less noticeable in Hodgson's porcupine.

The old story that porcupines are able to eject their quills at an enemy is quite untrue: when alarmed or angry, however, they can erect their quills, using subcutaneous muscles which lie below a thick layer of fat, at the same time deliberately rattling the hollow

*The name 'porcupine' – derived from the French *porcépin* ('spiny pig') – was probably given to the animal because of its grunting pig-like voice, and because its flesh is said to taste like pork.

†G. B. Corbet and L. A. Jones (1965) describe three species of crested porcupines, i.e. *indica, cristata* and *africaeaustralis*.

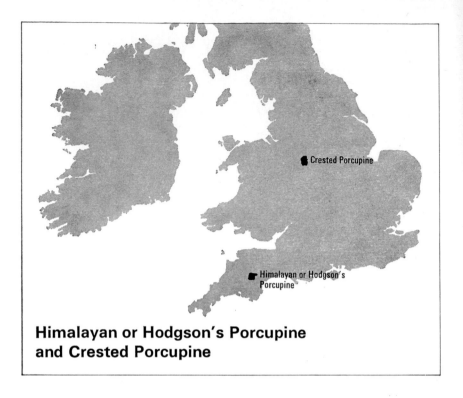

Crested Porcupine

Himalayan or Hodgson's Porcupine

**Himalayan or Hodgson's Porcupine
and Crested Porcupine**

ones at the tip of the tail: if this proves unsuccessful in frightening away an intruder, the porcupine may run sharply backwards in order to drive his quills into his opponent's body. The quills on the tail of Hodgson's porcupine are larger and more open than in any other members of the family *Hystricidae*.

The old-world porcupines are nocturnal in their habits and – unlike the American family which are adept tree-climbers – are purely terrestrial. Although the European variety is a basically solitary animal, those of India are more gregarious. Their diet is entirely vegetarian, consisting largely of roots, except in cultivated areas where they cause considerable damage to farm and horticultural crops. In Europe porcupines mate early in the year, and in late spring or early summer from two to four young are born in a nest formed of leaves, grasses and root-fibres, which is usually built in an old fox's earth, a badger's sett, in a natural cavity, or in a self-made hole. In Africa the hole of the aardvark (*Orycteropus afer*) is some-times used. In India, where porcupines frequently make their nest in long vegetation above ground, between two and four young are

normally produced, although the presence on the female of six teats (placed high up on the sides of the body as in the coypu) suggests the possibility of more multiple births. In captivity a gestation period of around 100 days has been recorded, and females may be ready to mate at the age of fourteen months.

In 1969 a pair of Himalayan or Hodgson's porcupines escaped from the Pine Valley Wildlife Park, about 2 miles (3 km) north of Oke-hampton in Devon,* which had acquired the animals direct from Calcutta. Unfortunately the escape was not reported at the time; it was thus not until 1971 – when a number of quills were found and damage to conifers by barking around the base of the trunk was first reported – that it was realized that a new alien mammal was at large in the English countryside. A survey carried out between 18 and 23 June 1973 disclosed that approximately 15 per cent (valued at about £300) of a 10-acre (4-hectare) plantation of young Norway spruce at Folly Gate, 2 miles (3 km) north of Okehampton, had been freshly attacked: the Forestry Commission estimated the cost of damage in their nearby plantation in Springett's Wood at between £500 and £1000.

*For information on porcupines in England I am indebted to Dr L. M. Gosling of the Ministry of Agriculture, Fisheries and Food's Coypu Research Laboratory in Norwich, where an intensive study is being made of these two members of the hystri-comorph group of rodents.

In June or July 1971 an adult porcupine was killed by a terrier owned by a badger-digger on Forestry Commission property at Bogtown near Northlew, some 6 miles (10 km) north-west of Oke-hampton: this remains the most westerly record in Devon. In November 1973 the corpse of a sub-adult – which had almost certainly been born in the wild and had probably died in the pre-ceding August – was found in the mouth of a badger's sett at Risdon, north of Okehampton in the valley of a tributary of the river Okement. Another was seen crossing a road between Whiddon Down and Throwleigh, 7 miles (11 km) south-east of Okehampton, within the borders of the Dartmoor National Park: this is so far the most southerly and easterly record in Devon. The most northerly evidence of porcupines in the county is provided by a number of quills found on Lock's Hill, south of Dolton, some 10 miles (16 km) north of Okehampton. Mr J. Davey and Mr R. Lancaster of the Ministry of Agriculture carried out a survey as far south as and including Fernworthy Forest inside the Dartmoor National Park, without finding any further trace of porcupines in this area.

The principal range of feral porcupines in Devon extends over an area of about 2 square miles in the shape of a letter 'h', from Oaklands in the south through Hook, Abbeyford, North, Springett's and Parsonage Woods, past Risdon, Folly Gate and Inwardleigh, as far north as Hayes Barton south of Jacobstowe. Here the animals appear to be largely confined to conifer plantations near the river Okement and one of its small tributaries. There thus appears to be some form of dependence on water, although this may be mainly for drinking purposes. The conifer plantations include Norwegian and sitka spruce, Scots pine, Japanese and common larch, and silver fir.

At Risdon the habitat consists mainly of a twelve-year-old planta-tion of Norway spruce on the side of a steep bank, where the under-growth consists principally of brambles, bracken, gorse and birch saplings. At the foot of the slope is a 40 yard (36 m) wide stretch of level ground intersected by a steeply banked minor tributary of the Okement. Here the canopy is of oak, silver birch and sallow, with an undergrowth of ferns and wood falsebrome. Dr Gosling discovered three burrows among the spruce trees each containing quills, which indicated recent occupation. Mr Davey found a number of entrances to badger setts which seemed to have been in use for a short time, and an extensive and recently occupied sett, with tunnels through much of the plantation, appeared to have been taken over by the porcupines: no sign of the former inhabitants has been seen since the arrival of the latter species. A complex system of runs about 7 to 8 in. (18 to 20 cm) wide and containing porcupine quills traverses the thick undergrowth, both vertically and horizontally, some of which were probably made by badgers and some possibly by porcupines.

During his survey of the Risdon plantation Dr Gosling found evidence that the porcupines were feeding extensively on young Norway spruce. A number of trees had been attacked near the base, and both the exposed wood and nearby pieces of bark bore clear incisor marks: the bark had been cut just above ground level, probably with the lower incisors, and then gripped between the teeth and ripped upwards and sometimes sideways: this appeared to be repeated several times, and fragments of bark measuring up to 11 in. (28 cm) by 3 in. (8 cm) had been removed. This requires far greater strength than can be achieved by a man using both hands. Some trees appeared to have been attacked on more than one occasion, the main objective apparently being the inner cambium layer, although small pieces of bark may also have been eaten. From the fact that a number of different sized incisor marks were observed, it was evident that the porcupines had been breeding in the wild. About 15 per cent of this plantation was damaged to a greater or lesser extent, some 54 per cent of the trees attacked being less than a quarter barked and only 8 per cent ring-barked. Other food consumed by porcupines at Risdon includes an exposed silver-birch root, the rhizomes of the fern *Polystichum setiferum* and those of the male fern (*Dryopteris felix-mas*), the large tuberous roots of black bryony (*Tamus communis*), and possibly the roots of the primrose (*Primula vulgaris*) and the hogweed (*Heracleum sphondylium*).

In Abbeyford Wood a small number of spruce were found to be damaged; these, however, had mostly been attacked on their surface roots, possibly because the trees were much older and larger than those at Risdon. In a mixed plantation of sitka spruce and Scots pine at Hayes Barton, five spruce trees had been damaged, one of which had been completely ring barked. Elsewhere, porcupines are reported to have fed on the bark of Japanese and common larch and silver fir, and on potatoes and swedes.*

---

*In India and Kenya porcupines cause considerable damage to potato crops: in India they also consume the cambium tissue of tree bark, sweet potatoes (yams), and other fleshy roots. In Malaysia the bark of oil palms is often eaten.

The present population of Hodgson's porcupines at large in Devon is unknown; it has been calculated by projection that by late 1974 the number could theoretically have been about a dozen or more (given that in England the animal has no natural predators), although in fact it is probably considerably less.

Evidence from this country and from abroad indicates that, even in small numbers, porcupines can cause considerable damage to conifers and root crops. (They have such immensely strong incisors and jaws that in Africa and India they are reported to be able to gnaw the tusks of dead elephants.) It is clear that if the population here were to increase in number and range and to become firmly established, porcupines could cause even greater damage than that done by coypus. In England they also pose a distinct threat to badgers, whom they have ejected from age-old setts.

As a result the Ministry of Agriculture started a campaign in June 1973 to try to live-trap the Devon porcupines, with the object of transferring them for study to the Coypu Research Laboratory at Norwich. Up to two dozen badger- and coypu-traps were set after a period of pre-baiting, but it was not until January 1974 that the first success was achieved, when an adult female weighing over 24 lb (11 kg) was captured: on 27 March following an adult 22 lb (10 kg) male was caught. In the late summer of 1974 after an assumed gestation period of ninety-one days (calculated by the elapse of time between observed copulation and birth) a male Hodgson's porcupine was born in Norwich: unfortunately, after a few weeks the mother accidentally killed it: in September 1975 a second young porcupine was born, which in February 1976 was reported to be thriving. In captivity the favourite food of Hogdson's porcupine appears to be sugar-beet, parsnips, apples, oranges and carrots. This suggests that the animal could prove a serious threat to beet crops if it ever became established ferally in districts, e.g. East Anglia, where this root is grown extensively.

Between 19 and 23 August 1974 a further intensive examination was made of the Okehampton area, where Mr Davey and other Ministry officials were actively engaged in surveying and trapping. Although there was no recent evidence of damage by porcupines, this was not considered surprising, as most tree-barking occurs in the spring. A number of areas of damage were noticed which had not been observed before, and the ravages at Risdon were much worse than in 1973. The Forestry Commission reported possible fresh barking since the capture of the male in the preceding March, with apparent incisor marks of two different sizes. Although unconfirmed sightings were reported at some distance from the main area of infestation, it was not until March 1975 that a feral porcupine was seen again in the Okehampton district: a further unconfirmed sighting was reported in the following month.

In the summer of 1972 a pair of crested porcupines – which had been bought as a breeding pair – escaped from the botanical gardens at Alton Towers in Staffordshire, some 12 miles (19 km) east of Stoke-on-Trent, in the valley of the river Churnet – incidentally, only about 16 miles (26 km) south of the Peak District wallabies.*

On 6 and 7 June 1974 Dr Gosling visited Cote Farm on the Farley Hall estate about 1 mile (1·6 km) north-west of Alton Towers, where the escaped porcupines were reported regularly to have taken stored grain. One of these animals had previously been caught at the farm and returned to Alton Towers, from which it promptly re-escaped. The hilly terrain on the Farley estate is plentifully supplied with conifer plantations intersected by minor tributaries of the Churnet, thus providing an ideal habitat for the porcupines, somewhat similar to that colonized in Devon. The range of reported sightings in Staffordshire appears to cover an area of about 4 square miles between Ramshorn, Farley and Alton Towers, but does not extend south of the Churnet. At least one plantation of larch was found to have been attacked by porcupines. Here, although the damage resembled that caused by the Hodgson's porcupines in Devon, the crested porcupines had barked mature trees with a diameter of 8–10 in. (20–25 cm), instead of the younger trees damaged in Devon, thereby directly affecting the final crop. After a more extensive survey in the spring of 1975, more damaged larches were discovered, but no other varieties of trees had been attacked. The crested porcupine is a heavier and generally larger animal than the Himalayan variety, reaching up to 45 lb (20 kg) in weight and measuring some 3 ft (91 cm) in length. Thus, although the damage caused by both species is similar in appearance, that caused by the

*See pages 42–4.

crested porcupine is greater in extent, and barking sometimes reaches a height of 18 in. (45 cm) up the trunk. Of approximately thirty damaged trees in an area of some 2¼ acres (about 1 hectare), two were completely ring barked and several others more than 50 per cent barked. As in Devon, a number of trees had been damaged more than once, and the method of attack appeared to be the same although more extensive areas of bark had been removed.

Between 5 and 7 August 1974 Dr Gosling re-visited the Alton Towers/Farley Hall area, where he found that about 4 per cent of the trees examined had been damaged. It was also clear that porcupines were attacking the most economically valuable trees in the district, thereby proving themselves a serious threat to forestry interests.

In January 1974 a porcupine was seen on an exposed grassland ridge on the north-western edge of Wilsford and Charlton Downs in the north of Salisbury Plain, east-south-east of the village of Market Lavington in Wiltshire. The same or a second animal was seen in this area two nights later. On the night of 23 October following, a motorist reported having seen a porcupine crossing a road between Alton Priors and Wilcot, some 8 miles (13 km) north-east of Market Lavington.

On 17 and 18 March 1975 Dr Gosling visited the district of these three sightings, which covers about 7 square miles: he found no suitable porcupine habitat within 2 or 3 miles (3–5 km) of Market Lavington, and concluded that the animal was travelling through the area when it was observed. A search of Tawsmead Copse, consisting of mixed spruce and larch and situated about 150 yards (140 m) south of the Alton Priors/Wilcot Road, yielded no evidence of porcupines. Nor was there any evidence of damage by porcupines in the nearby *Cupressus* sp. plantation of Blackball Firs; however, a number of small conifer plantations in the neighbourhood planted with trees of different ages and interspersed with pools of standing water, provide a near-ideal habitat which could well support a porcupine population.

In 1971 the Animals Act (Eliz. II. cap xxii) came into force, the effective result of which has been summarized thus: 'Anyone who, for his own purposes, keeps on his land anything likely to do mischief if it escapes, is answerable for all the damage which is the natural consequence of its escape. The liability does not depend on negligence: all the victim has to prove is that the mischievous thing was kept there, escaped, caused him damage, and it avails the defendant nothing to show that he took all possible care or even that he did not know the thing was dangerous.'*

The appearance on the statute book of this Act came just after the escape of porcupines in Devon, and just before that of those in Staffordshire: these were the first alien animals to become established in the wild in Britain after the Act became law. *Pace* those misguided people who welcome the naturalization in Britain of any exotic animal – however harmful it may be to the environment and the native fauna and flora – it might prove a salutary example if action were taken under this Act against those who keep in captivity, but cannot control, potentially dangerous pests. Porcupines, like a number of other equally harmful animals, are engaging and endearing creatures, but is is all too clear what havoc they could wreak if they were allowed to become established in the countryside.

Should attempts at live-trapping porcupines fail, it might conceivably be necessary to resort to killing them, however repugnant this might be. In India porcupines are controlled by poisoning, by fencing and by gassing in their burrows with Cyanogas: in Malaysia they are shot and trapped, and are prevented from attacking oil-palms by painting the trunks with chemical repellents. Of these methods, gassing is considered to be the most suitable in England: the local danger to badgers by gassing† should not be great, because most have already been evicted by the porcupines; any local risk involved would be offset by the prevention of a more widespread reduction in their numbers, which might well occur if the population

*P. Sieghart, 'Biohazards and the Law . . .', *Nature*, 251, p. 182, 1974. *The Animals Act 1971, and Common Law* (*Rylands* v. *Fletcher*) Eliz: ch. 22, pp. 1–7.

†A Government amendment to a Private Members' Bill, introduced by Mr Peter Hardy, M.P., permits the Ministry of Agriculture and licensed individuals to gas badgers in those areas, *e.g.* parts of south-western England, where they are suspected carriers of bovine tuberculosis. An article entitled 'Tuberculosis in wild badgers in Gloucestershire' in the *Veterinary Record* (14 December, 1974), concludes: 'However, badger infection is probably concentrated in isolated pockets of the population, and any indiscriminate slaughter of the species in the hope of removing the reservoir of infection would be quite unjustified. Further work is needed to determine the most reliable means of identifying infected badger communities.' See also *Oryx* 13(2), p. 114, 1975: 13(3), pp. 240–2, 1976.

of porcupines were permitted to increase. The presence in the countryside of free-living feral porcupines is a typical example of the lack of foresight in permitting incompetent people to keep in captivity such potentially harmful, if charming, animals.

14. Polecat-Ferret
(Mustela putorius furo)

The polecat-ferret is, as its name implies, a cross between the wild (European) polecat (*Mustela putorius*) and the domestic ferret (*Putorius furo*). There has been – and indeed still is – considerable controversy among systematists concerning the origin of the domestic ferret: all agree that it is a domesticated albino descendant of the polecat, but whether its ancestor is the European polecat or the Asiatic species *M. eversmanni* remains uncertain.

The European polecat is today our rarest carnivore, being confined almost exclusively to the woods and hills of Wales and parts of the extreme west of England. Its long fur, which is blackish with an almost purple sheen apart from the whitish cheeks, is changed annually in early May. The male is about 30 in. (76 cm) long, including an 8 in. (20 cm) tail, and weighs about $2\frac{3}{4}$ lb (1200 g): the female is about 20 in. (50 cm) in length plus a tail measuring about 6 in. (15 cm), and weighs approximately $1\frac{3}{4}$ lb (800 g). Four or five young are born in April or May after a gestation period of some 40 days. The polecat's foetid smell, which comes from its anal glands, has given it its scientific name (derived from the Latin *puteo*, meaning 'to stink'), and the English alternatives, foumart ('foul marten'), fitchet and fitchew.

The ferret, which is yellowish-white in colour with the pink eye of the true albino, is a smaller animal than the polecat, measuring about 14 in. (35 cm) long with a 5 in. (13 cm) tail. The albino colour has been deliberately achieved by selective breeding in order that the animals may be more easily seen in twilight, and because it has been found that albino wild animals are more easily trained and domesticated than those with a natural coloration. The male (buck, dog or hob) is marginally larger than the female (doe, bitch or jill). These fierce and vicious little carnivores are more prolific than polecats, producing between six and nine youug in two litters annually.

I

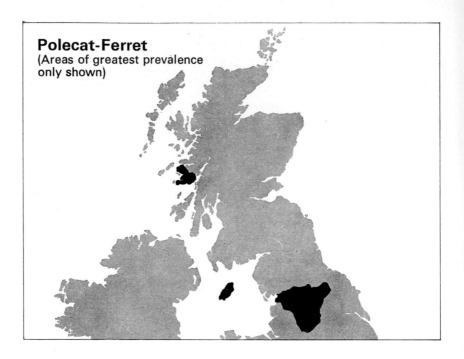

Polecat-Ferret
(Areas of greatest prevalence
only shown)

The ferret was well known in domestication to the ancient Greeks and Romans: the Greek historian Strabo (born *c.* 63 B.C.) states in his *Geography* that it was imported to Spain from North Africa, and the Roman, Pliny (*c.* 23–79 A.D.), describes in his *Naturalis Historia* how, under the name *viverra*, it was used for hunting rabbits. (The English name ferret is not derived from *viverra*, as might perhaps be expected, but from the French *furet*, which is itself derived from the Latin *fur*, meaning a thief or robber.) The exact date of the introduction of the ferret to Britain is uncertain, but it has been known in England since the late thirteenth century: it is therefore possible that it was first brought here during the reign of one of the middle or later Plantagenet kings, to help in providing sport and to control the rabbits introduced by his forebears some one hundred years earlier. In 1272 the capture of rabbits with ferrets at Waleton is mentioned by Thorold Rogers in his *History of Agriculture and Prices in England*, and ten years later Richard le Forester was paid 3s. 6d. for catching rabbits and keeping ferrets for the King at Rhuddlan Castle in Flintshire.*

*See page 71.

The polecat-ferrets of Britain are the result either of escaped albino ferrets breeding with wild polecats, or are escaped dark or particoloured animals which man has produced deliberately by crossing the domestic albino ferret with the wild polecat. Polecat-ferrets resemble the polecat in colour but are usually rather smaller in size. This constant interbreeding means that there may well be no fullblooded wild polecats left in Britain today; it also helps to explain why it is frequently almost impossible to differentiate between a true polecat and a polecat-ferret.

On the island of Mull, where the polecat has never been indigenous, ferrets and polecats were both introduced as domestic animals in about 1933 or 1934, and soon escaped. They bred freely in the wild, and before long were firmly established as a pest throughout the island, living on rabbits, ground-nesting wild birds and domestic fowls. In 1944 a trapper at Oskamull near Ulva Ferry killed twenty-two large feral polecat-ferrets with thick shaggy coats, and a number of others were caught in various parts of the island: it is probable that some still remain on Mull today. Seton Gordon recalls that polecat-ferrets were set free on the island of Harris in the Outer Hebrides to control the rabbit population: they were also, in Mr Gordon's opinion, responsible for exterminating the ptarmigan (*Lagopus mutus*) on the hill of Clisham – its last haunt in the Outer Isles.

'Polecats' – almost invariably in fact escaped or feral polecat-ferrets – are constantly being reported from many parts of Britain, (particularly from northern England) including country districts in close proximity to urban areas.

15. North American Mink
(Mustela vison)

The North American mink ranges over the greater part of that continent except the south-western United States and the extreme Arctic north.

The European representative of the group (*Mustela lutreola*), sometimes known on the continent as the nertz or sumpf-otter, is slightly smaller than but otherwise similar to the American species. Its habitat is France and eastern Europe, where it is found in Poland, Finland and most of Russia west of the Urals.

The mink is distinguished from other members of the genus by its semi-aquatic habits, and by the partial webbing of the toes which are also peculiar in possessing no long hairs between their naked pads. Like the martens but unlike the weasels the mink has a long and rather bushy tail, which is equal to half the length of the head and body. The ears are small and hardly appear above the general level of the fur. The coat consists of a dense, soft and matted underfur mixed with long, stiff and glossy hairs. In colour the mink varies from light, dull yellowish-brown to deep chocolate-brown, the usual tint being rich dark-brown, very little paler below than above. The tail is always decidedly blackish, and very occasionally may be tipped with white. The chin is invariably white, although the extent of the white area is subject to much individual variation, and there may also be small irregular-shaped patches of white on the under-parts. A fully grown mink measures from 15 to 18 in. (38 to 45 cm) in length with a tail about another 8 to 9 in. (20 to 23 cm) long. A large ranch mink may weigh from 6 to 7 lb (about 3 kg): a wild mink will weigh about half this amount.

In its habits the mink is amphibious, and it is therefore only found where water is abundant. Its food consists of various aquatic creatures, such as waterbirds; frogs; fish (including salmon and trout, but mainly cyprinids such as minnows, gudgeon, roach,

North American Mink
(Principal areas only shown)

rudd, carp, barbel, tench, bream, and chub, and percids, *e.g.* the perch); fresh-water molluscs (especially crayfish; some land birds (including domestic poultry); brown rats; mice (especially the wood mouse); voles, (especially the field vole and the water vole); rabbits; hares; eggs and insects. In America, where it is regarded as a voracious predator second only to the aptly named glutton or wolverine (*Gulo luscus*), the mink also preys, especially in winter, on the muskrat (*Ondatra zibethicus*). Mink are good climbers and have been seen in trees up to 50 ft (15 m) from the ground: they frequently pursue their prey entirely by scent, and hunt both by night and day.

The nest of the mink is situated either in a hole in the river or lake bank, or in a hollow log, and is usually well lined with feathers. In both America and Britain the young, numbering from four to six in a litter, are usually born early in May and stay with the female until the following autumn. The scent produced by the anal glands, which is characteristic of all members of the weasel family, is highly developed in the mink, and no other animal, with the exceptions of the skunk (*Mephitis mephitis*), the polecat (*Mustela putorius*) and the ferret (*Putorius furo*) possesses such a powerful smell.

The American mink, which on the continent has been bred commercially in Scandinavia, Finland, Germany and Russia, was first imported to fur-farms in Britain in 1929, and escapes were soon being reported from places as far apart as Morecambe Bay, Lancashire, Stratford-on-Avon, Warwickshire and Pawston Hill, Berwickshire. In the 1950s the industry began to expand considerably, and large numbers of American mink (as many as 700 in a single shipment) were imported from America and Scandinavia.

The first unconfirmed rumours of escaped mink breeding in the wild occurred shortly after the war; they came from Singleton near Blackpool in Lancashire, where there were a number of large fur-farms. One farm specialized in selling pairs of mink to prospective breeders, and many were sold to private householders by whom they were inadequately housed in boxes and cages which proved far from escape-proof. As a result, some did escape and caused considerable depredations among local poultry farmers; they did not, however, become established.

In 1953 a water-bailiff reported that a stretch of the river Teign in the area of the Teign Valley around Chudleigh and Bovey Tracey in Devon was devoid of rats because of a local colony of mink. These were thought to be descendants of the survivors of about a dozen mink which had escaped from a farm at Bovey Tracey, which closed down at about that time. In the same year a male was shot on the Teign neat Moretonhampstead, having probably followed the river up from Chudleigh and Bovey Tracey. In 1956 a female with some young was seen playing in an old mill leat in the same area; two others were shot after some ducks had been killed, also near Moretonhampstead, while yet another was killed in the Teign. There were several reports by fishermen of further mink seen at large in the Teign as far up river as Chagford, and in November 1957 Dr Ian Linn of the University of Exeter announced at a meeting of the Mammal Society that they were breeding. Feral mink were also reported at about this time from the river Avon in Hampshire, probably having escaped from a farm in that district, and at Teignmouth. In April 1958 a pregnant female containing six young was trapped on the river Deben near Woodbridge in east Suffolk, and there were reports of feral mink from Yorkshire. Fitter records that Professor Wynne-Edwards had two mink – from different areas of Scotland well away from mink farms – brought to him for identification in Aberdeen.

In the late 1950s and early 1960s wild mink began to extend their range considerably, and were reported from South Brent on the Devon Avon; from the Cilgerran Gorge–Newcastle Emlyn stretch of the river Teifi on the Carmarthenshire/Cardiganshire border; from the Fishguard/Letterston area of Pembrokeshire in south-west Wales; and in 1961 from Throof, Bisterne, Somerley, Kingston, Fordingbridge and Bodenham on the Hampshire and Wiltshire Avon and the rivers Itchen and Test. Reports of feral mink were also received from the rivers East Cleddau and Taf on the Pembroke-shire/Carmarthenshire border; from the Stour in Dorset, the Avon in Gloucestershire, the Wharfe and Lune in Lancashire and York-shire, and from the Dee and Tay in Aberdeenshire and Perthshire. In the middle and late 1960s mink increased in numbers in Devon, Lancashire, Wiltshire, Hampshire, Dorset and west Wales, and first occurred in Sussex on the Cuckmere and Ouse (where 24 were killed in 1964–5, and 30 in 1965–6), and in Somerset, Kent, Northampton-shire and Northumberland. In Buckinghamshire two mink were trapped on Wilstone reservoir near Tring in 1962; one was killed at Wigginton, also near Tring, in 1965, and in the same year another was shot on the river Chess near Latimer: in 1966 a single mink was seen on the Thames at Medmenham and another killed a tufted duck on Weston Turville reservoir in January 1967. By the late 1960s mink were being reported from the rivers Taw, Dart, Exe, Axe, Coly, Culm, Tavy, Mole, Torridge, East and West Webburn and Yeo, and from much of Dartmoor. In 1967 they were observed on the rivers Tiddy, Hayle and Tressillian (at St Clement) in Cornwall, and one was found dead on Goss Moor. In September 1968 a colony killed seventy-four ducks on the banks of the river Looe.* In 1975 a mink was seen pursuing an adult moorhen which it captured and killed, on Marazion Marsh near Penzance.

By 1963 only Aberdeenshire, Perthshire and Kirkcudbrightshire in Scotland reported mink present in any numbers; since then they have occurred throughout Scotland, especially in Ayrshire, Stirlingshire, Banffshire, Mid-Lothian, Roxburghshire, Peeblesshire, Lanarkshire, Kincardineshire, Fife, Selkirkshire, Dumfriesshire and Morayshire.

In Ireland the first fur-farms were set up between 1950 and 1953, and by 1960 some forty breeders were established. On 18 September 1965

*F. A. Turk, 'Notes on Cornish Mammals', New Series, Bull. Corn. Nat. Trust, 1968–9.

an order was brought in by the Irish Government forbidding the keeping of mink without a licence; this prohibition was extended to Ulster on 16 January 1968. Most of the mink imported to Ireland were from Scandinavia, supplemented by stock from North America and Britain.

The following table, adapted from Deane and O'Gorman, gives particulars of early reports of feral mink in Ireland.

Date	Location	Remarks
1961	Omagh, (near) Co. Tyrone.	30 escaped: some colonized river Strule: by 1967 had spread to R. Mourne and Glenelly: by 1968 to Victoria Bridge, 15 miles (24 km) west of Omagh.
1962	Dunsinea, Co. Dublin	1 killed in spring.
5 December 1963	Mount Merrion, Co. Dublin.	1 killed
1964	R. Dodder, near Tallaght Co. Dublin.	2 seen.
4 January 1965	Tynan Abbey, R. Tynan, Co. Armagh.	1 male shot by Sir Norman Stronge, Bt (Ulster Museum).
24 September 1965	R. Connswater (Newtownards Road, East Belfast), Co. Down.	1 nearly black mink killed (Ulster Museum).
1965	Grand Canal, near Stradbally, Co. Leix	1 seen in autumn.
July 1966	Carrickfergus, Co. Antrim.	1 seen.
14 August 1966	Deramore Estate, south Belfast, Co. Down.	1 female captured and escaped on lake.
1966	Ballycurry, Co. Wicklow.	Several escaped from a farm.
4 May 1967	R. Comeragh, above Drumkeena Bridge, Co. Kerry.	1 seen.
1967	Castledermot, Co. Kildare.	2 escaped from a farm.
April 1968	R. Margie, Ballycastle, Co. Antrim.	1 shot.
19 May 1968	Lissanoure, Flushwater, Co. Antrim.	1 trapped.

Date	Location	Remarks
May 1968	Ashford, Co. Wicklow.	1 seen.
June 1968	R. Finn (between Stone-bridge and Rosslea) Co. Fermanagh.	1 seen.
18 September 1968	Brookeborough, Co. Antrim.	1 killed on road.
September 1968	Dublin–Navan road.	1 seen.
	R. Boyne, Oldbridge Co. Meath.	?
11 October 1968	Finnebruge, R. Quoile, Co. Down.	1 male killed on road (Ulster Museum).
10 December 1968	Glanleam, Valencia Island, Co. Kerry.	1 shot.
3 January 1969	Dunshaughlin (3 miles/ 5 km.)	1 black female shot.
February 1969	Rathangan, Co. Kildare.	1 shot.

Today feral mink in Ireland are breeding in some numbers in Co. Tyrone and have spread throughout the watershed of the rivers Dodder and Liffey, Co. Dublin.

Feral mink are present in significant numbers in England today on the rivers Exe and Teign (Devon); Avon (Hampshire and Wiltshire); Wylye and Beaulieu (Wiltshire); Stour (Dorset); Wharfe and Lune (Yorkshire and Lancashire); Wyre (Lancashire); and the Cuckmere and Ouse (Sussex). In Wales they are found on the Teifi (Carmarthenshire and Cardiganshire), and on the Cych, Ceri, Dulas, and East and West Cleddau and Taf (Pembrokeshire and Carmarthenshire). In Scotland mink occur on the Tweed, Gala Water, Teviot and Kale Water (Roxburghshire, Selkirkshire and Berwickshire); Forth and Teith (Perthshire and Stirlingshire); at Boquhan (Forth), Blane, Earlsburn reservoir, Balfron (Selkirkshire and Stirlingshire); on the Tay, Almond and Earn (Perthshire); Dee, Don, Deveron and north and south Ugie and Cruden Water (Banffshire and Aberdeenshire); Spey (Banffshire and Morayshire); Doon (Ayrshire); Urr Water, Dalbeattie (Kirkcudbrightshire); and at Dalkeith, Gorebridge and Penicuick (Midlothian).

Between 1953 and 1967 a total of 2888 feral mink were caught in the wild in England, 195 in Wales and 1256 in Scotland – a total of 4339 animals.

On the continent feral mink are well established in Norway, Sweden, Denmark and Finland, and in Iceland where they have caused considerable damage to domestic poultry, game-birds, fisheries and wild water-birds; at Lake Thingvalla in Iceland, for example, they have virtually exterminated the waterfowl population – both ducks and waders. Some 18000–20000 wild mink are destroyed annually in Sweden and about 14000 in Norway, which gives some indication of the problems which could arise in Britain if such ruthless killers are not strictly controlled.

To achieve this end the Government in 1962 took action through the Mink (Importation and Keeping) Order, made under the Destructive Imported Animals Act (1932), and issued a series of regulations for prospective mink-breeders. The Ministry of Agriculture intensified its campaign – which began to be phased out in 1970 – to limit the number of mink in those areas where they were already well established, and to attempt to prevent their further encroachment into previously uncolonized districts. As a result, the total population of mink in Britain is probably fairly stable, although there continues to be some increase of range.

Mr Harry V. Thompson of the Ministry's Pest Infestation Control Laboratory Field Research Station is of the opinion that mink will neither be exterminated in Britain (as was the muskrat in 1934), nor that they will ever be materially reduced in their present range and confined, as is the more sedentary coypu in East Anglia, to a restricted area. In North America adult mink regularly migrate over considerable distances in winter and spring during the mating season, while the young disperse in the autumn, and there seems no reason why they should not do so here: mink are also adaptable to different habitats and climatic conditions, and are less easy to see and trap than are coypus.

Escaped ranch mink may survive well in the wild, as they are mostly a cross between the large and fertile 'Alaska' strain and the superior-pelted 'Quebec' strain. Ranch mink come in a variety of colour mutants – beige, blue, cream and white for example – and in the trade are known variously as Aleutian, Palomino, Pastel, Pearl, Platinum and Topaz. The offspring of feral mink usually revert to the standard dark-brown coloration. When mink first escaped from captivity it was hoped that they would never become sufficiently numerous to breed in the wild; the species was of doubtful fecundity, as the female only ovulates after sexual stimulation and is receptive to the male for a period of a few days only from the end of February to the beginning of April. This theory was, however, soon disproved, and the animal's fertility confirmed when up to six young were produced per litter.

It is tempting to regard the mink – the only alien non-domesticated carnivore to succeed in firmly establishing itself in Britain in historic times – as an interesting addition to the British fauna. The mink does, however, present a very real threat to many species of native wildlife. It has, for instance, been suggested that there is some correlation between the presence of mink and absence of otters in various parts of the country through a competition for food. In the 1967 hunting season the first mink to be caught by otterhounds was killed near Kendal in Westmorland, and many more have been killed since then by packs in other counties. Reports recently received from Sweden suggest that otters will not breed where mink abound. It has already been shown what harm can be done to native birdlife, especially waterfowl such as moorhens, coots, little grebes and ducks, which have largely disappeared from some mink-infested areas, and it is known that the presence of mink can have an inhibiting effect on migrant waterfowl.

Recent investigations carried out by the Universities of Edinburgh and Exeter into the diet of feral mink in their areas, have confirmed that one of the principal reasons for the success of the mink in becoming established in Britain is that it has filled a broadly-based feeding niche previously unoccupied by any other carnivore. Because of the comprehensive nature of the mink's diet, with its almost unique combination of aquatic and terrestrial prey, the animal is not a serious compctititor for food with any other carnivore, and is more difficult to catch than an animal with a less catholic menu. Together with the coypu and grey squirrel, the mink provides concrete evidence of the danger in bringing to this country an animal which is so well adapted to life in Britain and is, moreover, so potentially dangerous to our native wildlife.

16. Domestic Cat

(Felis catus)

The domestic cat is almost certainly not descended directly from the European wild cat (*Felis silvestris*), but from a North African species, *F. libyca*, which is, however, usually now considered to be a race of *silvestris*.

The wide range of colour, size and weight to which the domestic cat is subject is well known. The European wild cat appears to breed only once a year: the Scottish subspecies *F. s. grampia* normally produces two litters of kittens a year – in May and August – with occasionally a third litter in December or January. The domestic cat usually breeds twice a year, although sometimes – especially in the wild – this number is increased to three or even four litters annually. The multiple breeding season in the Scottish wild cat has been taken to show the close relationship it bears to the domestic cat, with which it interbreeds freely in the Highlands. By nature the cat is a purely carnivorous and non-gregarious animal, preferring – except during the breeding season – to live and hunt alone.

According to Mivart, 'there can be no question as to the Cat having been domesticated in Europe before the Christian era. There are signs that it was domesticated among the people of the Bronze Age period'. The animal was first introduced to Britain at some time prior to the Middle Ages, originally to help in the control of vermin and as a companion to man. The earliest remains of the domestic – as opposed to the wild – cat were discovered in a midden in the Roman city of Silchester in Hampshire.* The first known written evidence of the domestic cat in Britain is found in a code of laws which has been attributed to the tenth-century Welsh prince, Hywel the Good:

Whoever shall kill a cat which guards a barn of a King or shall take it

*W. H. St John Hope, and G. E. Fox, 'Excavations on the site of the Roman City of Silchester, Hampshire', *Archaeologia*, I, pp. 87–112, 1899; 60, pp. 149–68, 1905.

stealthily, its head is to be held downwards on a clean level floor, and its tail is to be held upwards; and after that, wheat is to be poured about it until the tip of its tail is hidden [and that is its worth]. Another cat is four legal pence in value [the same amount as for a sheep].*

Over the years many domestic cats have been deliberately turned out by their owners, who no longer wished or could afford to keep them, and others have themselves chosen to lead a feral rather than a domesticated existence. Writing in 1898 that great naturalist, W. H. Hudson, believed that of an estimated total of some 400000 domestic cats in London, no fewer than 80000 to 100000 – or approximately one quarter – were leading a feral existence. In 1944 Colin Matheson estimated that there were about 6600 feral cats in Cardiff, out of a total population of some 23500.

There have been a number of instances of deliberate introductions of cats to various parts of the British Isles. About a dozen were set free on St Kilda in 1930, but by the following year only three remmained. Two of these – both females – were shot, and as one proved to be pregnant the third must have been a male: in their reports on the

*A. A. Wade-Evans, *Welsh Medieval Law*, pp. 226–7, Oxford, 1909; cited by Matheson.

status of the St Kilda house mouse (*Mus musculus muralis*), T. H. Harrisson and J. A. Moy-Thomas* were of the opinion that feral domestic cats on the island must have been a contributory factor in reducing the population of this rodent.

In the 1890s cats were set free on the island of Noss in the Shetlands to control the population of rats: fresh blood was introduced from time to time, which may have helped in preventing the reversion to the true 'wild' tabby colour. Cats were also introduced at various times to the island of South Havra and to Holm of Melby in the Shetlands. According to Seton Gordon, domestic cats were once set free on an island in the Hebrides to control the rabbits which had themselves been introduced a few years previously. In Ireland some three dozen domestic cats were turned down on Great Saltee Island off Ballyteige Bay in Co. Wexford in 1950, again with the object of rabbit control.

Few urban areas are today without their stock of free-living commensal domestic cats. What is perhaps less well known is that many rural districts also support flourishing populations of feral cats which live on rats, mice, rabbits and small birds. In the wild the descendants of domestic cats often increase considerably in size, and frequently become as fierce as the true wild cat. As has already been mentioned, in the Highlands of Scotland escaped domestic cats frequently breed with the Scottish wild cat, which has led to a considerable variation in size and marking of the latter species.

*Nature, 129, p. 131, 1932, and *Jour. Anim. Ecol.*, 2, pp. 109–15, 1933.

17. Chinese or Reeves's Muntjac
(Muntiacus reevesi)

Indian Muntjac
(Muntiacus muntjak)

Two species of muntjac* – the Chinese or Reeves's and the Indian – have been introduced to England; both varieties have escaped from captivity and may have inter-bred in the wild.

The present-day home of the Chinese or Reeves's muntjac is the thickly forested hilly areas of southern China and the island of Taiwan: it is named after John Reeves (1774–1856), who in 1812 was appointed assistant – later rising to chief – inspector of tea for the East India Company in Canton. In 1817 Reeves was elected a Fellow of the Royal Society and of the Linnean Society.

This variety is very small, standing only some 16–18 in. (41 to 46 cm) high at the shoulder, and weighing barely 25 lb (11 kg). The colour of the coat is a rich chestnut-red speckled with grey in summer, changing to a deep brown in winter, and is darker above than beneath. The chin, throat and underside of the tail, which is some 6 in. (15 cm) long, are white. When the animal is alarmed the tail is erected and the white hairs fan out. The antlers, which are scarcely 2½–3 in. (6–8 cm) long, converge slightly and end in a single spike, with otherwise only a small brow tine: each antler grows from a long skin-covered pedicle – along the front of which is a line of dark hair – which extends as a ridge on each side of the face. In the doe the face line ends as a tuft of dark hair in the position of the pedicle of the buck. The facial depression formed by these ridges encloses a pair of glands, in the shape of two folds of skin which

*The name muntjac is derived from the Sunda Islands' dialect, meaning 'springing' or 'graceful'.

K

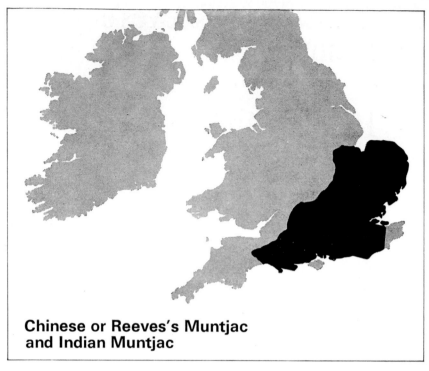

**Chinese or Reeves's Muntjac
and Indian Muntjac**

form a 'V': from this feature muntjac are sometimes known as 'rib-faced deer'. The upper canine teeth of the buck are enlarged and project from the mouth as small tusks, each about 1 in. (2·5 cm) long, which are used in fighting – especially dogs to which muntjac appear to possess a marked antipathy – in preference to the antlers. Muntjac are also sometimes referred to as 'barking deer', from the loud, hoarse and resonant bark which is uttered when the animal is alarmed and during the mating season. They also 'chatter' with rage by grinding their molars, and on occasion produce a strange 'clicking' sound of unknown origin.

The Indian race – which is said to reveal the presence of tigers by barking and is known in Hindustani, from its voice, as the *Kakar* – is a native of India, Ceylon, Burma and Thailand; it is slightly larger and heavier than its Chinese counterpart and has a rather darker pelage; its antlers measure between $3\frac{1}{2}$ and 5 in. (9 and 13 cm) in length. *M. reevesi* × *M. muntjak* is mid-way in size and weight between the two races, and has a slightly redder coat than the former variety.

Muntjac are a very ancient breed which was established in Europe

in prehistoric times: they are not gregarious, and are normally found either singly on in pairs. They are essentialy a forest-dwelling species, and being extremely shy and secretive like plenty of thick cover such as bramble, gorse, and rhododendron bushes, through which they are adept at making their way at high speed: they run in a curious and characteristic 'hunched' position, with the head held low and the hindquarters elevated. The principal food of muntjac consists of brambles and grass, to which in autumn are added acorns, chestnuts and apples. There is no special breeding season, and the spotted fawns may be seen at all times of the year.

Around the turn of the century the 11th Duke of Bedford introduced some Indian muntjac to his collection at Woburn, Bedfordshire,* where for a number of years they thrived both within the confines of the park and in the surrounding woodland. They were later replaced by the rarer and less aggressive Chinese or Reeves's muntjac, which also flourished both inside and outside the park. Both races may have interbred in the wild, and during the past seventy-five years have extended their range considerably beyond Woburn; they are now

*It has been suggested that the Duke may have brought the original stock from Rambouillet, south-west of Paris, when he was there in 1872. He also introduced some muntjac to his estate at Cairnsmore, Kirkcudbrightshire, but so far as is known none escaped.

For much of the history of deer in Britain I am indebted to G. K. Whitehead's monograph (see Bibliography).

firmly established as breeding animals in many areas of central England.

The Woburn stock was supplemented in the wild by escapees from Whipsnade, which between 1929 and 1931 was presented with a total of eighteen Indian muntjac by the Duke of Bedford, and during the 1930s by escapees from the collection of Major A. Pam at Wormley Bury, Broxbourne, Hertfordshire.

In Bedfordshire the first feral muntjac was observed at Wrest Park, some 7 miles (11 km) east of Woburn, in 1922; another was seen a few years later in Ashridge Park, Hertfordshire, 12 miles (19 km) south of Woburn. Around Brackley and in Salcey Forest and Yardley Chase on the Buckinghamshire/Northamptonshire border muntjac are reported to have been at large since 1933–4: the woods around Rockingham, north of Kettering, were colonized before the last war, presumably from this stock. They were well established in Ampthill Forest, Bedfordshire, and in the woods around Luton at the time of the 1938 deer census. In about 1937 a muntjac was reported at Kelmarsh near Kettering, Northamptonshire, which may have escaped from the collection of Mr H. J. Stevens at Croyland Abbey, Wellingborough. The species had penetrated to eastern Warwickshire before 1941, when a number were seen in Walton Wood near Kineton. In February of that year a young buck was shot in Needwood Forest on Duchy of Lancaster property, south of Sudbury and Marchington, in Staffordshire. In 1934 muntjac were seen in Ryton Woods, 15 miles (24 km) south of Kineton, and in 1947 at Bannerhill Farm, 3 miles (5 km) north-west of Warwick and in Chesterton and Oakley Woods, north of Kineton. By 1951 the animal had colonized Wellsbourne Wood, some 3 miles (5 km) west of Kineton.

In 1940 or 1941 a deer which was probably a muntjac was shot at Parham, some 3 miles (5 km) south-east of Framlingham in east Suffolk, and in 1941 or 1942 two more were killed in East Anglia, one near Colchester and the other near Easthorpe about 6 miles (10 km) west of Colchester in Essex. When the 1948 deer census was undertaken, muntjac had extended their range northward and westward to the Forestry Commission property at Hazelborough, Northamptonshire, and one was reported from as far off as Matlock in Derbyshire. In May 1949 a muntjac was reported to have been seen near Bolderwood in the New Forest and others were reported in the previous year in Alice Holt Forest, Hampshire, although these reports remained unconfirmed.

By the early 1950s muntjac were well established in the woods around Whittlebury and Silverstone between Northampton and Buckingham, and were reported from the Valley of the Ouse (east of Northampton), Overstone (5 miles/8 km north-east of North-ampton), Knightley Wood near Daventry, and from Kelmarsh and Weekley near Kettering.

In 1952 the slots* of muntjac were first observed in the Forest of Arden in Warwickshire. In the same year came the first reports of muntjac in Oxfordshire, where they were noticed in woods at Stoke Lyne near Bicester.

In the following year three muntjac were reported in the Breckland forests of Suffolk, and a further pair were seen at Leiston in the east of the county near Aldeburgh. Also in 1953 muntjac tracks were first observed in Charnwood Forest, Leicestershire, and in April one was shot at Bayford near Hertford.

In 1954 there were two further reports of muntjac in Hertfordshire – one from Hoddesdon and the other from Leavesden – and a third was seen near Warwick. On 13 February 1954, a muntjac was shot near Kilworth Hall, half-way between Rugby and Market Har-borough on the Leicestershire/Northamptonshire border: this animal may well have been the one which escaped in December 1953 from Mr Stevens' collection at Croyland Abbey, a little over 20 miles (32 km) south-east of Kilworth.

In 1955 a muntjac was seen on the southern outskirts of Birming-ham, and in May of the year following a buck was shot at Bear Place, Twyford, Berkshire, home of Lord Remnant, 5 miles (8 km) east of Reading.

The slots of muntjac were first noticed in Gloucestershire in November 1958, when they were reported from the northern edge of the 400 acres (161 hectares) of Withington Woods – well known for its fallow deer heads – some 9 miles (14 km) south-east of Cheltenham. On 30 January 1960, a muntjac buck was shot by Air Commodore White at Fairford, approximately 10 miles (16 km) south-east of Withington: on 16 November of the same year a second buck was caught in a snare at Stanway near Snowshill some 22 miles

*Footprints.

(35 km) due north of Fairford: on the 28th of the following month a pair of muntjac were seen in a field of kale at Stanway, and on New Year's Eve another was seen during a shoot at Dumbleton, 6 miles (10 km) to the west. In parts of Huntingdonshire feral muntjac appear to have been breeding freely since at least 1959.

Isolated early records of muntjac have come from Walsingham, Norfolk, where a Reeves's muntjac was shot on 4 October 1951: at Enfield, Middlesex (1958), where a buck was caught and later set free in Epping Forest in Essex: in Hatley Wood, Cambridgeshire: at Lullingstone Castle Park, 6 miles (10 km) south of Dartford in Kent: at Melbury Osmond, near Evershot, where a young buck was killed in the autumn of 1962, and at Llanvair Grange, Abergavenny, Monmouthshire, where a buck was discovered dead in the same year. Three more were seen in east Dorset in 1968. As early as 1963 muntjac were reported to be present in Tackley Wood, west of the river Cherwell in Oxfordshire.

Today feral muntjac are generally distributed in suitable woodland areas of central England as far west as the Welsh Marches, east to Norfolk and Suffolk, and north to Derbyshire. They are particularly numerous in Bedfordshire (especially near Ampthill and Luton), Buckinghamshire (principally in the triangle formed by Bletchley, Buckingham and Aylesbury), Hertfordshire (where they are particularly numerous around Hitchin, Ashridge, Hemel Hempstead, St Albans, Knebworth and Bishop's Stortford), Northamptonshire (in the woods around Whittlebury and Silverstone), Oxfordshire (where they have been established in the Chiltern Hills around Watlington since 1958), and to a lesser extent in Essex and southern Leicestershire.

Muntjac increased their range and population noticeably during the last war when, like so many other animals, they were less frequently persecuted by man, and in recent years there has been little diminution in the rate of expansion of both range and numbers.

The conspicuous success of the muntjac in becoming naturalized has been attributed to a number of causes: being unobtrusive, secretive, and causing only limited damage, it draws scant attention to its presence, and its antlers provide only a meagre trophy for the sportsman. Muntjac commenced their colonization at a time when large estates were in a state of decline, and when Forestry Commission planting was providing many suitable habitats. During the two world wars pest control officers and sportsmen were forbidden entry to ammunition dumps, training grounds and other restricted areas, where muntjac thrived. Their primitive antlers and tusks are features of a 'territorial' rather than a 'herd' animal, and the territorial organization which tends to banish young bucks has been another cause of the muntjac's successful expansion. Muntjac cast and re-grow their antlers at all seasons; in most deer the growth of new antlers is directly linked to the production of male sex-hormones, but muntjac bucks are potent throughout the year; this virility, coupled with their comparative freedom from internal parasites combines to ensure the health and continuation of the species.

Muntjac can cause a certain amount of damage to young hardwood plantations and to both flower and vegetable gardens, but otherwise appear to do little harm to man.

The following table, compiled from the distribution records sent by members* to the British Deer Society, gives particulars of the spread of muntjac deer between 1964 and 1975.

Year	County	District/Remarks
1964	Berkshire	A very small number in Windsor Forest and in Swinley Forest, Ascot: occasionally killed on roads.
1965	Middlesex	? Bushy Park (November): West Ham (November).
1966	Berkshire	Highclere (Boxing Day).
	Gloucestershire	Chedworth and Withington Woods Cheltenham (since 1963).

*Members who supplied information were: D. Holyoak, A. Hill, A. Rogers, I. Newton, Earl of Stradbroke, J. Robinson, D. and N. Chapman, J. E. Paton, E. E. Tupper (*via* Diana E. Brown), Jean Cobb, F. Marshall, F. J. T. Page, Mrs A. Hammond (*via* F. J. T. Page), P. H. Carne, I. C. N. Alcock, J. Goldsmith, Lieut.-Col. R. H. A. Cockburn, D. Ireland, Earl of Cranbrook, W. H. Payn, L. J. Petyt, K. MacArthur, Maj. K. C. G. Morrison, M. Savage, Beatrice Gillam (*via* R. Prior), Brig. H. V. Vaughan, L. MacNally, C. K. I. Littleboy, D. Hart-Davis, S. R. Worsfold, R. N. Sanders, R. Viccari, H. J. Horswell (*via* I. C. N. Alcock), Capt. R. Casement, M. T. Horwood, Margaret Ralph, Maj. MacEwan and Mrs June Atkinson (*via* R. Prior), Mrs S. J. Patrick, L. J. Warner, Dr A. McDiarmid, W. L. Baron, J. Essex Davies, M. Evans, S. Pouncey, A. Gresham Cooke, B. Reynard, D. Percy, F. Thurlow, R. J. King, Diana Brown, W. R. Wright, M. Hill, K. Howell, M. Clark, P. McManus, A. M. T. Davis, K. R. Duff, R. A. Harris, D. Talbot, D. J. Cross, R. Eastham, G. K. Whitehead, E. A. Ellis, Mrs S. Taylor, W. G. Teagle, D. Davis, W. Wilkes, P. Delap, E. S. Philip, I. R. Beanes, Mrs R. Clarke-Hall, D. Wood, J. Gassman, G. Darwall, J. Beckett.

Year	County	District/Remarks
	Hampshire	In area of Liss.
	Surrey	Buck captured on Barnes Common on 6 January; photographed and released in Richmond Park.
	Sussex	Borden Wood, Rogate: Harting Combe (March/April): ? Kingham Wood (near Borden): Watergate Hanger, Compton.
	Wiltshire	Warren Plantation, Ash Hill and Bewly Common, Sandy Lane: Green Lane Wood, Trowbridge: Chisbury Wood, Froxfield: Savernake.
1967	Buckinghamshire	Reported as resident in south Chiltern Hills – especially in Hambledon valley – for past 8–10 years.
	Devon	Widworthy, Honiton (March).
	Hampshire	Crondall (January): Liphook: ? Fernhurst.
	Middlesex	Slots seen at Enfield, Grange Park, Potters Bar and Northaw.
	Norfolk	Buck shot and doe seen in Wensum valley, 20 October.
	Sussex	Beacon Hill, South Harting: North Marden Down: Oakreeds Wood: a pair still in Borden Woods in August.
	Wiltshire	Bradon Forest, Chippenham.
1968	Berkshire	Stratfield Saye, Reading.
	Dorset	Studland (23 April): Rempston, Purbeck (25 April).
	Hampshire	Bramshill Forest.
	Suffolk	West Stow Forest, Bury St Edmunds.
	Surrey	Milford Common.
	Sussex	Iping, Midhurst (February and March).
	Wiltshire	Biss Wood, Trowbridge.
1969	Denbighshire	One reported at Pontfadog in the Ceiriog Valley.
	Dorset	Cranborne (January).
	Surrey	Putney: Hammersmith.
	Warwickshire	Buck captured at Clifton, Rugby (7 March).
1970	Berkshire	One seen and photographed in Windsor Forest (May/June).
	Monmouthshire	Llanfair.

Year	County	District/Remarks
	Oxfordshire	Wotton Underwood, Thame.
	Surrey	Three seen in Dunley Hill area, Dorking.
	Sussex	One ? buck seen (August) at Kilnwood, near Faygate, Horsham. Slots seen in Setley Common woods, Hoyle, Midhurst (April).
1971	Buckinghamshire	Cryers Hill, High Wycombe.
	Dorset	? Chetnole Withy Bed.
	Hertfordshire	Fairly widespread around Watford, Croxley Green and Rickmansworth, especially in Beechengrove, Lees, Whippendell and Chipperfield Woods, and on allotments and in gardens in Watford.
	Northamptonshire	Fermyn Woods and Wakerley Great Wood in Rockingham Forest.
	Oxfordshire	Nettlebed Woods: at least six in Hambledon valley near Henley: Luxters Farm and Bix, Stonor.
	Rutland	Slots seen in Rutland Wood: first recording for county and most northerly district to date.
	Suffolk	Ousden, Newmarket: Bradfield Great Wood, Bury St Edmunds.
	Surrey	One seen between Cobham and Byfleet on A3 road (December).
	Warwickshire	Bilton near Rugby.
	Wiltshire	Two seen in Maiden Bradley Wood: droppings seen 8 April in Potterne Woods, Devizes.
1972	Berkshire	Buck caught at Lockinge near Wantage. One buck seen at West Byfleet.
	Buckinghamshire	Buck and doe seen in Stowe Park (just prior to the Game Fair): also in Tingewick area.
	Cornwall	Slots, droppings and sighting of an animal at Mithian (7 miles/11 km. northwest of Truro) in February.*
	Derbyshire	A young buck visited a tennis-court on Normanton Recreation Ground on 28

*F. A. Turk, 'Notes on Cornish Mammals', New Series, *Bull. Corn. Nat. Trust*, No. 13, 1972.

Year	County	District/Remarks
		February: it was discovered dead in a nearby street on the following day.
	Devon	Bishopsteignton churchyard, Teignmouth, Christmas Eve.
	Dorset	Cerne Abbas and Evershot: four seen around Studland Bay.
	Herefordshire	A buck caught near Leominster was later released in Queen's Wood, Dinmore: a doe was reported from Wellington Woods.
	Surrey	One buck seen at West Byfleet.
	Sussex	Signs observed near Steyning: ? one seen on 17 October crossing a ride in Rewell Wood near Arundel. ? One seen on 6 October in woods near Rogate.
	Warwickshire	Buck seen on Brailes Hill, Shipston-on-Stour, 28 July.
1973	Hampshire	One seen and numerous slots observed in Hartley Wood, Hartley Wespall, Basingstoke.
	Norfolk	Drayton and ? Sparham, Norwich. Cranwich, Brandon. Great Hockham, Thetford. South Wootton and Hillingdon, King's Lynn.
	Oxfordshire	Fairly plentiful in Shabbington Wood and Bernwood Forest on border with Buckinghamshire: Wychwood Forest, Charlbury.
	Somerset	Wanstrow, Frome.
	Yorkshire.	One on beach at Flamborough, 14 January.
1974	Cumberland	One at Wedholm Flow, Lawrenceholm near Carlisle.
	Devon	Some (introduced from Woburn) escaped in Dawlish area in 1971; about six around Ashcombe; about four near Newton Abbot.
	Essex	Hatfield Forest; Epping Forest; Walden Forest; Weald Park; Honeywood Forest, Halstead.
	Hampshire	One in Micheldever Wood, Winchester, 2 February. One in Hammer Wood, between Petersfield and Midhurst.
	Norfolk	Swaffam end of Thetford Chase and

Year	County	District/Remarks
		especially in King's Forest between Thetford and Bury St Edmunds.
	Oxfordshire	Small number presumed to be in Wytham and Bagley Woods in west of county south of the Thames.
	Suffolk	Bradfield Woods; King's Forest; Mildenhall woods.

18. Fallow Deer
(Dama dama)

The Fallow deer is a native of southern Europe, where its range extends from Spain and Portugal eastwards along the border of the Mediterranean to Asia Minor and Iran.

The typical summer pelage of fallow deer, which is assumed in May, is reddish-fawn ('fallow') dappled with numerous white spots; in October this gives way to a uniform dark-greyish brown above, without any spots, and greyish-white beneath. A line of black hairs extends back from the neck to the tail where it divides to outline and emphasize the caudal disc, which remains white at all seasons. A number of colour variations exist, those most frequently encountered being; black, as in, for example, the long-established herds of the New and Epping Forests and Bringwood and Cannock Chases; menil, where the spots remain throughout the year; and white. Harris and Duff refer to a variety with long 'shaggy' hair which first appeared in the 1960s in the 4500 acre (1800 hectare) Mortimer Forest near Ludlow in Shropshire.

The fallow buck grows antlers which are cast annually in April and May, and are re-grown and clean of velvet by late August: they have brow and trez points but lack the bez, and the top is expanded into a palmated plate on the hind edge of which are a number of small points, known as 'spellers', 'spillers', or 'snags'. In many places, such as the New Forest, the antlers are weak and the amount of palmation small when compared with well-fed park stock.

Fallow are basically gregarious animals – more so when living in parks than in the wild – and spend most of the year in single-sex herds. The rut takes place in October, when the bucks gather harems of does. The two sexes stay together for the winter, but re-form in separate herds by the spring. The fawns, which are usually spotted like the adults, are born between the end of May and early July, twins being rare. Fallow are forest-dwelling deer, spending most of the day

lying in thick cover and normally only venturing into the open at dawn and dusk to feed. A fallow buck stands about 3 ft (90 cm) at the shoulder, and weighs an average of 225 lb (102 kg); the doe measures a few inches shorter and weighs some 65 lb (29 kg) less.

The status of the fallow deer in Britain as an indigenous or introduced species has long been in doubt. The idea that it might at one time have been indigenous usually rests on the evidence of remains of *D. dama* or a closely related species known as the Clacton fallow (*D. clactoniana*) or Brown's fallow (*Cervus browni*) which have been discovered in Pleistocene deposits at Swanscombe, Kent, and Clacton, Essex. Following the discovery in Joint Mitnor Cave, Buckfastleigh, Devon, of the remains of a species of fallow deer with rather smaller antlers than *D. clactoniana,* Dr A. Sutcliffe suggested that *D. dama* may have existed for a brief time in England during the Last Interglacial period of the Upper Pleistocene, around 150000 years B.C., long after *D. clactoniana* had become extinct. Other remains of fallow deer have been discovered in Eastern Torrs Quarry Cave, Yealmpton, and

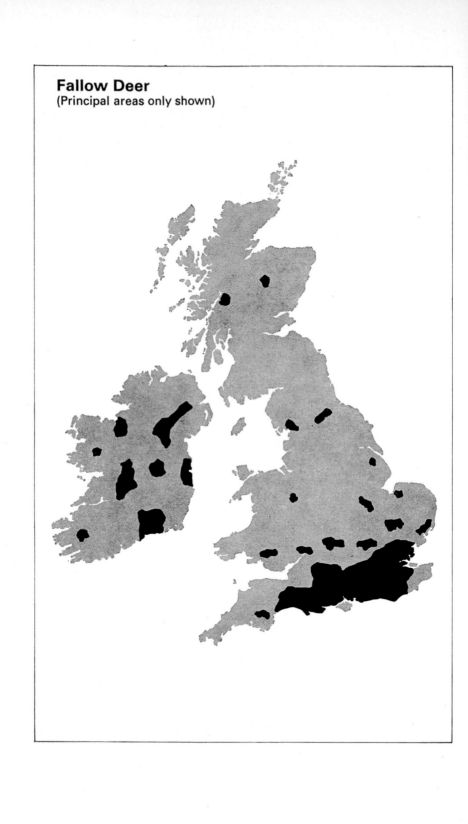

Fallow Deer
(Principal areas only shown)

Tornewton Cave, Torbryan, Devon; in Milton Hill Cave, Wells, Somerset; in Hoe Grange Cave, Brassington, Derbyshire; at Barrington, Cambridge, and in the deposits of the Upper Floodplain Terrace of the Thames. An ancestor of *D. clactoniana* – the smaller *D. nestii* or *D. savini* – may have existed during the Antepenultimate Interglacial period of the Lower Pleistocene, some 550000 years B.C.

All these deposits, however, are recognized as being Interglacial, and no remains of fallow in Britain have been found since the end of the Ice Age around 8000 B.C. As Dr G. B. Corbet* points out, 'it's absence amongst the abundant deer remains at the Mesolithic site of Star Carr, Yorkshire, seems significant.'† According to Fitter, it is still possible 'that the fallow did reach this country when it was warming up after the recession of the ice, but before the severing of the North Sea land bridge, some time in the seventh or eighth millennium B.C., leaving only small populations in limited areas to survive the climatic deterioration of the first millennium'.

On the assumption, however, that man rather than nature is responsible for the presence of fallow deer in Britain today, their origin has usually been ascribed to the Romans or, as J. G. Millais states, to the Bronze or Iron Age Phoenicians:

The Phoenicians were great sailors and traded with Britain for many years before the Roman Conquest, bringing with them all the first things of artistic beauty or use that our savage ancestors had seen: it is more likely that these daring sailors should have brought and bartered strange wild animals which were at that time common on their coasts, than the more practical Italian warriors.‡

Alternatively, it is possible that fallow may have been reintroduced by the Romans after the Phoenician stock died out. One writer has even gone so far as to suggest that neither the Phoenicians nor the Romans, but the Gauls, were responsible for the introduction of fallow deer to Britain; the evidence given, however, hardly bears close examination.

*The Distribution of Mammals in Historic Times. Systematics Association Special Volume No. 6, *The Changing Flora and Fauna of Britain*, pp. 179–202, 1974.
†F. C. Fraser and J. E. King, 'Preliminary Report on Excavations at Star Carr, Seamer, Scarborough, Yorkshire, (Second Season, 1950),' *Proc. Prehist. Soc.*, 1950. *Excavations at Star Carr, an early mesolithic site at Seamer, near Scarborough, Yorkshire* (J. G. D. Clark, ed.), pp. 70–95, C.U.P., 1954.
‡*The Mammals of Great Britain and Ireland*, 1906.

It is now generally agreed that fallow deer were well established in Britain in Roman times: according to G. Jennison, in approximately 238 A.D. the Emperor Gordian I exhibited in Rome 'according to the list in the *Augustan History* . . . 200 stags of the fallow deer (*cervi palmati*) including some from Britain'.* Abbot Aelfric 'the Grammarian' refers to fallow as 'bucks' in his *Colloquium* written at Cernel (Cerne Abbas in Dorset), and Domesday Book (1086) records them in no fewer than thirty-one parks in southern England. By the early seventeenth century there were more than 700 parks in England enclosing 'more fallow in a single English county than in the whole of Europe. Every English gentleman of £500 to £1000 rent by the year hath a park for them, enclosed by pales of wood for 2 to 3 miles [3 to 5 km] compass'. During the Civil Wars the fences of many deer-parks fell into disrepair, which resulted in the escape of numerous fallow deer into the surrounding countryside. The eighteenth century saw a revival of the fashion of keeping deer in parks, and by the close of the nineteenth century there were almost 400 parks in England holding an estimated 71 000 fallow deer. The two world wars of the present century resulted in the break-up of many deer-parks – comparatively few of which survive today – and the escape of further considerable numbers of fallow deer into the surrounding woods.

Just as the status of the species in Britain has long been doubtful, so has the true origin of the black variety. Many authorities have held that the black fallow deer is the earlier strain, and that the spotted variety was introduced from southern Europe at a later date. Fitter considers 'the black fallow to be a domesticated form that has become stabilized in a small number of true-breeding feral herds'. According to Harting, black fallow were first noted in Windsor Forest in 1465, but on the continent they are nowhere indigenous. In 1611 the King of Denmark, Frederick II, sent some black fallow from Sweden or Norway to James I and his Danish-born wife Princess Anne, at Dalkeith Palace, Midlothian: they were subsequently transferred to Theobalds in Hertfordshire at a cost of £30 5s. 1d., and later still were liberated in Epping Forest and Enfield Chase. Writing of the fallow of Epping Forest, Harting states:

Animals for show and pleasure in ancient Rome, Manchester University Press, 1937.

they have held their own, in spite of all difficulties, and have strangely preserved their ancient character in regard to size and colour. Locally they are referred to as 'The Old Forest Breed', and are comparatively small in size, of a uniformly dark brown colour, with very attenuated antlers – peculiarities which have no doubt been brought about by continued isolation without the admixture of any fresh stock for many generations.

The following tables, adapted and condensed from Whitehead, give particulars of the principal areas in the British Isles where herds of feral fallow deer may be seen today:

England

County	Distribution	Origin
Bedfordshire/ Cambridgeshire.	Potton, Biggleswade: Hatley; perhaps small numbers elsewhere.	Probably wanderers from over the border in Hertfordshire.
Berkshire.	Widely but thinly distributed.	In Windsor Forest since thirteenth century. Present stock escapes from Hall Place, Aldermaston, Hampstead, Welford and Littlecote Parks.
Buckinghamshire.	Chiltern Hills; Bledlow; Vale of Aylesbury; Whittlewood Forest.	In Brill Forest since at least 1229. Present stock escapes from Ashridge Park and estates of Lords Portman and Burnham.
Cornwall.	Tetcott.	Mentioned by Norden in his *Survey of Cornwall* (written around 1584; published 1728). Present stock park escapes.
Derbyshire.	Duffield; Ambergate; Kedleston; Locko; Sponden; Calke Abbey.	In Duffield Forest since fourteenth century. Present stock park escapes.
Devon.	In south-east, along Dorset border.	On Dartmoor in seventeenth century.

L

County	Distribution	Origin
		Present stock escapes from Shute, Bicton, Cready, Whiddon and Werrington Parks.
Dorset.	Widely distributed, especially in west.	In Stock Gaylard Park since 1248. Present stock escapes from Sherborne, Hooke, Melbury, Hyde House and possibly Charborough Parks.
Durham.	Barnard Castle.	In Teesdale Forest in 1538. Present stock possibly escapes from Marwood or Langley Chase.
Essex.	Epping Forest; Bishop's Stortford; Hatfield Forest; Waltham; Elsenham; Stanstead; High Roding: White Roding; Berners Roding; Stondon Massey and Kelvedon Common.	In Epping Forest since fifteenth century. Present stock elsewhere escapes from Langleys, Easton Lodge, Dunmow, Chelmsford and Weald Parks.
Gloucestershire.	Widely distributed.	In Kingswood Forest since before 1325. Present stock escapes from Lydney, Over, Great Barrington, Brockhampton, and other parks.
Hampshire.	Widely distributed, especially in the New Forest and around Winchester.	Fallow deer were probably first introduced in Britain to the New Forest: they were certainly well established there by the twelfth century. Present stock escapes from Alder-maston, Hackwood,

County	Distribution	Origin
		Hurstbourne, Hursley, and Uppark Parks.
Herefordshire.	Wyastone Leys, Whitchurch: Kentchurch Court; between Kinnersley and Bodenham, especially at Wormsley and Dinmore. Mary Knowl; Richards Castle; Haugh Wood; Mortimer Forest.	Henry VIII hunted wild fallow deer in Bringwood Chase. Present stock escapes from Moccas, Garnstone (black variety), Hampton Court, Ludford, Haye, and Moor Parks.
Hertfordshire.	Watton-at-Stone; Bramfield: Hatfield; King's Walden; Lawrence End; Ashridge; Berkhamsted, Royston.	Escapes from Woodhall (in the blizzard of March 1916), Hatfield, Ashridge (1928), and King's Walden Parks.
Kent.	Widely distributed.	Escapes from Knole, Surrenden-Dering (Mr Walter Winans), hilham, Mersham Hatch, and Waldershare Parks.
Lancashire.	Cark-in-Cartmel, Furness.	In Croxteth Park since 1348. Present (menil) stock escapes from Holker Hall Park.
Leicestershire.	Charnwood Forest: between Ashby-de-la-Zouch and Melbourne; Donington (small stock).	In Leicester Forest (west of R. Soar) since at least 1313. Present stock escapes from Staunton Harold, Calke Abbey and Donington Hall Parks.
Lincolnshire.	Very small stock in south-east.	In various parks for over 600 years. Present stock escapes from Grimsthorpe, Burghley and Brocklesbury Parks.
Monmouthshire.	In most of the woods around Monmouth, on	Probably escapes from Wyastone Leys and

County	Distribution	Origin
	both sides of R. Wye, especially in Beaulieu and Lady Park Woods and Redding Enclosure towards Staunton in the the east; a few to the west at Hendre and White Hill.	Hendre Parks.
Norfolk.	Widely but thinly distributed.	Fallow in Cossey Park mentioned in Domesday Book (1086). Present stock escapes from Blickling, Dudwick (twelve black fallow in 1942), Sandringham, Houghton, and Holkham Parks.
Northamptonshire.	Whittlebury (Wakefield Lawn and Wicken Wood); Rockingham Forest; Badby Wood; Salcey Forest; Castor Hanglands.	Fallow possibly present in Rockingham Forest in eleventh and twelfth centuries, and certainly by early thirteenth century. Present stock supplemented by escapes from various parks, *e.g.* Milton and Fawsley.
Northumberland.	Chillingham Park, Wooler.	Fallow in county in seventeenth century. Present stock park escapes.
Nottinghamshire.	Only in west (old Sherwood Forest).	Established by thirteenth century. Present stock supplemented by escapes from Welbeck, Thoresby and Rufford Parks.
Oxfordshire.	Wychwood Forest; Tangley Woods; Waterperry; Studley; Stanton; Chiltern Hills, from Goring to Whipsnade, especially	In Wychwood Forest by 1226. Present stock escapes from Ditchley, Cornbury, Great Barrington, Glympton, Shotover, Stonor, and

County	Distribution	Origin
	around Stonor, in the Hambledon Valley, and at Watlington and Bledlow.	Nuneham Parks – some of which obtained their original supply from the ancient stock in Wychwood Forest. Fallow are recorded in the park at Stonor as early as the fourteenth century.
Rutland.	In woods between Pickworth and South Witham, especially in Morkery Wood.	In Rutland Forest (Leefielde), where they probably died out in the seventeenth century, since before 1269. Present stock probably escapes from Exton and other parks.
Shropshire.	Mortimer Forest (about 250 in 1973); Wyre Forest; Ludlow (from Bromfield to Croft Castle); Bringwood Chase; Loton Park, Aldbury.	In Wyre Forest since thirteenth century. Present stock escapes from Mawley Hall, (about 1880), Cleobury Mortimer, Longnor; Loton; and Attingham Parks.
Somerset.	Brendon, Croydon and Staple Hills.	In Petherton Forest by 1315. Present stock probably escapes from Dunster, Nettlecombe, Combe Sydenham, and Hatch Court Parks.
Staffordshire.	Needwood Forest; Cannock Chase.	In Cannock Chase since at least 1271, and in Needwood Forest since before 1313.
Suffolk.	Holbrook Wood, Ipswich: Blyford and Sotherton, Southwold; Livermere and Ickworth, Bury St Edmunds.	Escapes from Woolverstone Park, Ipswich; Henham Park (1914), Southwold; Livermere Hall and Ickworth, Bury St Edmunds.

County	Distribution	Origin
Surrey.	Along Sussex Border, especially in Charleshill (Elstead), and Farnham areas.	Probably in Stocha or Stoke Park mentioned in Domesday Book and Guildford Park enclosed around 1160; certainly in Surrey Forest by 1327. Present stock mostly escapes from Witley Park, Godalming, and Farnham Park.
Sussex.	Widespread, especially in west, in Worth, Tilgate and St Leonard's Forests and in Goodwood area. Also around Netherfield, Dallington, Mountfield and Ashburnham near Battle, and in Ashdown Forest.	Eridge Park mentioned in Domesday Book probably contained fallow; Ashdown Forest (Lancaster Great Park) certainly did in reign of James I. Present stock escapes from Ashburnham, Brightling, Buckhurst, Arundel, Petworth, Parham, and Cowdray Parks.
Warwickshire.	Small numbers at Ragley, Alcester: a few elsewhere.	Possibly in Feckenham Forest in 1300. Present stock escapes from Ragley Hall, and possibly Charlecote, Alscot and Ettington Parks.
Westmorland/ Cumberland.	A few at Pooley Bridge; possibly a small number elsewhere	In Leven's Park by 1360. Present stock escapes from Dalemain and Lowther Parks.
Wiltshire.	Mainly in Savernake Forest and Cranborne Chase; small numbers elsewhere.	In Savernake Forest by twelfth century. Present stock supplemented by park escapes.
Worcestershire.	Possibly very small stock near Halesowen.	In Feckenham Forest by 1225. Present stock (if any) escapes from Hagley Park.

County	Distribution	Origin
Yorkshire.	Mainly in the southern part of the North Riding between the Hambleton Hills and Allerston Low, especially in the Antofts Woods at Helmsley between Thirsk and Pickering, and Boltby Wood, Ampleforth, Murton and Ghyll.	In Pickering Forest since at least 1289. Present stock escapes from Duncombe Park, Helmsley, and possibly from Castle Howard and Scampston Park, Knapton.

Recent reports of wild fallow deer in new districts in England include between twenty and thirty – both black and brown varieties – in Lullingston Park, Kent, in 1970. In 1972 about 100 were reported from Bramshill Forest, Hampshire, most of which were white, but including the black and common varieties. Some fallow present in the same year on the Wiston Estate near Steyning in Sussex were probably descended from escapees from nearby Parham Park. In 1974 twenty fallow were seen in Allerton Woods in Yorkshire, and the (presumed) descendants of escaped stock from Duncombe Park were observed in Foresty Commission plantations west of Helmsley. Outliers from Studley Royal were reported in the same year in the Laver Valley. Two bucks were seen on 9 August 1974 in Marley Wood, near Lulworth in Dorset – an entirely new locality.

In Scotland there is a tradition that the monks first brought fallow to Kildalton on the Isle of Islay in about 900 A.D. They are first mentioned as being present on the mainland in 1283 when Professor Cosmo Innes (1860) records that 'an allowance was earmarked in the accounts of the King's Chamberlain for mowing and carrying hay and litter for the use in winter of the Fallow Deer' in the King's Park at Stirling, which had been enclosed by Alexander III in 1263. Fallow are known to have been on Inchmurrin and Inchcailloch Islands at the southern end of Loch Lomond since very early times: in 1326 the owner of the islands, David Graham, signed an agreement allowing King Robert the Bruce to use them as a hunting ground. In

1424 fallow first came under the protection of Scottish law when it was decreed that those who 'slay deare, that is to say harte, hynde [red deer], doe [fallow doe], and rae [roe]' should be fined 40s. and their lairds £10. In about 1539 James V and Mary of Guise 'hunted Fallow Deer on the Lomonds' while staying at Falkland Palace in Fife, and fallow were among the game at the tinchel* organized in Glen Tilt on the Braes of Atholl by the Earl of Atholl for Mary Queen of Scots in 1564. Twenty-three years later legislation was enacted which made 'the slayers and schutters of hart, hinde, doe, roe, haires, cunninges and other beasts' liable to the same penalties as the stealers of horses or oxen. As previously mentioned, the black fallow sent by the King of Denmark to James I in 1611 were the first of that variety to be seen in Scotland.

By the middle of the seventeenth century fallow deer were established on a number of forests in central Scotland. The *Wardlaw Chronicle* of 1642 records that '. . . a gallant, noble convoy, well appointed and envyed by many' went hunting in the Forest of Killin in mid-Perthshire where they found 'fallow-deer hunting to their mind. . . .'

During the eighteenth century there were a number of fresh introductions of fallow deer to Scotland, *e.g.* to Ross-shire by 1729 and Dumfriesshire in 1780. As in the rest of Britain, a number of deer-parks were enclosed in Scotland during the nineteenth century from which there were in course of time the usual escapes.

Scotland

County	Distribution	Origin
Argyllshire.	Inverary; Mull of Kintyre: Islands of Mull (Glenforsa; Ben More); Islay (Kildalton); and Scarba.	Introduced to mainland probably in nineteenth century by 8th Duke of Argyll. To Mull in 1868 by Colonel Greenhill Gardyne. To Islay either by monks around 900 A.D. or in fourteenth century by 'John the Good'.
Banffshire.	Arndilly House woods, Craigellachie.	?

*Deer-drive.

County	Distribution	Origin
Caithness.	Berriedale.	White fallow introduced to Berriedale from Welbeck Abbey Park, Nottinghamshire, by 6th Duke of Portland about 1900.
Dumfriesshire.	Raehills, Lockerbie.	Twelve introduced to Raehills, 1780. Present stock survivors of those killed off in 1898.
Dunbartonshire.	Loch Lomondside (between 70 and 90 in 1973).	Present on Inchmurrin and Inchcailloch Islands since very early times, and well established before 1326: reintroduced to former in 1530.
Inverness-shire.	Loch Ness at Balmacaan.	Present since around turn of the century.
Morayshire.	Spey Valley from Fochabers to Craigellachie.	Since eighteenth century.
Perthshire.	Dunkeld (Craigie Barns and Drumbouie Woods), Bolfracks, Aberfeldy.	James V and Mary of Guise 'hunted Fallow Deer on the Lomonds' about 1539. Introduced to Dunkeld Forest by Duke of Atholl early in nineteenth century.
Ross and Cromarty.	Achnalt/Garve; Corriemoillie; Kinlochluichart; Lochrosque; Scatwell.	Well established by 1729. Scatwell stock are escapes from Lord Seaforth's park at Coul.
Sutherland.	Dunrobin.	Introduced at Dornoch about 1840. Dunrobin herd may have originated from Welbeck Park.

In recent years there have been reports of fallow deer at large in some of the woods of Galloway, Kirkcudbrightshire, which are presumably interlopers from over the border in Dumfriesshire. In

February 1966 a black fallow buck was seen near Loch Lintrathen in Angus, some 15 miles (24 km) from the nearest known herd at Dunkeld. In 1970 at least twelve white fallow were living at Berriedale and around Langwell Water in Caithness. There are at present an estimated 2000 to 3000 wild fallow deer at large in Scotland.

Since fallow deer are mentioned in neither *Topographia Hibernica* (1183–6) by Giraldus Cambrensis nor in the *Polychronicon* of Ranulf Higden (*c*. 1299–*c*. 1364), it has been assumed by some authors that they may not have arrived in Ireland until the fourteenth century: Fitter, however, states that they were 'evidently introduced by the Normans for hunting, as there were some in woods in Munster by the twelfth century', and Mooney suggests that they 'possibly came with the Normans, with the introduction into the Glencree [Co. Wicklow] Royal Park in 1244. One interesting record tells of a gift of twelve fallow deer in 1296 from the Royal Forest at Glencree to Eustace le Poer, ancestor of the Powers of Curraghmore in County Waterford.' Harting quotes from the *Description of Ireland* (1599–1603) by Fynes Moryson, Secretary to Lord Mountjoy, Lord Deputy of Ireland: 'The Earl of Ormond, in Munster, and the Earl of Kildare, in Leinster, had each of them a small park enclosed for Fallow-Deer . . . they have also about Ophalia and Wexford, and in some parts of Munster, some Fallow-Deer scattered in the woods'; this establishes the existence of feral fallow deer in Ireland as early as the beginning of the seventeenth century. Harting also refers to a letter written by Lord Deputy Stafford to the Archbishop of Canterbury in May 1638, in which he describes hunting fallow deer at Coshawe, Co. Galway.

In 1772 John Rutty wrote of 'the *Cervus platyceros*, the Buck, or Fallow-deer, whose horns are palmated', which was by then quite common in Co. Dublin. During the nineteenth century a number of parks throughout Ireland were enclosed which contained fallow deer, some of which escaped into the nearby woods.

Between 1920 and 1922, when many estates were illegally taken over by the thugs of the Irish Republican Army, a number of deer-fences were broken down through lack of maintenance: in at least one instance – at Charleville Castle, Co. Offaly – the owner opened his gates and deliberately let his herd escape rather than be slaughtered by the murderous Sinn Feiners. As a result, by the mid-1920s herds of wild fallow deer were firmly established in nearly every county in Ireland, where by 1970 the total population outside deer-parks was estimated to number about 2500.

Ireland

County	Distribution	Origin
Antrim.	Randalstown.	Escapes from Shane's Castle Park, about 1933. Dark-coloured type.
Clare.	Tuamgraney; Newmarket-on-Fergus.	Escapes from Dromoland Castle and Raheen.
Down.	Seaforde.	Escapes from park, about 1933.
Fermanagh.	Colebroke; Florence Court; Crom Castle.	Escapes from Florence Court, Crom Castle and Kiltiernay House Park, Kesh (1941).
Galway.	Loughrea; Portumna; Athenry; Ballygar; Clonbrock and Woodford.	Mentioned by Lord Deputy Stafford in 1638. Present stock park escapes.
Kerry.	Killarney.	Introduced to Kenmare mid eighteenth century.
Leitrim and Mayo.	Clooncormick; Bloomfield Park, Claremorris.	Escapes from Bloomfield Park.
Leix.	Ballyfin, Mountrath; Emo.	Escapes from Emo Park.
Limerick.	Glenstal.	Escapes from Glenstal Park, Murroe, about 1935.
Monaghan.	Clones; Glaslough.	Escapes from Hilton Park.
Offaly.	Tullamore; Kinitty.	Escapes from Charleville Forest Park; Kinitty Castle; Bin Castle: Knockdin Castle, and Thomastoun Park.
Roscommon.	Boyle.	Escapes (about 1910) from Rockingham Park (enclosed around 1800).
Sligo.	Collooney; Lough Gill.	Escapes from Markree Castle Park.

County	Distribution	Origin.
Tipperary.	Glen of Aherlow; Bansha; Suir Valley; Dundrum; Clonmeland; Glogheen; Nenagh.	Escapes from Ballinacourty Park; Dundrum House (Earl of Montalt), in about 1918, where they were introduced in mid-eighteenth century: Gurteen Park, 1890s: Castle Lough, about 1920.
Tyrone.	Caledon (between 300 and 500 in 1974); Baronscourt.	Park escapes.
Waterford.	Portlaw, Suir Valley; Kilsheelan; Cappoquin; Dungarvan, and Nier.	Escapes in about 1909 from Curraghmore Park, where they were first introduced from Glencree, Co. Wicklow in 1926.
Wicklow.	Carrick Mountains; Aughrim; Rathnew; Rathdrum; Ballinglen; Luggala; Glendalough, and Glenmalur.	Introduced to Glencree Royal Park, according to Mooney, in 1244. Peter Delap records that the feral fallow in Co. Wicklow are descended from stock at Powerscourt and Ballycurry (near Ashford) Parks.

Wild fallow deer also occur in small numbers in parts of Co. Carlow, Co. Cavan, Co. Donegal, Co. Down, Co. Dublin (where they are mentioned by John Rutty in 1772), Co. Kildare, Co. Louth, Co. Meath, and Co. Westmeath, and perhaps in Co. Armagh and Co. Londonderry. In 1975 between twenty and thirty fallow were reported to be living a precarious existence on the Marquess of Dufferin and Ava's estate at Clandeboye on the outskirts of Bangor, Co. Down, where there had been only between four and six deer in 1964.

Although fallow deer are not indigenous to Wales, they appear to have been there for at least 700 years. Whitehead quotes Dr Jackson who found at Dyserth Castle, Flintshire, a large distal fragment of a tibia and a calcaneum (dating from *c.* 1250) which 'agree very closely with similar bones of the Fallow Deer and can . . . be reasonably referred to this species'.* By the sixteenth century fallow were running free in the Royal Forest of Snowdon, as is shown by the following warrant quoted by Whitehead from Pennant:†

These are to require you to delyver to my friend Maurice Wynne, Gent. or to the bringer hereof in his name, one of my fee stags or bucks of this season, due to me out of the queens majestys forest of Snowdon:
. . . From Cardigan, the 14th August 1561.
   Yr. loving friend, H. Sydney. To my very loving friende John Vaughan, forrester of the queens forest of Snowdon, in the counties of Anglesey, Merioneth and Carnarvon. . . .

There were also by this date a number of deer-parks established in Wales, *e.g.* that at St Dinothes (St Donat's Castle) in Glamorgan referred to by John Leland‡ in 1540 and that at Baron Hill, Beaumaris, on the Isle of Anglesey, where, according to Pennant, Sir Richard Bulkeley (1533–1621) 'kept two parks well stored with . . . Fallow deer'. In Pembrokeshire, George Owen, quoted by Whitehead, states that by 1603 fallow were enclosed 'in two small parkes onlye, and not in anye fforest or chase, and the nomber very fewe'.§

As in the rest of the British Isles, a number of deer-parks were formed in Wales especially during the nineteenth century, from which there were the usual escapes, mainly after the two world wars. The present range of feral fallow deer in Wales is not extensive nor is the population large, but it is being augmented to a certain degree by wanderers from over the border in Herefordshire and Shropshire; in the latter county fallow have escaped from Loton Park, Shrewsbury, some of which have found their way on to Breidden Hill in Montgomeryshire, where they have been joined by escapees from Powis Castle, Welshpool. In 1974 a solitary buck was seen in the hitherto deerless Gwydyr Forest in Snowdonia.

Archaeologia Cambrensis, pp. 77–82, 1915.
†*Tours in Wales*, 1772.
‡*Itinerary in Wales of John Leland in or About 1536–39.*
§*The Description of Pembrokeshire.*

Wales

County	Distribution	Origin
Anglesey.	Woods surrounding Bodorgan, Aberfraw.	Sir Richard Bulkeley (1533–1621) 'kept two parks well stored with . . . Fallow deer' at Baron Hill, Beaumaris.
Carmarthenshire.	Woods around Golden Grove, Llandilo.	Escapes from Golden Grove Park.
Denbighshire.	Chirk; Wynnstay; Abergele; Henllan.	Escapes from Wynnstay Park.
Flintshire.	St Asaph; Kinmel.	Llannerch Park.
Glamorgan.	Port Talbot.	At St Dinothes (St Donat's Castle) in 1540. Introduced to Dunraven, according to Matheson, from Clearwell Court, Gloucestershire, by Thomas Wyndham around 1796.
Merionethshire.	Woods around Nannau Park, Dollgellau (about 100 in 1971); in Forestry Commission property at Coed-y-Brenin (about 30 in 1971), as far west as Bontddu in Dovey Valley and north to Trawsfynydd.	Escapes from Nannau Park where originally introduced between 1770 and 1820.
Montgomeryshire.	Breidden Hill, Welshpool.	Escaped from Powis Castle in spring of 1947: others escaped from Loton Park, Shrewsbury.

19. Japanese Sika Deer

(Cervus nippon)

The home of the Japanese sika deer* is the dense forests in the hilly regions of the main Japanese islands, and around Vladivostok in the Maritime Province of Siberia.

The winter coat of the adult Japanese sika stag is a uniform dark greyish-brown, that of the hind being somewhat lighter; in summer the pelage changes to a warm reddish brown liberally covered with yellowish-white spots. The tail, which is almost as long as that of the fallow deer, is white underneath with a faint black line above. The hairs on the extensive white caudal patch, which is edged above and on both sides with black, become erect and are fanned out as in the north American prongbuck (*Antilocapra americana*) when the animal is alarmed. The prominent skin-glands below the hock on the back legs are visible throughout the year. The antlers of the stag are of the red-deer form, but invariably lack the bez tine and seldom number more than four points each: they are cast in the spring and the new set is clear of velvet by late August or early September. A Japanese sika stag weighs about 120 lb (54 kg) and a hind about 40 lb (18 kg) less, although weights are subject to considerable variation in different parts of the country: both sexes stand about 32 in. (82 cm) at the shoulder.

Sika deer are a mainly woodland species, only emerging from cover at dawn and dusk to feed; in Dorset – one of their principal strongholds in England – they mainly graze on grass, moss, leaves, sedges, ivy, and heather, and also consume in autumn acorns, chestnuts and fungi. In the New Forest they prefer to browse on both

*Three races or subspecies of sika deer are represented in the wild in Britain: the Japanese (*C. n. nippon*); the Manchurian (*C. n. manchuricus*), and the Formosan (*C. n. taiouanus*), all of which have been known to interbreed. In addition, herds of Kerama or black sika (*C. n. keramae*), Chinese sika (*C. n. kopschi*), and Dybowski's or Pekin sika (*C. n. hortulorum*) have been kept in captivity at Woburn, Whipsnade, and Surrenden-Dering Parks.

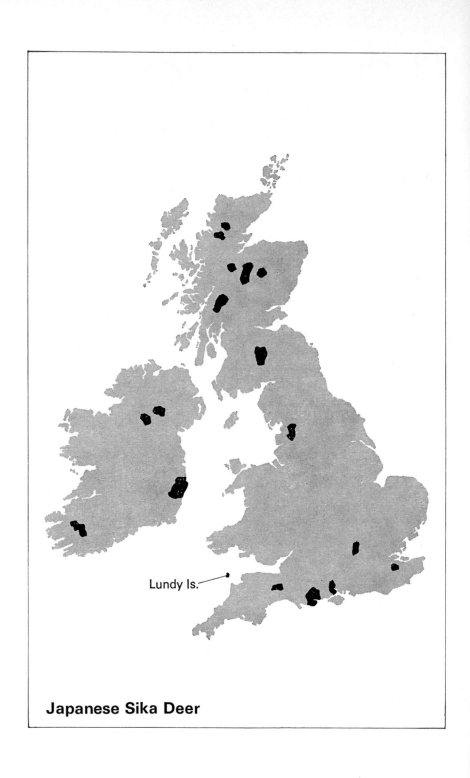

Lundy Is.

Japanese Sika Deer

broad-leaved and coniferous trees. The rut takes place in September and October when each stag gathers together five or six hinds; both sexes remain together until the following March, when the stags depart to grow their antlers. The calves – brown in colour spotted with white – are born in May and June. During the rut the stag utters a strange whistle sometimes terminating in a harsh scream (corresponding to the roar of the red stag and the groan of the fallow buck), while the hind sometimes makes a flute-like, whistling sound. Sika are perhaps the most pugnacious and potentially dangerous of British deer, frequently holding their ground and appearing to make a direct and aggressive challenge when confronted by man. The prominent and angular facial bones give them a bad-tempered and even fierce expression. In Dorset and the New Forest they seem to cause little timber damage, apart from minor cases of fraying.*

Japanese sika deer were first introduced to England in 1860, when a pair were presented to the Zoological Society of London for their collection in Regent's Park. In the same year a stag and three hinds were obtained from the German-born animal dealer Johann Christian Carl (Charles) Jamrach (1815–91), who had premises in

*Damage to timber caused by cleaning velvet from antlers.

M

Ratcliff Highway (now St George Street) in the East End opposite the entrance to the London Docks, by the then Viscount Powerscourt and were brought to his home at Enniskerry, Co. Wicklow: this park became the source of supply for the original stocking of a number of parks in England and Scotland, the first of which were probably those at Waddesdon in Buckinghamshire in about 1874 and at Tulliallan in Fife, around 1870. Since then some thirty estates have preserved Japanese sika deer, a number of which have escaped to form feral herds in the surrounding countryside.

The following tables, adapted and consensed from Whitehead, give particulars of when and where Japanese sika deer were first introduced to deer-parks in the British Isles.

England

County	Estate	Date
Buckinghamshire.	Waddesdon Park (from Powerscourt, Co. Wicklow).	c.1874.
Dorset.	Melbury Park.	c. 1880.
Hampshire.	Hursley Park, Winchester.	c. 1885.
Essex.	Weald Park, Brentwood.	c. 1890.
Kent.	Knole Park, Sevenoaks.	c. 1890.
Rutland.	Exton Park, Oakham.	c. 1890.
Sussex.	Leonardslee, Horsham.	1892.
Dorset.	Brownsea Island, Poole Harbour.	1896.
Northamptonshire.	Whittlebury Park, Towcester.	Late nineteenth century.
Surrey.	Park Hatch Park, Shillinglee.	Late nineteenth century.
Westmorland.	Rigmaden Park, Kirkby Lonsdale.	Late nineteenth century.
Bedfordshire.	Woburn Park.	c. 1900.
Norfolk.	Melton Constable Park, Fakenham.	c. 1900.
Kent.	Surrenden-Dering Park (by Mr Walter Winans).	Early twentieth century.
Yorkshire.	Park Nook, Bolton-by-Bowland (by Lord	1904–6.

County	Estate	Date
	Ribblesdale and Capt. Peter and Capt. Alec Ormerod, for the Ribblesdale Buckhounds).	
Dorset.	Hyde Park, Wareham (by Major Radclyffe).	Early twentieth century.
Gloucestershire.	Great Barrington Park, Burford.	Early twentieth century, and about 1928.
Northumberland.	Hulne Park, Alnwick.	Early twentieth century.
Nottinghamshire.	Rainworth Lodge, Mansfield (by Mr Joseph Whitaker).	Early twentieth century.
Oxfordshire.	Crowsley Park, Henley (from the Mull of Kintyre).	Early twentieth century.
Oxfordshire.	Fawley Court, Henley (by Mr Austin Mackenzie).	Early twentieth century.
Oxfordshire.	Nuneham Park, Abingdon.	Early twentieth century.
Sussex.	West Grinstead Park, Horsham.	Early twentieth century.
Westmorland.	Lowther Park, Penrith.	Before 1914.
Shropshire.	Weston Park, Shifnal.	1925.
Devon.	Lundy Island (by Mr Martin Harman from Surrenden-Dering).	1927 or 1929.
Kent.	Mersham Hatch Park, Ashford.	1930s.
Buckinghamshire.	Langley and Ashridge Parks.	Between the wars.
Sussex.	Arundel Castle.	Between the wars.
Sussex.	Eridge Park, Tunbridge Wells.	?
Sussex.	St Leonard's Park, Horsham.	?
Wiltshire.	Rushmore Park, Tollard Royal, Cranborne Chase.	?
Yorkshire.	Allerton Park, Knaresborough.	?

Feral herds of Japanese sika deer are at large today in a number of English counties, mainly in the south. Here they are most common on the heathlands of south-east Dorset in Wareham Woods and around Puddletown, extending their range to the limits of the Bagshot and Reading Beds, and in the south of the New Forest in Hampshire. They appear to be more successful in establishing themselves on the acid soil of the Tertiary deposits rather than on such rich or chalky land as that around Melbury and Tollard Royal, Cranborne Chase. A small herd is established in Tyneham Wood about half a mile (800 m) from the Bagshot Beds on the Wealden. They also occur in small numbers between Beaminster and Sherborne having escaped from Melbury Park. Some of the sika introduced to Brownsea Island in 1896 escaped by swimming across Poole Harbour – reputedly on the day of their introduction – to the Isle of Purbeck, where they became established. When Hyde Park was requisitioned by the War Office in 1939 a number of sika escaped to join up with those already on the Purbeck peninsula: these deer were the ancestors of the present-day feral sika of south-east Dorset.

In 1948 a Government committee first conceived the idea of establishing a National Park on the Purbeck peninsula, parts of which have been used by the army since 1916 as a tank gunnery school. In 1971 the Defence Lands Committee, under the chairmanship of Lord Nugent of Guildford, was set up to review the current land holdings of the army in Britain. In July 1973 the Nugent Committee published its findings, which recommended, *inter alia*, that the Royal Armoured Corps gunnery school should be moved from Purbeck to Castlemartin in south Pembrokeshire. On 29 August 1974, however, the Defence Secretary, Mr Roy Mason, announced in a White Paper the Government's rejection of the Nugent Report, largely on the grounds of cost. At the present time commercial pressures on Purbeck are being added to those of the military: oil has recently been discovered and is currently being extracted at Steeple and Kimmeridge, and potentially valuable deposits of rich china clay have been found on the Arne peninsula. Were the Purbeck National Park to become a reality it would be both the first designated National Park in lowland England, and also the first within one hundred miles (160 km) of London. It would undoubtedly prove of immense benefit to the Japanese sika on the peninsula – which today form the largest group of the species in Britain – as well as to other varieties of animals and plants at present leading a highly precarious existence. Only by providing the protection which would result from the creation of a National Park can the survival of the fauna and flora of the Purbeck area be ensured. A small step forward was taken when in February 1975 it was announced that a working party (set up as the result of the 1974 White Paper), under the chairmanship of Brigadier Roy Redgrave, Commandant of the R.A.C. centre at Bovington, had recommended that the public should be granted limited access to the Lulworth Cove area, at an estimated cost of £100000 *per annum*. This proposal opened up the 6 mile (10 km) stretch of coast between Kimmeridge Bay and Lulworth Cove, thus making it possible for the first time in fifty years to walk the entire length of the ancient Dorset Ridgeway from Ballard Point near Swanage to Abbotsbury Castle at the western end of the Chesil Bank.

In 1904 a pair of sika escaped from the collection of Lord Montagu at Beaulieu, in Hampshire, into the neighbouring Ashen Wood, where they were joined in the following year by a deliberately liberated pair: these four deer were the ancestors of the present New Forest stock, which mainly inhabits the Beaulieu/Brockenhurst district, and the Beaulieu woods west of the Beaulieu river: in 1967 they were estimated to number some sixty head.

The sika occasionally seen in Cranborne Chase in Wiltshire are probably descended from escapees from Rushmore Park, Tollard Royal.

In Devon and Somerset the escaped sika from Pixton Park, which were at large in the Barle Valley and in the woods to the west of Haddon Hill southwards along the Valley of the Exe in the mid-1960s, appear now to be extinct. There is however a herd of about twenty-five sika on Lundy Island in the Bristol Channel where they were originally introduced by Mr Martin Harman in 1927 or 1929.

In southern Oxfordshire and Buckinghamshire a small number of sika frequent the woods of the Chiltern Hills around Stonor, Fawley and Crowsley in the Henley area, and the Hambledon valley.

The feral sika living on Forestry Commission land between Charing and Challock in Kent are the descendants of those which escaped from Surrenden-Dering Park at the outbreak of war, while those around Sevenoaks owe their origin to the herd in Knole Park. The sika at large in the St Leonard's Forest district of Sussex originate from those kept in Leonardslee and West Grinstead Parks since the early years of the century, while the small stock sometimes to be seen in the Shillinglee/Chiddingfold district on the Surrey/Sussex border are descended from some which escaped from Park Hatch Park in 1939.

In Yorkshire and Lancashire the sika which frequent Pendle Forest and Ribblesdale, between Bolton-by-Bowland and Ribchester, are the descendants of those turned out in Park Nook between 1904 and 1906 by Lord Ribblesdale and Capts. Peter and Alec Ormerod of Wyresdale Park, Garstang, for hunting by the Ribblesdale Buckhounds. The Bowland sika today range from Ribblesdale to about 1000 ft (305 m) in Gisburn Forest: they cover an area of about 9 miles (14 km) by 8 miles (13 km) bounded on the west by Dunsop Bridge, Browsholme and Bashall Eaves; on the east by a line between the northern limit of Gisburn Forest and the village of Gisburn, and on the south by the A.59 Skipton to Preston road. In 1971 there were estimated to be between sixty and seventy in this region.

It has been known for some time that hybridization between sika and red deer occurs in north-western England. John Heaton, late Secretary of the Lunesdale and Oxenholme Staghounds, who seems to have been the first authority to draw attention to this fact, believed that sika from Rigmaden interbred with red deer from Middleton Fell. As there were no sika in the Rigmaden deer park after the outbreak of the First World War, when the deer fence fell into disrepair and the deer could no longer be artificially fed, the animals mentioned must have either already been hybrids or have interbred with red deer soon after leaving Rigmaden. The earliest recorded hybrid stag appears to have been one which was roused in Barbon Wood in the Lune Valley by the Ribblesdale Buckhounds in 1940. A number of hybrids have been seen or shot since that time in the same area. It seems possible that in the years after the First World War some of the sika which had been liberated in Bowland Forest by Lord Ribblesdale between 1904 and 1906 may have wandered northward towards Rigmaden. If this movement did take place, it may well be that these deer are at least in part responsible for the hybrid animals which exist today.

As pointed out by Lowe and Gardiner, the centre of this hybrid herd has moved westward since the Second World War, perhaps because of the increase in afforestation and the policy of not shooting red deer in the area of Cartmel Fell some 8 miles (13 km) south-west of Kendal at the southern end of Windermere where the herd now appears to be centred; outliers are sometimes to be found on Gummer's How in the north, on Whitbarrow Scar in the east, and as far south as Witherslack and the valley of the river Winster.

Lowe and Gardiner came to the conclusion that virtually all hybridization in the past appears to have taken place with park red deer. They point out, however, that in recent years a number of hybrid stags have been seen and shot further away from the Cartmel Fell district: these include two to the east of Thirlmere 16 miles (26 km) to the north; one in Holker Park near Cark-in-Cartmel 8 miles (13 km) to the south; and one near Hawkshead at the northern end of Esthwaite Water in Grizedale Forest 6 miles (10 km) to the north-west on the far side of Windermere. It thus appears that the hybrid animals are continuing to expand their range to the north, south and west, where they are now mixing with native red deer populations. With the possible exceptions of the two stags from Thirlmere, however, there is as yet no evidence of hybridization with native red deer stock, but it will be interesting to see whether this takes place in the future.

Scotland

County	Estate	Date
Fife.	Tulliallan (by a daughter of Admiral Lord Keith).	c. 1870.
Ross and Cromarty.	Achanalt, Garve; Loch-rosque, Achnasheen (by Sir Arthur Bignold from Powerscourt: 1 stag, 4 hinds).	c. 1887 or 1889.
Argyllshire.	Carradale, Mull of Kintyre (by Mr Austin Mackenzie from Fawley Court, Buckinghamshire).	c. 1893.
Sutherland.	Rosehall Park (by Mr W. Ewing Gilmour).	Late nineteenth century.

County	Estate	Date
Inverness-shire.	Glenmazeran and Glenkyllachy (by Mr William D. Mackenzie from Fawley Court, Buckinghamshire).	c. 1900.
Inverness-shire.	Aldourie Castle, by Inverness (by Colonel E. G. Frazer Tytler from Rosehall, Sutherland).	c. 1900.
Peeblesshire.	Dawyck Park, Stobo. (by Mr F. R. S. Balfour, from Japan).	1908.
Ross and Cromarty.	Coulin, Wester Ross.	c. 1919.
Caithness.	Berriedale (by the Duke of Portland from Welbeck Abbey, Nottinghamshire).	c. 1920. and c. 1930.
Ross and Cromarty.	Rosehaugh, Black Isle (by Mr Douglas Fletcher).	Early twentieth century.
Angus.	Kinnaird Castle, Brechin (by the Earl of Southesk).	?

A number of places in Scotland today support herds of feral Japanese sika deer. Those which escaped from Mr A. N. Balfour's collection in Dawyck Park, Peeblesshire, in 1912, are the ancestors of what is at the present time probably the largest Scottish population of free-running sika, which still mainly frequent the 4000 acres (1618 hectares) or so of the Dawyck estate, although in recent years small numbers have crossed the border into Dumfriesshire. A one hundred acre (40 hectare) wood outside the deer-park at Kinnaird Castle maintains a small herd, while from Tulliallan in Fife sika have wandered over the border into the Devilla Woods near Alloa in Clackmannanshire. In Argyllshire sika which escaped from Mr Austin Mackenzie's estate at Carradale have colonized Saddell, Glen Lussa and Torrisdale to the south of Carradale, and Cour, Crossaig and Claonaig, near Skipness to the north. They are also met with in the Forestry Commission woods around Achanaglachach in South Knapdale and as far north as Poltalloch north of the Crinan Canal. In recent years sika have also been reported south of Campbeltown and from some estates on the west Kintyre coast.

Further north, in Inverness-shire, sika are found in the Great Glen on both sides of Loch Ness and the Caledonian Canal: to the west they frequent Balmacaan and Creag-nan-Eun; to the east escapees from Colonel Frazer Tytler's introduction from Rosehall, Sutherland to Aldourie Castle about 7 miles (11 km) south of Inverness, have spread along the east side of Loch Ness through Glencoe, Dell, Fort Augustus and Culachy, south to Aberchalder near Invergarry. Away from the loch-side Japanese sika have been reported on Corriegarth near Gorthleck and at Flichity, Strathnairn.

In Ross and Cromarty Japanese sika which escaped from Sir Arthur Bignold's collection at Achanalt, Lochrosque, and from Coulin are still to be found in the former area and around Strathbran, between Inchbae and Garbat, northward to Kinlochluichart, Stratvaich, Amat Lodge and Alladale in Strathcarron. In Sutherland and Caithness the descendants of sika which escaped from Mr W. Ewing Gilmour's Rosehall Park when the deer-fence was breached during the last war, have spread to Achany, Glencassley and Glenrossal and neighbouring estates.

Ireland

County	Estate	Date
Co. Wicklow.	Powerscourt Park, Enniskerry (by Viscount Powerscourt from the animal-dealer, Jamrach).	1860.
Co. Kerry.	Kenmare, Muckross (from Powerscourt Park).	c. 1865.
Co. Fermanagh.	Colebrooke Park (from Powerscourt Park).	1870.
Co. Tyrone.	Baronscourt Park (from Colebrooke Park, Co. Fermanagh).	1891–2.
Co. Down.	Castlewellan Park (from Powerscourt Park, Co. Wicklow).	?
Co. Limerick.	Glenstal Park (from Powerscourt Park).	?

As previously stated, the Japanese sika deer introduced to Powers-court in 1860 were – jointly with the pair presented to the Zoological Society of London in the same year – the first of the species to be seen in the British Isles; Powerscourt, where by 1884 there were over one hundred head of sika, became the most important single source of supply of the species to other collections in England and Scotland as well as in Ireland.

The three sika deer – a stag and two hinds – introduced to the Kenmare estate around 1865 were set free in the woods surrounding Muckross Lake, and by about 1935 were said to have increased to between 3000 and 5000 head. Today they are still found in large numbers – over 500 in 1972 – on the 100000 acres (40467 hectares) or so of the Kenmare estate, including the deer-forests of Derrycunnihy and Glena, and in the Bourn Vincent Memorial Park at Muckross, as well as in the State-owned forest of Killarney: from Co. Kerry Japanese sika have wandered over the border into Co. Cork where they have taken up residence in the woods of Glengariff. The species is common in most of the woods of the eastern Wicklow Mountains – some 40000 acres (16186 hectares) – especially in the Glencree area and around Enniskerry, where in 1972 there were about 300. From here they have wandered north-west some 12 miles (19 km) across the river Dodder, to Saggart, Co. Dublin.

The five hinds and one stag brought from Powerscourt to Cole-brooke in Co. Fermanagh in 1870 by Sir Victor Brooke had increased to about 300 in 1891: from Colebrook they swam to Inismore Island on Lower Lough Erne, and by 1902 had crossed to Tempo and the woods around Crom Castle: by the 1950s they had reached Ballinglen and the Glen of Imaal, 30 miles (48 km) from Powers-court. Between 1885 and 1897 sika/red deer hybrids were reported from Colebrooke; today there are estimated to be between 150 and 160 sika in this area.

A number of the sika at Baronscourt Park in Co. Tyrone broke through the deer-fence in 1896 and became naturalized on Bessy Bell Hill, where they now number about 250: according to the Duke of Abercorn they are all of the *nippon* race. Sika presumably from Baronscourt appeared in Lislap Forest at Gortin in 1962 and at Seskinore, Knockmany Hill and in Glengannce Forest in 1968.

The Japanese sika of Ireland were almost exterminated by the severe winter of 1962–3: since then, however, they have staged a dramatic recovery, and there is now estimated to be some 2500 of the species throughout the country.

In Wales, a herd of some two dozen Japanese sika deer were kept in Vaynol Park, near Bangor in Caernarvonshire between about 1900 and 1950. In recent years a small number were introduced to Llannerch Park, near St Asaph in Denbighshire, but so far as is known no escapes took place from either collection, and there are today no wild Japanese sika deer at large in the Principality.

The following table, compiled from the distribution records sent by members* to the British Deer Society, gives particulars of the spread of Japanese sika deer in the decade between 1966 and 1975.

Year	County	District/Remarks
1966	Somerset	One seen at Hawkcombe, presumably descended from escapees from Pixton Park.
1967	Dorset	Deer from Wareham first appeared in vicinity of Charborough Park, Wimborne.
1968	Hampshire	In woods bordering Setley Common, Brockenhurst.
	Ross-shire.	Reported from Alladale Forest.
1969	Kent	Reported around Ashford and Sevenoaks; escapees from Merstham Hatch (Lord Brabourne) and Knole Park.
1970	Caithness	About 20 at Berriedale and around Langwell Water.
	Dorset	Reported from High Wood, Wimborne. A young 6-pointer – an escapee from Melbury Park – seen in autumn.
	Sussex	Burgess Hill district.
1971	Dorset	Three stags jumped into Sherborne Park, having escaped from Melbury Park. One stag seen at Up Cerne near Dorchester. Small numbers reported to have crossed the Purbeck Hills from the heathland around Poole Harbour. Woods in former Admiralty property at Holton Heath near Wareham colonized by sika, which have spread east to Holton Point and north-west to Poole Harbour. About 20 on estate outside Melbury Park.

*See note at foot of page 151.

Year	County	District/Remarks
	Dumfriesshire	Three stags reported around Moffat.
	Ross-shire	First record (a knobber*) from Torridon (July); nearest known stock at Achnasheen, 20 miles (32 km) to the east.
1972	Dorset	Reported from two fresh areas in the Poole Basin; Combe Wood, Wool, and Yellowham Wood, Dorchester. Also in woodland and on heathland west to Owermoigne and Warmwell, Dorchester.
1973	Dorset	Two in February on Brownsea Island.
	Inverness-shire	One at Knoydart which had presumably travelled *via* Loch Ness and Glengarry.
	Ross-shire	A decline in numbers reported from Amat Forest following re-afforestation and fencing by the Forestry Commission. Still on Alladale, Kincardine, in Easter Ross, colonized from Rosehall, Sutherland.
1974	Angus	Maintaining numbers and may be increasing at Kinnaird.
	Dorset	Isolated records from west of county. Appears to be increasing around Melbury Park and on Lundy Island. Estimated to number upwards of 1000 in the Poole Basin.
	Fermanagh	150 at Colebrooke in February: annual cull of 20–25.
	Ross-shire	Appears to be increasing in north-east and as far south as Amat Lodge.
	Somerset	Isolated records.
	Sutherland	Appears to be increasing around Rosehall and neighbourhood.
	Yorkshire	Two escapees from Allerton Park seen in woods near Knaresborough.
1975	Cornwall	A single animal seen in September near the Cheshire Home at Marazion.

*Second-year male.

20. Chinese Water-Deer
(*Hydropotes inermis*)

The home of the Chinese water-deer is the tall reedbeds and coarse grassland fringing the banks of a number of rivers in China and Korea.

The summer pelage – grown between April and June – of the adult Chinese water-deer is a light foxy reddish brown; this changes in winter to a dull sandy brown indistinctly flecked with grey. The hair on the body is rather harsh and loose, in order to contend with the severe climate of the animal's homeland. Both sexes stand about 20 in. (50 cm) high; a buck weighs between 25 and 30 lb (11 and 16 kg) and a doe some 5 lb (2–3 kg) less – approximately the same size and weight as an Indian muntjac. Neither the buck nor the doe of the Chinese water-deer grows any antlers,* but the former is provided with a pair of scimitar-shaped tusks, approximately $2\frac{3}{4}$–3 in. (7–8 cm) long, which protrude from the upper jaw. (In both these features the species resembles the unrelated musk deer (*Moschus moschiferous*) of the Himalayas.) The rut normally takes place in November and December and, most unusually among deer, between two and six fawns – a dullish brown or red with two ill-defined rows of yellowish-white spots – are born in the following May or June.

Chinese water-deer are, like the muntjac, a territorial rather than a herd species, and are seldom seen other than singly or in pairs; also in the manner of muntjac, they move at a great pace, usually progressing in a series of quick leaps. In England they frequent woodland and downland as well as the marshes and reed-beds which they favour in the east. Their voice has been described as a high-pitched bark or yelp, amounting almost to a wail.

The prolificacy of the Chinese water-deer doe, coupled with the absence of antlers and presence of tusks in the buck, tends to suggest

---

*Its scientific name means 'weaponless water-drinker.'

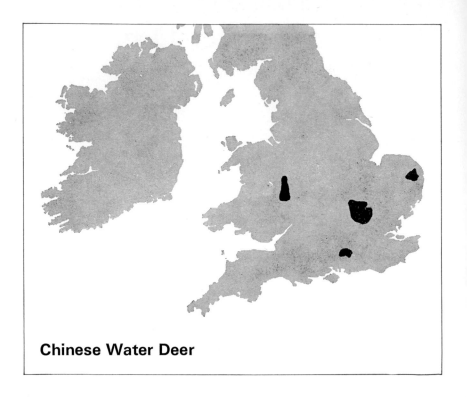

Chinese Water Deer

that the species is, like the muntjac, the survivor of a very ancient breed. In most or all of the early Tertiary deposits the male deer lack antlers but possess long upper-jaw tusks. When antlers became fully developed as efficient weapons, the need for and possession of tusks gradually disappeared. The muntjac, in which both antlers and tusks are of a somewhat rudimentary nature, is a good example of the transitional stage in this development.

Compared with the muntjac, our knowledge of feral Chinese water-deer in England is meagre indeed. At about the same time that he introduced muntjac to Woburn Park, the 11th Duke of Bedford added a pair of Chinese water-deer to his collection, both of which however died shortly afterwards. A few years later the duke made a second attempt to establish these deer in the park, which proved more successful, and thanks to their fecundity they soon formed a breeding herd. During the war a few managed to escape, and rapidly became naturalized in the surrounding woods, mainly in the south

and on the eastern side of the park around Ampthill and Flitwick. Their numbers were augmented by escapees from Whipsnade, which the duke had presented between 1929 and 1931 with thirty-two animals which by 1934 had increased to some two hundred head. At about this time a small number were despatched from Woburn to the collection of Mr Alfred Ezra, at Foxwarren Park, Cobham in Surrey, where they thrived but did not long survive the war.

In 1944 the Duke of Bedford sent some Chinese water-deer to two Hampshire estates – Leckford Abbas near Stockbridge, and Farleigh Wallop near Basingstoke, home of the Earl of Portsmouth – from both of which they appear to have escaped. In November 1945 Mr Brian Vesey-Fitzgerald shot one to the north of Farnham in Surrey, and a few years later they were reported from Silchester Common some 10 miles (16 km) south-west of Reading on the Hampshire/ Berkshire border. As Fitter points out, it seems possible that the two reports of muntjac in Hampshire in 1948 and 1949 really referred to Chinese water-deer. In 1949 several water-deer escaped from Farleigh, and in October 1953 one was shot at Preston Candover a few miles south of Farleigh Wallop: two months later Lord Portsmouth sent six of his water-deer to Cwmllecoediog, Aber Angell, near Machynlleth in western Montgomeryshire, where none survived for long.

In 1945 came the first report of water-deer from Buckinghamshire, and in 1950 from Northamptonshire and Oxfordshire. In the latter year two pairs were sent from Woburn to Walcot Park, Lydbury North, near Bishops Castle in Shropshire, home of Mr Noel Stevens. A considerable herd had been built up by the time Walcot was sold in 1956, when a number escaped and became naturalized in the surrounding countryside. Mr Stevens transferred about twenty to his new home, Hope Court, Hope Bagot, near Ludlow, where they soon more than doubled their number.

Also in 1950, six water-deer from Woburn were introduced to Studley Royal Park, near Ripon in Yorkshire, where three soon escaped. In 1952 a further two pairs were sent from Woburn and on 27 May of that year one was found seriously injured near Harrogate, about 10 miles (16 km) south of Ripon. By early 1954 only three or four remained at Studley Royal – all of them outside the park fence.

In 1961 a water-deer was reported from Souldern near Bicester in Oxfordshire, and in the same year an animal which was probably of this species was seen in reed-beds fringing Weston Turville Reservoir in Buckinghamshire.

Two Chinese water-deer seen at Swim Coots and Ling's Mill near Hickling in east Norfolk in 1958 were found to have escaped from a private collection at nearby Stalham. In the following year two escaped from the recently re-stocked Studley Royal Park. In 1970 a Chinese water-deer was shot in Marston Thrift Wood, 4 miles (6 km) north of Woburn, and another was found dead on a road at Wisbech in Cambridgeshire near the border with Norfolk and Lincolnshire. Two years later several were seen at Hickling Broad northeast of Norwich, and the species was reported as 'well-established and self-perpetuating' between Woburn and Whipsnade, while small numbers were apparently living a feral existence outside Walcot Park and Hope Court in Shropshire.

At the present time Chinese water-deer are naturalized in suitable areas of Bedfordshire, Buckinghamshire, Hampshire, Northamptonshire, the Norfolk Broads and Shropshire, and outliers are seen from time to time in neighbouring counties. In spite of its fecundity the species is not so widely spread nor so firmly established as the muntjac. One reason for this may be its vulnerability to predators – both carnivores and raptors – as the fawns weigh barely 1 lb (448 g) at birth. Because of its heavy coat, so necessary to its survival in the east, it is also prone to dehydration and heat-exhaustion in a more temperate climate. Should it become more widely naturalized it is likely to cause no worry to man as it is a grazer rather than a browser and does little if any damage to trees, shrubs, or crops.

N

21. Wild Goat
(Capra hircus)

The domesticated goat (from which is derived the wild – or, more properly, feral – goat of the British Isles)* is descended, with some cross-breeding with other races (of which there are many), from the Persian *pasang* ('rock-footed') or Grecian ibex (*Capra hircus aegagrus*). The range of the *pasang* is extensive; it is found on a number of islands in the Aegean Sea, eastwards to the Caucasus, Asia Minor and Iran.

The Grecian ibex or *pasang* (known also as the Bezoar) is an extremely active animal; it frequents rocky districts where it is notably sure-footed, making enormous leaps with unerring accuracy. The colour of the upper parts is brownish grey in winter, changing in summer to yellowish or rufous-brown, while the underparts are generally yellowish-white at all seasons. The adult male grows a short blackish beard: he stands about 3 ft (91 cm) at the withers, and possesses scimitar-shaped horns which measure from 40 to 50 in. (101 to 127 cm) in length.

Professor James Ritchie has stated that there is no evidence to show that the goat was ever indigenous to Britain. It was introduced to this country probably during the Neolithic era when the British Isles were still joined to the continental land-mass, and together with the sheep and ox was one of the earliest domesticated animals to be imported to these shores.

Whitehead cites a number of examples of evidence of goat husbandry in Britain within historic times. In 1229 when he was at Stamford, Henry III received a number of petitioners who claimed that Hugh de Neville, Keeper of the Forest of Rockingham, had refused to allow them to graze their goats in Clifford Forest. Between 1323 and 1324 no fewer than fifty-six people were tried by the

*For much of the information on wild goats in the British Isles I am indebted to G. K. Whitehead's monograph (see Bibliography).

justices of Epping for illegally grazing goats within the forest. The inhabitants of Broughton in Amounderness, Lancashire, claimed for themselves at the Forest Eyre* of 1334–6 the right, dating from time immemorial, of common pasturage for all their domestic animals except goats in Fulwood Forest. In about 1460 Syr Dafydd Trefor of the Isle of Anglesey wrote a poem asking for a quantity of goats to be sent to him from Snowdonia: an island neighbour, Gruffydd ab Tudur af Howell, in a second poem, objected to the proposed introduction on the grounds that the goats would damage Anglesey's forestry and agriculture. At a swanimote† held in 1479 in the Forest of Wyresdale, eight goatherds were arraigned for permitting their charges – some of which belonged to the Prioress of Seton who was fined 4d. for the offence – to enter the forest. A survey of 1615 reported that 'Sheepe and goates, most pernitious cattle, intolerable in a forest, make a far greater show than his Majesties' Game' in the appropriately named Kingswood Forest in Gloucestershire. As evidence that by the seventeenth

*Itinerant Court.

†Forest Assembly held triennially in accordance with a Forest Charter of 1217, originally to enable forest officials to superintend the clearance of the forest of pigs, cattle, sheep and goats.

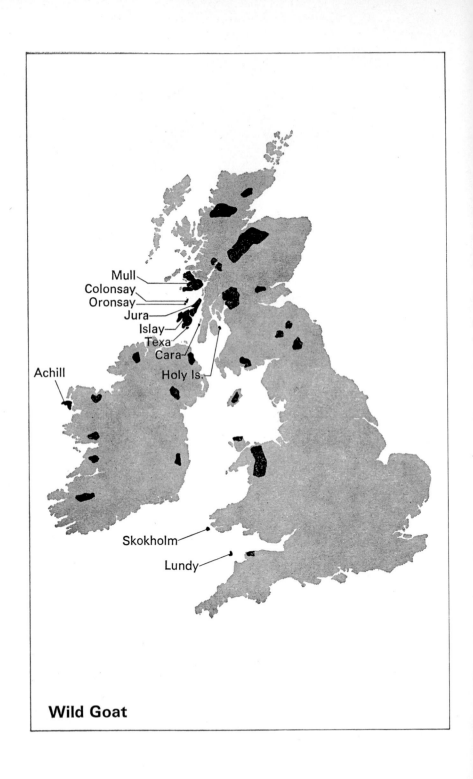

Mull
Colonsay
Oronsay
Jura
Islay
Texa
Cara
Holy Is.
Achill
Skokholm
Lundy

Wild Goat

century feral goats were already established in many districts, John Manwood wrote

there be some wilde beastes . . . that so long as they are remaining within the bounds of the Forest, the hunting of them is punishable by the laws of the Forest, such are wilde Goats, Hares and Connies.*

The wild goats in Britain today are in many cases the descendants of ancestors which had been wild for centuries, which were themselves the offspring of feral domestic stock. As one of the earliest of domesticated 'utility' animals, the uses made of goats since very early times have been many. Their skins have been much used in the manufacture of 'morocco' leather and in the making of parchment – a reminder of which is the paper known today as 'Goatskin Parchment'. Whitehead quotes Thomas Pennant who wrote that goatskin is

well adapted for the glove manufactory, especially that of the kid; abroad it is dressed and made into stockings, bed-ticks, bolsters, bed-hangings, sheets and even shirts. In the army it covers the horseman's arms and carries the foot-soldier's provisions. It takes a dye better than any other skin, it was formerly much used for hangings in the houses of people of fortune being susceptible of the richest colours; and when flowered and ornamented with gold and silver, became an elegant and superb furniture.†

In Ireland, according to John Rutty, kidskin was much used for making fans and firescreens.

Goat-horn is used for innumerable purposes, while Pennant records that in Caernarvonshire wild goats were hunted on the hills in autumn to provide tallow for candles.

The hair of goats was much in demand in the eighteenth century in the manufacture of wigs, and has also been extensively used in rope-making: domesticated Angora and Cashmere goats have from time to time been kept in Britain for their wool – known respectively as mohair and cashmere.

Whitehead quotes Pennant on the subject of goat-meat in Wales:

the haunches of the goat are frequently salted and dried, and supply all the uses of bacon: this by the natives is called *Coch y wden* or hung venison . . . the meat of a splayed goat of six or seven years old (which is called

*A Treatise of the Laws of the Forest, 1665.
†Tours in Wales, 1772.

Hyfr) is reckoned the best; being generally very sweet and fat. This makes an excellent pasty; goes under the name of rock venison, and is little inferior to that of the deer.*

In Ireland, according to Rutty, again quoted by Whitehead:

the flesh of the male goat, castrated and fed, makes a good venison, and lately kids are reared about the mountainous part of the south of Dublin for the delicacy of the flesh preferable to that of the lamb: for this purpose they are taken into the house presently after they are dropt, *viz*: before they have tasted their mother's milk, and suckled by ewes and fed with cow's milk spouted into their mouths.†

The milk of goats has for long been popular, both medicinally and as a valuable source of nourishment; the cheeses from Cheddar in Somerset were originally made exclusively from goats' and sheeps' milk.

Many years ago goats were owned by crofters much more commonly than they are today; they probably proved, however, the least useful of the domesticated animals, and with the decline of lactation and as they grazed further afield when the herbage coarsened in late summer, they became lost and were eventually allowed to run wild. Many, too, were turned loose when the crofters were evicted from their holdings. Pennant recorded that the goat was 'the most local of our domestic animals, confining itself to mountainous parts of these islands'.‡

In Wales Merionethshire and Snowdonia are today the principal strongholds of wild goats; here they occur mainly on Tryfan, the Glyders, Moelwyn, and the Rhinogs.

In *Summer Road to Wales* the author, an observant naturalist, thus describes his first sighting of the wild goats of Rhinog Fawr:

I was absolutely astounded at what I saw. They reminded me of pygmy bisons. Each animal, as far as I could see, had a very dark sienna-brown head, neck and forequarters, and their long hair was draped like a

---

*Tours in Wales, 1772.
†An Essay Towards a Natural History of the County of Dublin, 1772.
‡A Tour in Scotland & Voyage to the Hebrides, 1772, (Vol. I; 1776).

yak's . . . these creatures seemed to be nearly double the size of domestic goats.*

The goats found on the Great Orme, off the Caernarvonshire coast, are not truly wild but are the descendants of Cashmeres imported from Windsor Park at about the turn of the century to provide regimental mascots for the Royal Welch Fusiliers.

The majority of Welsh wild goats today are the descendants of ancient semi-domesticated herds which were allowed their freedom on the hills throughout the greater part of the year: it is also possible that some originate from escapees from herds of Irish goats which were at one time driven through Wales *en route* to sales in England: one of the last such drives took place in the autumn of 1891, when three hundred animals driven by three men, three boys and five dogs, travelled from Cardigan to Kent. Truly wild goats first appeared in Wales in the early 1770s, when Pennant refers to them on the Rhinogs; thirty years later Bachwy in Radnorshire was mentioned as being one of their few strongholds in south Wales. George Barrow described a herd which he saw between Festiniog and Beddgelert:

they were beautiful creatures . . . white and black, with long silky hair and long upright horns; they were of large size and very different from the common race.†

In England there were goats on Lundy Island in the eighteenth century, but these had disappeared by 1914; more were introduced by Mr Martin Harman in 1926, which soon increased to some two hundred: by 1956, however, their numbers had been reduced to about fifty, and by 1975 to around thirty. There have long been wild goats on the Lynton Rocks in Devon, while those on the Isle of Man may originate from some introduced by lighthouse-keepers between 1818 and 1875. A number of flourishing herds of goats exist on the Cheviot hills in Northumberland.

Whitehead traces the history of England's most celebrated herd of semi-feral goats at Bagot's Park, Blithfield Hall, Uttoxeter, Staffordshire – ancestral home of the Bagot family – where they have existed for some six hundred years.

In about 1380 the then representative of the family, Sir John Bagot, first adopted as his crest a goat's head; this has ever since remained a part of the Bagot arms: it appears on the graves of Lewis

*Nicholas Kaye, 1964.
† *Wild Wales*, 1854.

Bagot (1461–1534) and his son Thomas (1504–41) in Blithfield church, and resting on a tilting helm, hangs above the tomb of Richard Bagot (1552–96) and his wife. The colouring of the animals at Bagot is unique among British goats, the head, neck and shoulders being black, the rest of the body white: both sexes are long-haired and horned, and a billy stands about 30 in. (76 cm) at the shoulder. The home of these goats is in Valais, Switzerland – especially along the valley of the Rhône – where they are known as Glacier or Saddle or Schwarzhals goats. How they first came to Bagot remains uncertain; one theory is that they were brought over from Normandy at the time of the Norman Conquest; a former Lord Bagot believed that they were presented to an ancestor, John Bagot, by Edward II; the late Lord Bagot was of the opinion that the goats did not arrive until the late fourteenth century; alternatively, Sir John Bagot may have brought them back with him on his return to England in about 1387 from an expedition to Europe with the Duke of Lancaster (John of Gaunt), or they may have been presented to him directly by Richard II.

In 1939, when the herd numbered about eighty, the War Agricultural Executive Committee granted the then Lord Bagot permission to maintain the herd at a maximum of sixty: in the late 1940s the number rose to one hundred; in 1954 the herd was reduced again to about sixty by culling and transplanting. In 1957 the Forestry Commission assumed responsibility for the woodland at Bagot, and the goats which since 1937 had freely roamed the 2000 acres (809 hectares) of wood and park, were restricted to a 110 acre (44 hectare) enclosure. After Lord Bagot's death in 1961 all but a dozen were sold to Mr Robin Bagot of Levens Park, Kendal, where a small herd survives today. A herd of about twenty is still kept at Blithfield where, tradition maintains, some goats must remain in order to ensure the continuance of the Bagot family.

In Scotland wild goats are thinly but widely distributed throughout the Highlands and Islands, as well as in a number of Lowland and Border counties.

A tribe of large white goats on the forest of Mamlorn in Perthshire is mentioned in a report on the rebellion of Hew Murray from 1661–79, when it was said that 'he and his men would never be subdued while they could get a goat on Creag Mhor'. Millais mentions an old established tribe on the Morar hills, Inverness-shire, and others in Wester Ross, Strathpeffer, Rothiemurchus (where Seton Gordon saw them on Creag Leth Choin up to the outbreak of the First World War), and on the Perthshire hills from Glen Lyon to Rohallion. Goats are mentioned as being common at Altyre and Douchfour and in Glen Moriston and Glen Urquhart, Inverness-shire in 1835.

There were at one time several tribes of wild goats on the dangerous Black Isle cliffs in Ross and Cromarty, which were probably the descendants of those seen there by Charles St John when he lived in Scotland between 1837 and 1853: he also saw goats near Bridge of Dulsie on the river Findhorn, which he described as being 'long-horned, half-wild, and with shaggy hair and long, venerable beards'. In the Cairngorm region goats were to be found on the slopes of Lochnagar, Slochd (from where they disappeared in 1947), and in Glen Callater, Braemar, and at Abriachan, Balmacaan and Ben-nachie, Perthshire. A story of about 1829 relates that 'the only novelty from the hill this season is the murder of half-a-dozen goats by an English party in Atholl, who had mistaken their prey for deer'. In this same forest wild goats were part of the game driven to Mary Queen of Scots during her visit to Blair Atholl in 1564.

A fine bearded head in the British Museum (Natural History) marked 'Scotland, 1894, presented by Sir Donald Currie' is thought to be from a tribe then established at Schiehallion in Perthshire. Loch Achray and the slopes of Ben Venue, Callander, were until recently frequented by the descendants of about forty goats mentioned by the poet Southey in 1819, about which he wrote:

The extirpation of wild beasts from this island is one of the best proofs of our advancing civilization: but in losing these wild animals, from which no danger could arise, the country loses one of its great charms.

There were also goats at Loch Lairig Eala, Glenogle, Glen Falloch and a score at Ledard near Loch Ard in 1875. A petition of 1763 by crofters of Dalclathick asked for a reduction in their rents as they were no longer allowed to keep their goats on the forest. In 1590 'Mad' Colin Campbell and his men encountered a herd of wild goats on Stuchd an Lochain in Glenlyon.

The goats at Inversnaid on Loch Lomond are reputed to possess pedigrees dating from the time of Robert the Bruce (1274–1329), who ordained that they should never be molested, and provided a sanctuary for them named Pollochthraw. According to legend, in

1306 the king was hiding in a cave when a tribe of goats lay down outside the entrance; his enemies, thinking that the animals would not go near a man, omitted to search the cave. Another version of the story is that the goats provided the king with sustenance and shelter in his time of need. Tradition records that goats once inhabited a yew-tree-grown island called Inch Longaig on Loch Lomond where they killed all the trees.

Millais mentions a long-established tribe of thirty to forty goats coloured black, grey and black-and-white, on Gruinard Forest in north-west Ross-shire, and another on Fisherfield Forest. Goats of local origin were liberated on the nothern side of Strath Beag near Dundonnell in Wester Ross in 1911, and soon began to flourish. A small tribe of some ten or twelve magnificent goats with long, thick, pure white coats and strong, wide-spreading horns on An Tellach near Dundonnell were introduced as recently as 1927: some of these were later reported from the north shore of Little Loch Broom. About forty similar goats were to be found at Kildonan, Sutherland and at one time almost a hundred were at Rhirevoch on Beinn Ghobhlach. There were about thirty on Dundonnell near Ullapool in 1928, when they were also reported from Loch Broom.

A story of about 1745 tells of goats on an island called Eilean nan Gobhar (Isle of Goats) on Loch Ailort, Inverness-shire. There were others at Benderloch, Easdale and Eorsa, and there were once over fifty on Aonach Mor or Innean Mor in Morven, Argyllshire, though these have long since disappeared; there were at one time about the same number on Aird-Ghobhar (Ardgour), and on Stob Ghabhar (Hill of Goats), Blackmount and near Uamh nan Gabhar (Cave of Goats) on Colonsay.

Goats are also well known in the Hebrides, and Martin (before 1697) says of the 'Isle of Harries' that goats were in the Isle of Lewis. He relates a story of the islanders about 'a couple of goats that did grow wild on the hills and after they had increased they were observed to bring forth their young twice a year'.* There were still a few at Dune Tower on Lewis in 1930. Harvie-Brown describes the goats of Harris as having fine heads and horns; he also records goats on Canna Island off the north-west coast of Rum where they had been introduced before 1793; on Ardinaddy before 1812; on Tiree and Coll, and on 'another small island near Oronsay'.

---

*A Description of The Western Islands of Scotland, 1703.

Pennant suggested in 1776 that in Skye 'goats might turn to good advantage if introduced to the few wooded parts of the island'. 'Skie' is also mentioned by Martin as 'producing' goats among the 'cattell' at the biannual fair. There are two fine white heads from Scalpay dated 1895 in the British Museum (Natural History). Seton Gordon reported goats above Suishnish in Sleat, and Fraser Darling saw them on Priest and Horse Islands, and on Sgurr nan Goibhrean (Peak of Goats) on Rum. There was formerly a mixed tribe on Eigg, which at one time included a billy imported from Holy Island at the beginning of the century. In 1922, about ten or twelve grey and black-and-white goats were introduced to Little Colonsay, west of Mull. In 1802 there were a few on Staffa Island, but these soon died. On Mull there are goats at Lochbuie where Seton Gordon reported that 'wild goats spring from ledge to ledge on the great rocks that tower above', and there were at one time some at Pennyghael. Between fifty and a hundred were once at Kilpatrick on Arran. The island of Ulva west of Mull had a tribe of about fifty which once marooned themselves on a tidal islet in Loch Tuarth; in trying to swim back through the wind and current between twenty and twenty-five were drowned.

Pennant saw about eighty goats on Jura in the 1770s: the Royal Scottish Museum at Edinburgh has a fine cream specimen shot there in 1931, and also a black-headed goat presented by Viscount Astor. There are three highly successful colonies on Islay: the oldest, at Kinnabus near Mull-of-Oa, dates 'from time immemorial'; the second is at Smaull near Kilchoman, where about seventy were introduced from Kinnabus in 1787; the third is at Gortantaoid on Loch Gruinart, where they were introduced from Mull and Jura at the beginning of the century.

In the Clyde area there are goats on the Mull of Kintyre and possibly still on Davaar Island. There were magnificent white goats (said to be the descendants of survivors of a wreck of the Spanish Armada) on Ailsa Craig from Pennant's time to 1925. The *Old Statistical Account of Scotland* (1791) states that Ailsa was uninhabited 'save for the wild goats', and in about 1836 the tenant of the Craig 'went there for a few days' shooting of the wild goats which abound'. Pennant saw goats on Arran in 1772, and there have been all-white animals on Holy Island for many years. On Bute there was a herd of about forty at Garroch Head which was destroyed in about 1938–9.

In the Lowlands there was a well-known tribe near St Mary's Loch in Selkirkshire (mentioned among others by Sir Walter Scott), with more at Saddleyoke, Brandlaw and Megget Water. Goats are also sometimes seen at Moffat Water, and about twenty were once at Ewes near Langholm, Dumfriesshire. There were half a dozen near Kells, New Galloway in 1912, and they were once to be found at Glenapp and Cairnharrow. On the Scottish side of the Cheviots small numbers of wild goats are found to the south of Jedburgh in Roxburghshire.

The feral goats of the British Isles are all of the 'wild' type; when goats escape from domestication they revert to this wild form within a few generations, and sometimes within ten years. They are usually large in size with long shaggy coats, and bear horns measuring from 40 to 45 in. (101 to 114 cm) in length; the normal colours are brown, grey, black-and-white, or pure white, these last being the finest specimens of all. (The Duke of Montrose wrote that in his opinion the pure white goats of Ailsa Craig, Holy Island and Creag Mhor are really the correspondents of the so-called 'wild' white cattle at Chillingham, Chartley and Cadzow which now exist solely as semi-feral herds in parks*.) Wild white goats were once much in demand to provide regimental drummers' aprons, and the billies' beards or 'tassels' formed the original sporrans of some Highland regiments. Fraser Darling believes that the influence on their nitrogen metabolism, which makes feral goats run increasingly to hair and horn until the standard of the wild goat is reached, is natural selection acting mainly through the climate.

Except during the rut, which takes place from early September to late October and is hastened by cold weather and shorter days, the billies keep themselves apart from the nannies and kids in separate herds in which there is invariably a strict 'pecking order'. The kids, which when young frequently fall prey to foxes and golden eagles, are born from the end of January to late March (when the prevailing weather ensures that only the strongest will survive) after a gestation period of some 150 days, and twins are not uncommon. In Britain feral goats are thought to live for about twelve years and weigh on average between 112 and 140 lb (51 and 63 kg). Goats are browsers rather than grazers, eating the leaves and twigs of shrubs

*Countryman, April 1932.

such as bramble, briar, gorse, ivy and heather, and coarse weeds such as thistles, docks and nettles, in preference to richer herbage. They also cause damage to trees by 'barking', especially in winter, ash, elm, hazel, holly, rowan, willow and yew. Given the opportunity, goats will eat considerable quantities of salt, and numbers on the islands and west coast of Scotland have been drowned while searching for seaweed, for they are poor swimmers.

The myths and superstitions concerning goats are innumerable and go back at least as far as Biblical times. In many country districts in Britain it is still commonly believed that goats help to prevent contagious abortion among cattle and *enzootic ataxia* ('staggers') in horses. In the eighteenth and nineteenth centuries goats' whey was regarded as possessing valuable medicinal properties, and whey-drinking centres were established in Leswalt, Wigtownshire; at Blairlogie, Stirlingshire; in the foothills of Clackmannanshire; on the Isle of Arran, and at Abergavenny, Monmouthshire.

In Merionethshire it was thought that the smell of goats acted as a barrier against infection – a belief which was even extended to the skin of a dead animal – and that the smell of the billy kept rats away. Many hill-farmers believe that if goats and sheep are kept together the former will feed in the more precipitous places, thus keeping the latter to safer ground with richer feeding. Country people regard goats as a better weather prophets than sheep, as they seek earlier refuge on low ground from winter storms.

In Scotland wild goats are reputedly adept at killing adders – hence the Gaelic adage *'Cleas-na gooiths githeadh na nathrack'* ('As the goat eateth the viper'). Until the seventeenth century it was the custom of the fishermen of the Inner Hebrides to hang a male goat in the rigging of their ships to produce a fair wind. The 'Ettrick Shepherd', James Hogg (1770–1835), claimed for goats the power to drive evil spirits from the mountain tops.

In Ireland the goat is at the centre of the 'Puck Fair' held annually from 10 to 12 August at Killorglin, Co. Kerry – a ceremony which dates back to at least 1613, at which time it was held at Lammastide – 1 August – to honour a male goat which is crowned first 'Puck King of the Fair' and then 'His Majesty Puck King of Ireland'. Whitehead quotes from Margaret A. Murray a description of the fair and the information that 'the name Puck is a derivative from the Slavonic word Bog, which means God'.*

* *The God of the Witches*, pp. 43–4, 1956.

Once goats have begun to run wild they take to the higher reaches of the hills above the peat-line at about 1700 ft, (518 m), where the drainage is good, for they dislike damp ground. They move about with great agility even on the most dangerous rocks, over which they seem to know all the possible routes. They scramble about deliberately and seldom appear to be in a hurry; during the breeding season the billy leads, with the kids next and the nannies bringing up the rear to prod the kids if they stray from the path. Parties normally consist of about half-a-dozen nannies with their kids, led by an adult billy, with some young billies on the outskirts. They are extremely shy and wary, and if approached wait until the intruder is hidden by the ground before disappearing with great speed. They cause considerable damage to trees, and are most unpopular with deer-stalkers as when alarmed they utter a snorting hiss through the nose, making a noise which carries as far as the bleat of a sheep: deer can sometimes hear them up to half a mile away if they are in a high corrie on a still day when the sound is carried and amplified.

At the present time wild goats are of only local distribution in England, Ireland and Wales, although they are perhaps more widespread in Wales than is commonly thought, and are widely but thinly distributed in the Highlands and Islands and in parts of the Lowlands in Scotland. There seems to be little spread or migration, but where they are found the animals appear to be holding their own.

The following table, adapted and condensed from Whitehead, gives particulars of areas supporting populations of wild goats in the British Isles today:

England

Location	Remarks
Devon. Lundy Island, Bristol Channel.	Present eighteenth century; extinct 1914; re-introduced 1926; about 90 in 1952; about 50 four years later: around 30 by 1975.
Devon. Lynton Rocks, Lynton.	Population are descendants of those introduced many years ago to the Valley of Rocks.
Isle of Man. Laxey, Glen Sulby, Glen-Snaefell area.	Small population.

Location	Remarks
Isle of Man. Calf of Man (island off south coast).	Population are believed to be the descendants of some introduced by lighthouse keepers 1818–75.
Lancashire. Coniston and Tilberthwaite.	Possibly some small tribes on fells.
Northumberland. Cheviot Hills and College Valley near Wooler.	Small tribes from Brigg Fell, Hareshaw, to Nunwick Moor; Catcleugh and Cottonshope, Redesdale, and Harthope Linn on south slope of Cheviot. Herd of mostly blue-grey goats north of Cheviot, east of College burn, introduced about 1860; Hindhope; Kielderhead Moor; Bewcastle and Roan Fells. About 30 in 1971. Goats of the Cheviots may be descended from stock liberated by the monks on Lindisfarne.
Northumberland. Thrunton Crag, near Alnwick.	About 3 or 4 south of Whittingham near Bridge of Aln in 1962.

Wales

Location	Remarks
Isle of Anglesey. Amlwch (north coast).	Small tribe reported 1961.
Caernarvonshire. Snowdonia.	Mostly between Llyn Ogwen Valley and Pass of Llanberis; Glyder Fawr, Glyder Fach, Tryfan; Y Garn, Elidyr Fawr, Y Foel Goch, Moel Perfedd and Snowdon. About 78 in 1964: considerably more by 1968.
Merionethshire. Craig-y-Benglog, north-east of Dolgelley.	Introduced about 1870: on eastern face of Rhobell Fawr and Dduallt to Wenallt rocks north of Dolgelley/Llanuwchllyn road.
Merionethshire. Moelwyn, west of Ffestiniog.	Century-old (or more) herd of mostly black and white goats; about 15–20 in 1972.

Location	Remarks
Merionethshire. Rhinogs, north of Dolgelley.	Near Llanbedr range over Craig-y-Saeth above Llyn Cwm-by-Chan, Rhinog Fawr, Rhinog Fach, and Foel Ddu. Around Cwm Nantcol; Coed y Rhygen; Arenigs; Cader Idris; Craig Aberyn. Sometimes on Y Llethr and Diphwys, Llawr Llech and Y Graigrddrwg. In 1957 five goats from Bagot Park were released on Rhinogs by Major F. Bennett. Total of about 50 in 1972. Numbers of wild goats in Wales were killed in 1957–8 to prevent the spread of foot-and-mouth disease, and recent afforestation exterminated some herds. Total Welsh population estimated to be about 300 in 1972.
Pembrokeshire. Isle of Skok-holm.	Small number reported in 1970.

Scotland

Location	Remarks
Argyllshire. Conaglen, north of Ardgour.	On Guisachans at head of Loch Shiel. About 20 at head of Loch Eil in 1958.
Argyllshire. Glencrippesdale, near Morven.	Two small tribes above Camas Saluch.
Argyllshire. Kilmalieu, near Ardgour.	About 30–40 between Meall Na Each and Druim Na Maodalaich. Mostly dark grey or black and white.
Argyllshire. Mull of Kintyre.	Pennant refers to goats on Mull of Kintyre in 1772. Today about 50 are around Rudha Duin Bhain, Uamh Ropa and Earadale in the south-west, and between Machrihanish and Carskey Bay. About 40 are on Carradale Point, north of Campbeltown.
Ayrshire. Kirrieroch Hill.	A few are on Kirrieroch Hill and Mullwharchar north of Merrick.
Dumfriesshire. Craigieburn, near Moffat.	About 50 are at Saddleyoke, Birkhill, and Roundstonefoot.

Location	Remarks
Dumfriesshire. Langholm.	About 20 since before 1930 around Tarraswater.
Inverness-shire. Glen Affric.	About 40 in 1959.
Inverness-shire. Dalmigavie, near Tomatin.	About 20 east of Glenkyllachy.
Inverness-shire. Dochfour.	Between Dochfour House and Abriachan; believed to be descendants of those belonging to the builders of the Caledonian Canal.
Inverness-shire. Coignafearn near Tomatin.	In 1968 about 150 goats in Monadhliath hills.
Inverness-shire. Kinveachy, near Tomatin.	Century-old herd; 26 in 1951.
Inverness-shire. Loch Hourn.	Small numbers around the loch, and on Glenquoich, Arnisdale, and Barrisdale.
Inverness-shire. North Morar.	About 20 on Sgor na h-Aide and Carn Mor.
Inverness-shire. Corrybrough near Tomatin.	Small numbers.
Inverness-shire. Cairn Kincraig and Creag a Chrocain.	A few on the Inverness-shire/Nairnshire boundary.
Kirkcudbrightshire.	Range over hills from Merrick and Rhinns of Kells in north to Cairnsmore of Fleet in south. Population estimated at about 750 in 1972.
Morayshire. Lochindorb.	On Lochindorb Moor around Glentarroch Rocks. About 30 in 1952.
Nairnshire. Valley of the Findhorn.	Goats have been in the Findhorn Valley for more than a century. Two main herds today range over Dunearn, Drynachan, Lochindorb and Dalrachne. About 40 north of river in 1968.
Perthshire. Ben Lomond.	Old established herd.
Ross and Cromarty. An Teallach, Dundonnell.	About 40, originally introduced in 1927.

o

Location	Remarks
Ross and Cromarty. Beinn Eighe, Meall a Ghuibhais.	About 50 in 1952.
Ross and Cromarty. Clunie and Glenshiel.	Small numbers.
Ross and Cromarty. Five Sisters, Kintail.	About 60 on Sgurr na Carnach and Sgurr na Ciste Duibhe opposite Glenshiel.
Ross and Cromarty. Glomach.	Small numbers.
Ross and Cromarty. Gruinard.	Long-established herd of about 50.
Ross and Cromarty. Inverinate and Dorusduain.	About 50 around A' Ghlas-bheinn and Beinn Fhada (Ben Attow), Kintail.
Ross and Cromarty. Kinlochewe.	On Letterewe side of Slioch and around Loch Garbhaig and Lochan Fada: about 50 in number.
Ross and Cromarty. Lael.	About 40 in 1959; appear to be increasing.
Ross and Cromarty. Leckmelm, near Ullapool.	Small numbers, and on Ardcharnich and Meall Dubh.
Ross and Cromarty. Letterewe.	Three herds: (a) about 15 on the southern slope of Slioch: (b) about 20 on the north and south sides of Beinn Airidh Charr: (c) about 30 on the south face of Mhaighdean and Sgurr ne Laocainn – all probably originated from Gruinard. Small tribe on Eilean Suthainn on Loch Maree.
Ross and Cromarty. Scoraig.	Small number.
Ross and Cromarty. Shieldaig.	About 56 on the north shore of Loch Torridon between Diabaig and Red Point.
Ross and Cromarty. Slattadale.	Small number of black billies and grey nannies.
Ross and Cromarty. Tollie, near Poolewe.	About 20 reported.
Roxburghshire. Jedburgh.	Between 10 and 30 on Forestry Commission land at Lethem (Wauchope), and occasionally at New Castleton.

Location	Remarks
Selkirkshire. Ettrick and Eskdale.	Some around Birkhill on the Dumfriesshire border, and at times on Herman Law, Bell Craig and Bodesbeck Law east of Ettrick Water. Goats have been in this area since before 1820.
Stirlingshire. Ben Lomond.	On the eastern flank of Rowardennan to beyond Inversnaid, mainly in the Ptarmigan Loch district; also on Craig Rostan. 70–100 in 1952; perhaps about double that number today.
Sutherland. Kinlochbervie.	Small tribe at Rimichie on the south shore of Loch Inchard. Few on the rocks above Laxford Bay.
Sutherland. Loch Inver, (? Soyea Island).	Small number reported 1952.

Scottish Islands

Argyllshire. Cara.	White goats have certainly been present from the eighteenth century and are reputedly descended from stock which came ashore from wrecked galleons of the Spanish Armada in 1588. About 20 today.
Argyllshire. Colonsay. Islay.	Over 100 around Balnahard and Machrins. About 70 mostly white goats in 1952 near Gortantaoid and Staoisha. Some 300 in 1960s at Kinnabus on Mull of Oa – probably oldest Islay herd. Between 40 and 60 in 1952 near Smaull and Braigo, introduced from Kinnabus in 1787. In 1964 there were twenty white goats on the Rhinns where they had been introduced ten years previously from Smaull.
Argyllshire. Jura.	About 50 near Craighouse, mostly white or black and white. Some 200 mostly dark brown goats are between Corpach Bay and Loch Tarbert. On Ardlussa there are perhaps 100 more.
Argyllshire. Little Colonsay.	A few reported.

Location	Remarks
Argyllshire. Mull	On the Ardmeanach peninsula, including Burgh, Tavool and part of Balmeanach. Between 150 and 200 between Carsaig Bay and Uisken. About 25 near Bunessan. Occasionally seen on Beinn-an-Aoinidh. Between 300 and 400 on the south and west shores of Loch Buie.
Argyllshire. Oronsay.	Herd of some 30 dark brown and black goats has been established for some years.
Texa (south-east of Islay).	About a dozen in 1964.
Holy Island (east of Arran).	Thirty-nine white goats in 1969.

David Mackenzie, quoted by Whitehead, postulates the theory that the white goats of the Scottish coast 'owe their origin to Scandinavian seafarers who frequented these shores, in whose homeland the white 'Telemark' goat has long been popular. . . . The Viking longboats . . . may have carried the white goat of Norway to the Western Isles.'

Ireland (Republic).

Location	Remarks
Co. Clare.	A few in the Burren Hills.
Co. Donegal.	A small number on Glenveagh, particularly near Creenarg.
Co. Dublin.	Goats are on the Howth peninsula, but are not truly feral.
Co. Galway.	About 100 recently between Leam and Costelloe. Small tribes in the mountains of northern Connemara, and perhaps along the coast and on the islands. A few are possibly around Lough Lettercraffroe. About 50 are near Kylemore Lake.
Co. Kerry.	Small numbers on Skellig Rocks in St Finan's Bay. Herd of 150 mostly brown goats on Mt Eagle and on the western slope of Brandon Hill has been established for over 60 years. Herd of white goats reported in 1950 from the Dingle peninsula, which has sheltered wild goats for

Location	Remarks
	almost 100 years. Small tribe in Torc and Mangerton hills.
Co. Mayo.	Goats on Achill Island since the mid-eighteenth century. Today there remain less than 100 animals. Fifteen on Curraun peninsula in 1957.
Co. Sligo.	Small numbers reported in the hills in recent years.
Co. Tipperary.	About 6 reported recently in Slieve-na-Mon area.
Co. Waterford.	A few in Comeragh hills.
Co. Wicklow.	Bray Head herd is not truly feral. In the Wicklow Mountains about 20 wild goats are around Luggala Lake, a few are at Glencree, and some 60 are around Glendalough.

Ireland (Northern)

Location	Remarks
Co. Antrim.	Small number (6 in 1959) are on Fair Head. About 10 are on Garron Point. A few (21 in 1959) are near Dunemencock, Glenarm. Approximately 20 on Rathlin Island cliffs are possibly descendants of some liberated in about 1760.
Co. Armagh.	Herd of about 20 are on Camlough; 1959 census by District Forestry Officer reported 16 from Killeary and Ballintemple; 23 from Shanroe; 20 from Carrickbroad, and 9 on Slieve Gullion.
Co. Fermanagh.	A few reported from islets on Lough Erne and around Lough Navar.

22. Soay Sheep
(Ovis aries)

The Soay sheep of the St Kilda group of islands – situated some 50 miles (80 km) due west of Harris in the Outer Hebrides – have been called by James Fisher 'relics, living fossils of the domestic sheep of a millennium ago'. They represent what is today the most primitive breed of domesticated sheep surviving in Europe, and probably bear as close a resemblance to the peat- or turbary-sheep of the late Neolithic period as any existing in the world today.

The earliest inhabitants of these rugged islands kept a breed regarded by Professor James Ritchie as a primitive stage in the domestication of the wild sheep of Europe, the mouflon (*Ovis musimon*), which is today confined to the mountainous areas of Corsica and Sardinia. It seems likely that this stock are the survivors of a species which in former times inhabited the lowland woodlands of the Mediterranean region. Soay sheep are probably the direct descendants of this early domesticated breed.

The date of the introduction of Soay sheep to St Kilda is unknown, but R. N. Campbell in *Island Survivors* . . . has done much to unravel the mystery.

The islands have been inhabited by man for the greater part of the Christian era, and possibly for some time previously. They were almost certainly settled well before the arrival of the Vikings, who were themselves replaced by a Gaelic-speaking people from the Western Isles when the latter were ceded to the Scottish crown in 1266. The name of St Kilda first appears in literature in an Icelandic saga written before 1249, which refers to the arrival of a storm-driven vessel in 1202.

Soay Sheep may have been introduced to St Kilda in prehistoric times. Dr J. M. Boyd points out that the bones of present-day Soays closely resemble those of a breed discovered in late Neolithic deposits (*e.g.* at Skara Brae on Orkney) and in late Bronze

Age/early Iron Age (*c.* sixth to first centuries, B.C.) sites at Jarlshof in the Shetlands, as well as in Romano-British settlements in Wiltshire and at Bar Hill, Dunbartonshire. There is evidence that late Stone Age/Bronze Age man used St Kilda in conjunction with Megalithic standing stones to provide a sight-line in his observation of celestial bodies

It is known that the island of North Rona – some 50 miles (80 km) north-west of Cape Wrath and approximately half-way between the Shetlands and St Kilda – was inhabited in pre-Viking times, around 300–400 A.D. This, therefore, is another possible time for the introduction of Soay sheep to the Western Isles. On the other hand, the domination of the Vikings over the Outer Isles, coupled with direct evidence of their occupation of St Kilda, and the close resemblance between the bones of Soay sheep and those of a breed found in Viking settlements in Greenland, all point to a possible Norse origin. If this is so, they might well have been brought first to the Orkneys* and Shetlands from Denmark and Scandinavia around 800 A.D., and thence to St Kilda.

*Possible remnants of this early introduction are to be found in the 'kelp' or seaweed-eating semi-wild Orkney sheep of North Ronaldshay. With the exploitation of North Sea oil, there is a risk that this colony might be exterminated by an accidental discharge of oil, while the increase in human traffic poses an additional threat from foot-and-mouth disease. With the view to establishing a separate colony of this ancient breed of sheep, the Rare Breeds Survival Trust has recently purchased the more

[*Footnote continues on page 216*

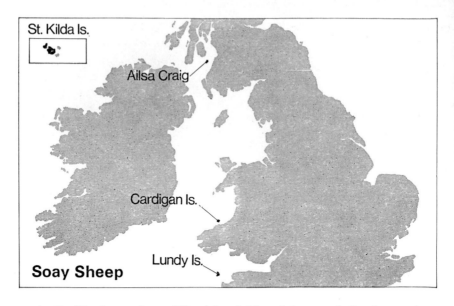

St. Kilda Is.

Ailsa Craig

Cardigan Is.

Lundy Is.

Soay Sheep

A. B. Taylor writes: 'The island [Soay] is noted for its ancient breed of sheep. The name (O.N. *Sauoa-ey* = Island of Sheep) does not provide proof that the Norsemen placed the breed on the island. It is more probable that they named the island from the sheep they found there.' As Campbell points out, of 130 principal Gaelic and/or Norse place-names on Hirta and Boreray (two of the islands in the group) only two refer indirectly to sheep, and twelve to other animals. This could be either because sheep were well established when the islands were first colonized by the Scottish Celts or Norsemen, or because the sheep arrived after the principal topographical features had been named.

Recent evidence, however, tends to lend support to the theory of a Viking origin for the Soay sheep of St Kilda. In a paper published in 1969 R. J. Berry* revealed that the St Kilda field mouse (*Apodemus*

sheltered 142 acre (56 hectare) island of Linga Holm in the Orkney group. A small flock of these sheep has also recently been given a home on the small Channel Island of Lihou – some 700 miles (1126 km) south of their Orcadian home – by Lieutenant-Colonel Patrick Wootton, a member of the R.B.S.T. The first Channel Island-bred Orkney lambs were born in mid-April 1975, the pregnant ewes having previously been transferred to Colonel Wootton's Guernsey estate, Les Rouvets, to protect the lambs from roving dogs. On North Ronaldshay the ewes normally give birth to single lambs only. It will be interesting to see whether they become more prolific as a result of the milder climate and richer feeding of the Channel Islands. ('The Orkney Sheep' *Country-Side*, 22(8), p. 377, 1974. 'Old British Livestock', John Vince (*Shire*, 1974). 'Land of the Seaweed Eaters', *Country Life*, p. 1187, 8 May 1975.

*See Bibliography of shrews, voles and field-mice.

sylvaticus hirtensis) more closely resembles Norwegian field mice than it does those of the British mainland, *A. sylvaticus*. In the same way, the now extinct St Kilda house mouse (*Mus musculus muralis*) was more like the species found in the Faeroe and Shetland islands than the typical British *M. m. domesticus*. The probability is that these mice were accidentally introduced from Scandinavia and Denmark to St Kilda in hay and other fodder which the Norsemen brought over to provide food for their previously imported domestic stock: among these would most probably have been Soay sheep, which later escaped from domestication and became feral. The present stock are therefore, as Dr Boyd remarks, 'living survivors of a domesticated breed of a thousand years ago'. Since then the stock of this prehistoric breed has remained intact on the island of Soay off the north-west tip of Hirta, the main island of the group, from which it is separated by a towering cliff and several hundred yards of tempestuous sea. Here they were seen by Hector Boece (1465?–1536) when he visited the islands in about 1520:

Beyond this island [Hirta] is yet another isle, but it is not inhabited by any people. In it are certain beasts, not far different from sheep, so wild that they cannot be taken except with a snare; their hair is long . . . neither like the wool of sheep or goat. . . . This last island is named Hirta, which in Gaelic is called a sheep; for in this island are great numbers of sheep, each one bigger than any male goat with horns longer and thicker than those of an ox, and with long tails hanging down to the earth.*

Fifty-one years later they sorely puzzled the Bishop of Ross, John Leslie (or Lesley; 1527–96), when he saw them during a pastoral visit to St Kilda:

Near here [Hirta] lies another island, uninhabited, where no cattle are found except some very wild, which whether to call them sheep or goats, or neither sheep nor goat, we know not. . . . They have neither wool like a sheep nor hair like a goat.†

Soay sheep have survived unchanged on St Kilda because they are geographically isolated, and because of the absence of development in the breed due to their small population and the lack of outside

---

Scotorum Historiae, 1527. Since Boece's time the sheep of Soay have become much smaller and have grown progressively shorter tails.
†*De Origine Moribus et rebus gestis Scotorum*, 1578.

influences. The stock on Soay – which has remained at about two hundred for centuries – may well be the direct descendants of those originally introduced hundreds, and conceivably thousands, of years ago.

In medieval times a four-horned breed, known as the 'St Kilda' or 'Hebridean' sheep, was introduced to Hirta, but not to Soay, from northern Europe, to improve the stock. It was this breed that Martin Martin described in *A Late Voyage to St. Kilda* (1698) following his visit to the islands in June of the preceding year, when the human population stood at around 180. In *The History of St. Kilda* (1764), Kenneth Macaulay described the same breed as being present when he was the minister on the islands in 1758, by which date smallpox had reduced the human inhabitants to about thirty. In about 1830 Scottish blackface sheep (which probably originated in the Pennines) were introduced to Hirta and Boreray, and remained the principal breed on those islands until the time of the human evacuation a hundred years later. During the first half of the nineteenth century the number of islanders rose again to around a hundred: thirty-six emigrated to Australia in 1856, after which the population remained at about seventy until the First World War. By 1928 the number of people had fallen to under forty, and the entire population of thirty-six was removed – at their own request – in 1930.

Soay sheep are small goat-like animals, with supple bodies and somewhat disproportionately long necks. They stand about 2 ft (61 cm) high at the shoulder, on rather thin and lanky legs: rams weigh around 75 lb (34 kg) and ewes about 20 lb (9 kg) less. There are two principal colour forms; one is dark chocolate-brown* above with an off-white underside and rump – all ewes of this colour being horned: this has led to the suggestion that this form may be descended from the Asiatic wild sheep (*Ovis vignei*), variously known as the urial, shapo or sha. The second colour form, which comprises about one-third of the present St Kilda population, is a pale oatmeal. The wool of the Soay sheep is close, fine and soft, but being intermixed with hair, has to be plucked rather than sheared every summer. Because of the colour variation in the wool of Soay sheep, it is probable that patterns for cloth were designed based on the natural

*This colour-phase was known to the St Kildans as *lachdann*, *cf.* the Manx *loghtan* for the snuff-brown native sheep of the Isle of Man. (K. Williamson and J. M. Boyd.)

colours of the wool – a practice which is still followed today in the northern isles. Alternatively, a kind of block pattern may have been employed, as used in the manufacture of Scottish clan tartans. For the early St Kildans, Soay sheep would have been a source of meat and milk as well as of wool. Two to three year old rams have high outswept horns: adult five year olds have heavy, dark horns which grow in a strong, thick, tight, downward-curving whorl. Many rams have thick, rough manes similar to those grown by true wild sheep. Over half the ewes have spiky horns which grow backward and out-ward. The rut takes place between early October and late November, after the rams have left their single-sex flocks in late September, and the lambs are born in the following April. Fleeces are shed between May and June. Like domesticated sheep, Soays are gregarious animals, but unlike domesticated breeds they tend to scatter rather than gather if approached by men and dogs. They are extremely agile and sure-footed, picking their way with an extraordinary sense of balance along seemingly impenetrable paths on their island home, grazing and moving freely on steep cliff-faces. In the evening they tend to assemble in small herds and move from the glens – to which they return in the morning – on to higher and steeper ground.

When in 1930 the inhabitants of St Kilda were evacuated to the Scottish mainland, they took with them between 500 and 600 black-face sheep from Hirta leaving the flocks of blackface on Boreray and Soays on Soay intact. In 1931 the 4th Marquess of Bute bought St Kilda from the Macleods of Macleod:* in the following year a number of the islanders returned to St Kilda, and on Lord Bute's instructions transferred 107 Soay sheep (20 rams; 44 ewes, 43 lambs, of which 22 were wethers and 21 ewes) from the 244 acre (97 hectare) island of Soay to the 1500 acre (600 hectare) island of Hirta. The St Kildans returned annually to tend the sheep on Boreray and Hirta until the outbreak of war in 1939, when the Soays on Hirta were said to number 500 head. In 1947 Fisher esti-mated the population at between 400 and 450, and Darling in the following year suggested that the total might be between 650 and 700.

In recent years comparatively little attention has been paid to the

*In 1957 the islands became the property of the National Trust for Scotland as a National Nature Reserve, under the administration of the Nature Conservancy.

Soay sheep remaining on the island of Soay, or to the feral blackface on the 190 acres (76 hectares) of Boreray. In July 1952 Dr Boyd, standing on the 693 ft (211 m) high hill of Cambir on Hirta, counted sixty Soay sheep on Soay, which is separated from the larger island by the several hundred yards wide Soay Sound. In the same year T. B. Bagenal saw a considerable number of blackface sheep grazing on the slopes of Boreray, where in the previous year J. Cunningham had counted 340.

Under the auspices of Dr Boyd a census of Soay sheep on Hirta was undertaken on 5 August 1952, and subsequently annually since 1955, for which figures taken from *Island Survivors* were as follows:

	Ewes	Lambs	Rams	Total
1952				1114
1955				710
1956				775
1957				971
1958				1099
1959	722	458	164	1344
1960	448	86	76	610
1961	417	417	76	910
1962	624	221	211	1056
1963	821	526	243	1590
1964	654	237	115	1006
1965	753	567	149	1469
1966	918	495	185	1598
1967	593	231	52	876
1968	571	415	118	1104
1969	770	357	70	1197
1970	633	351	78	1062
1971	855	753	175	1783
1972	747	600	207	1554
1973	809	482	109	1400
1974				1340
1975				1013

Counts of blackface on Boreray between 1951 and 1971 yielded the following totals:

	Total	
1951	340	
1956	365	
1959	430	(45 rams: 258 ewes and yearlings: 127 lambs)
1960	330	(25 rams: 200 ewes and yearlings: 105 lambs)
1963	413	(13 rams: 206 ewes and yearlings: 194 lambs)
1965	364	
1966	450	
1970	440	
1971	466	

All the blackface on Boreray are horned; the adults are mostly either all grey or all tan in colour, or grey with tan-coloured faces and legs; a few carry black fleeces. They differ from the domesticated breed in having short and rather bare tails, similar to those of Soay or Shetland sheep, and long necks.

Dr Boyd considers that the variation in the colour of the Soays of St Kilda cannot be attributed to interbreeding with those blackface which remained behind when the islands were depopulated in 1930; on the island of Soay, where there have never been any blackface, both dark and light colour forms of Soay sheep live today.

In 1955 Dr Boyd, standing on the cliffs of Cambir on Hirta, counted eighty-seven Soays on the island of Soay. He also counted some 150 feral blackface on Boreray: in both cases these were estimated to be about half the total stock. In 1966 and 1967 observers on Soay counted respectively 115 and between 140 and 160 light- and dark-coloured Soay sheep.

The 40 per cent fall in the population of Soays on Hirta between 1952 and 1955 may well have been due to the harsh winter of 1954–5. Any variation in population is unlikely to be caused by shortage of grazing: the quality of the pasturage on the maritime meadows of St Kilda is exceptionally high, due in part to the valuable supplies of guano deposited by the vast quantities of seabirds, including shags, razorbills, guillemots, puffins, fulmars, manx shearwaters, storm petrels, kittiwakes, and the largest gannet colony in the north Atlantic, numbering some 44000 pairs, which are drawn to the islands by the fish which abound in the plankton-rich sea. Late winter storms, which strike during the lambing season and when the resistance of the adults is lowest, are undoubtedly the worst threats which the Soays have to face. Their survival between January and April depends as much on the weather and feeding of the previous autumn as on current conditions. The ruined buildings of the islanders' abandoned village, and the seven hundred or so surrounding *cleitans*,* are of the greatest importance to the survival of the sheep on Hirta, providing valuable shelter from the fiercest winter weather. The *cleitans* are often chosen by the sheep as places in which to die.

A number of other British off-shore islands today contain, or have in the past held, feral populations of Soay sheep: these include Lundy Island in the Bristol Channel; Ailsa Craig off the coast of Ayrshire, and several islands off the Pembrokeshire coast.

In 1910 a breeding nucleus of Soays was transferred from St Kilda to Woburn, where the light-coloured variety was rigorously weeded out. In 1934 the Duke of Bedford sent two young rams and four ewes – all dark brown animals – to R. M. Lockley on the 242 acre (96 hectare) island of Skokholm: here they proved slow in establishing themselves, but had multiplied to about forty ten years later. After 1946 the flock was administered by the West Wales Field Society and Field Studies Council. On 18 April 1944 one ram and three ewes were sent from Woburn to the 40 acre (16 hectare) Cardigan Island in Cardigan Bay: Mr Ronald Stevens of Walcot Hall in Shropshire provided the same number, one of which, a ginger-coloured ram, was removed to the 21 acre (8·4 hectare) island of Middleholm in September 1945, together with a single ewe from Skokholm. By 1959 these had multiplied to a flock of about seventy largely ginger-coloured animals. On 21 July 1952 one ram and three ewes were introduced to the 20 acre (8 hectare) St Margaret's Island near Tenby, where by 1959 they had increased to between twenty and thirty. On Christmas night of that year the entire flock was killed by lightning. In 1958 some Soays were introduced to the island of Skomer: two years later there was a total of about forty on Skomer, Skokholm and Cardigan. In 1975 the only Welsh island to

*See page 78.

possess a flock of Soays was Cardigan, where in July there were about eighty sheep of which a high proportion were rams. The tendency is for the flock to increase, and when it exceeds about a hundred, the animals are culled.

The late Mr Martin Harman introduced some Soays to Lundy in about 1927, where by 1959 they numbered over eighty head. By 1975, when the island was on lease from the National Trust to the Landmark Trust, the flock had been reduced to about sixty dark-coloured sheep; these are culled from time to time: in 1973 a dozen rams were dispersed among a number of flocks on the mainland. The Soays on Ailsa Craig were introduced direct from St Kilda in the 1930s; in 1956 the herd consisted of fourteen light-coloured animals.

The wild Soay sheep of the St Kilda islands – and the islands themselves – are unique in a number of ways. The sheep live in a habitat totally isolated from outside influences, and in controlled conditions which afford admirable opportunities for scientific observation: there have been no removals or introductions of fresh blood for many years: there is a complete absence of ground predators and the few avian ones, such as hooded crows, ravens and greater and lesser black-backed gulls, take toll of only a small number of lambs and of the sick: the sheep are provided with an abundance of rich pasturage, for which there is no competition from other herbivores. Apart from deer and goats, Soay sheep are our sole surviving larger wild mammal – and the one which, moreover, probably most closely resembles its direct ancestors. They have been well called, by Williamson and Boyd, 'one of the faunal treasures of Britain'.

II Birds

23. Ring-Necked, Rose-Ringed or Green Parakeet

(Psittacula krameri)*

The ring-necked, rose-ringed, or green parakeet† (*paroquet* or *parraquet*) is found in Africa north of the Equator from Senegal and Nigeria to Sudan and Uganda, through Mauritius and the Seychelles eastward to Burma, India, Sri Lanka, Malaysia and south-western China. It has been introduced to Egypt, Iraq, Iran, Kuwait and Oman.

The male ring-necked parakeet is yellow and green in colour, and has a long and pointed tail. The throat is black, the eye pale yellow, the bill crimson, and the feet yellowish orange. Encircling the neck, the nape of which has a bluish bloom, is a half-collar of a delicate rose tinge. The female is similar in appearance, but lacks the black throat, blue nape and rose collar. Both sexes measure between 16 and 17 in. (35 and 38 cm) in length, including a 9 to 10 in. (23 to 25 cm) long tail.

In Africa the ring-necked parakeet frequents semi-arid country interspersed with patches of dense thorn-scrub. In India, however, according to Dr Jerdon, it favours

cultivated grounds and gardens, even in the barest and least wooded parts of the country, and it is habitually found about towns and villages, constantly perching on the house-tops. It is very destructive to most kinds of grain, as well as to fruit gardens. When the grains are cut and housed it feeds on the ground, on the stubble cornfields, also on meadows, picking up what grain it can; and now and then takes long flights, hunting for any tree that may be in fruit; and when it has made a discovery of one in fruit,

*The alternative scientific name *Palaeornis torquatus* ('ancient ring-necked bird') is derived from the tradition that the first specimen in Europe was brought back to Greece from the Punjab in north-western India by Alexander the Great in about 324 B.C.

†According to W. W. Skeat's *Etymological English Dictionary* (1879–82), the name parakeet is derived from the Spanish *periquito* or *perroqueto*, meaning a small parrot, the diminutive of *perico*, a parrot, which in turn may be a diminutive of the proper name, *Pedro*.

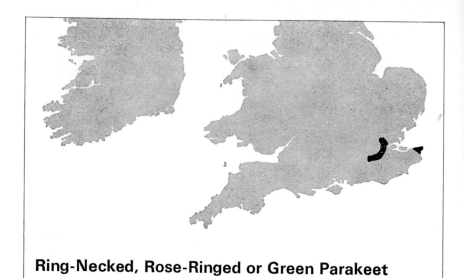

Ring-Necked, Rose-Ringed or Green Parakeet

circling round and swirling with outspread and down-pointing wings till it alights on the tree. It associates in flocks of various size, sometimes in vast numbers, and generally many hundreds roost together in some garden or grove. It breeds both in holes in trees, very commonly, in the south of India, in old buildings, pagodas, tombs, etc., and lays four white eggs. Its breeding season is from January to March. The ordinary flight is rapid, with repeated strokes of the wings, somewhat wavy laterally or arrowy. It has a harsh cry [kee-ak, kee-ak], which it always repeats when in flight, as well as at other times.

Being such colourful and decorative birds, members of the parrot tribe – *Psittacidae* – which includes parakeets – have always been extremely popular with aviculturists, from whose collections they have frequently escaped. Writing in the *Animal World* in about 1855 J. H. Gurney described how 'they have several times made nests [in the wild at Northrepps in Norfolk] and on five of these occasions the young have been brought to maturity'. Fitter records that in the early 1930s a number of feral green parakeets frequented gardens bordering Epping Forest for two or three years. A party of half a dozen or more was observed feeding on haws at Loughton in Essex, between Woodford and Epping, in the autumn of 1930; two years later, however, only a solitary male remained. In August 1899 a single South-American monk or quaker parakeet (*Myiopsitta monachus*) – which was frequently to be seen in the company of a flock

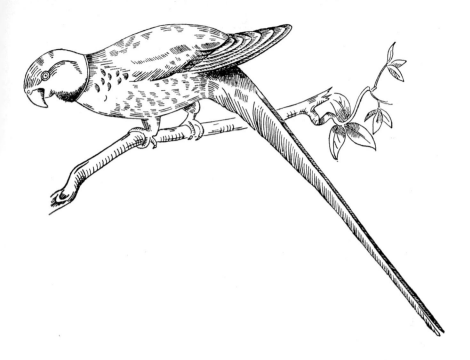

of starlings – was captured in the New Forest in Hampshire, where it was kept in semi-freedom until June 1902. In May 1900 a mate was procured for it, but although a nest was built on the roof of a nearby house in the following October, no young were reared. In October 1936 thirty-one quaker parakeets were turned out at Whipsnade, where they remained in complete freedom for a number of years, as they have also done from time to time at Woburn.

In 1969 a family party of ring-necked parakeets was observed at Southfleet, some 8 miles (13 km) west of Rochester in Kent. In the same year others were reported from the Northfleet, Gravesend and Shorne area. Today ring-necked parakeets are a familiar sight in the Gravesend area (where in 1975 one pair nested in the thatched roof of a farmhouse), from where they are spreading east to the Medway towns; they are also to be seen flying over the marshland villages (*e.g.* Cliffe) to the north. Between 1972 and 1973 a new colony was established based on Margate; in 1975 up to six pairs are believed to have nested successfully in Northdown Park (on the eastern outskirts of the town), where flocks of up to twenty-two parakeets have been counted, and birds are now seen regularly all over the Isle of Thanet south to Stodmarsh in the Stour Valley and as far west as Herne Bay. This broad distribution suggests widespread breeding, although this has so far not been confirmed.

In the spring of 1974 ring-necked parakeets were reported in the Cuckmere Valley in East Sussex, and in the area of Chichester Harbour in West Sussex. The warden of the neighbouring Pagham Harbour Bird Reserve believes that parakeets bred on the reserve in that year. In 1975 up to five 'green parrots' were reported from Hollingbury Park, Brighton, but there is as yet no evidence of feral breeding, and it has still to be confirmed that these birds were *krameri*.

Also in 1969 small groups of parakeets were seen in the neighbourhood of Bromley (Langley Park), Beckenham, Shirley (Monks Orchard) and Croydon, on the border between Surrey and Kent. In this area the birds were first proved to have nested in the wild near Croydon in 1971. In the spring of 1972 and 1973 a number were seen displaying and prospecting for nesting-sites in Langley Park, where a flock of eleven was observed in March of the latter year, which had increased to at least a dozen by the following June. Fifteen parakeets were counted in Langley Park on 5 September 1973, and there were at least twenty there three weeks later. From the middle of February 1974 a pair were noticed displaying in the Park, and in April the same or a different pair paid repeated visits to a large hole in an old elm tree: during the second week of May a single recently fledged juvenile was observed which was fed by both parents for several days. During the breeding season (which appears to extend from mid-February to late May) of 1975, at least two pairs successfully reared three young in the Park, where on 27 August a total of fifteen birds were seen. The same number were counted on 23 November and again in January 1976, suggesting that this colony is now well established and is reasonably resistant to winter conditions. The birds in Langley Park appear to range in small parties over a wide area by day to feed, returning to a communal roost in the park at night. They spend most of the time high up in the canopy of mature deciduous trees (especially ash, elm, horse-chestnut and oak), visiting open spaces principally in the early morning, where they are very wary and difficult to approach.

From 1973 there have been several reports of ring-necked parakeets from the Bexley area of north-east Kent, some 8 miles (13 km) east of Langley Park, mostly from near Tile Kiln Lane, Joyden's

Wood and Green Street Green, where breeding has not yet been recorded.

Thirteen miles (21 km) west of Langley Park, between Esher and Claygate in Surrey, a pair of ring-necked parakeets first appeared in 1970; they bred in the following year, and a family party of six was reported in February 1972: nesting was again confirmed at Esher in 1973 and 1974. Most of the records for Surrey have been from the Ashstead/Bookham/Chessington/Claygate/Esher/Surbiton district, but parakeets have also been seen at Dorking, Lightwater, Nonsuch Park, Send, Clandon Park near Guildford, Vann Lake and Wallington.

From 1972 onwards flocks of up to ten ring-necked parakeets have frequently been seen around Old Windsor in Berkshire, and up to seven near Wraysbury, Buckinghamshire, where there is, however, as yet no evidence of breeding in the wild. On 21 October 1975 a single bird was seen by the author at Winkfield in Berkshire, some 5 miles (8 km) south-west of the Old Windsor colony. A 'homing' flock at Marlow in Buckinghamshire, where the birds are from time to time allowed to fly from their aviary, is another potential source of feral parakeets in this district.

Another population of ring-necked parakeets has been established since at least 1971 north of the Thames around Woodford Green and Highams Park in south-west Essex: up to twenty-two birds have been counted in a single flock, and a pair was seen examining a hole in an oak tree in 1973.

At Herstmonceux in East Sussex two families of parakeets were seen in 1975, but breeding was not confirmed.

On 1 June 1975 a pair of parakeets was seen building a nest, which was reported as looking rather like that of a magpie, high up in an oak tree at Alfreton, some 10 miles (16 km) south of Chesterfield in Derbyshire. From the description of the nest it would appear that these birds were not ring-necked parakeets, which nest in holes, but monk or quaker parakeets. This seems to be the only record in recent years of feral parakeets breeding in the wild away from the London area.*

Three separate sources have been suggested as the origin of the feral parakeets of south-east England: they may have come from free-

*My thanks are due to Mr Robert Hudson for drawing my attention to this record.

flying homing colonies kept at semi-liberty by aviculturists, which have failed to return to their aviaries; they may be escapees from pet-shops or exotic bird-farms, such as that at Keston some 3½ miles (6 km) south of Bromley, since London is one of the main centres of commercial aviculture; or they may have been deliberately turned loose by members of ships' companies when they realized that importation would be delayed by a long and expensive period of quarantine. Whatever their origin may be, feral ring-necked parakeets are today a familiar sight to city-bound suburban commuters in a number of districts to the south, south-west and south-east of London.

In a letter to *British Birds* magazine in 1974 the writer suggested that the success of the ring-necked parakeet in becoming established in south-east England is due partly to the fact that, because it is cheaper than other species, larger numbers are imported, and partly because its readiness to steal food (including mixed seeds, dried fruits, nuts, coconut and, on at least one occasion, fat from a tit-feeder) put out for other birds helps it to survive in all but the harshest winters. 'It would not be difficult', the writer continues, 'to stop the nonsense (to put it at its lowest) of a parrot getting on the British and Irish Lists.' But what is nonsensical about a parrot on the British List? Its presence there would be no more incongruous on zoogeographical grounds than that of the Egyptian goose, the mandarin duck or the golden pheasant, and no one, surely, feels that their inclusion on the List is 'nonsense'?

In view, however, of the immense amount of damage which parakeets could cause to agricultural and horticultural crops, especially in a fruit-growing county such as Kent where they have been observed eating the buds and mature fruit of apples, pears and plums; because they commandeer for their own use the nesting holes of native species; and because of the not inconsiderable danger which they present to humans from psittacosis ('Parrot Fever') – a highly dangerous and sometimes fatal respiratory disease akin to pneumonia – an eradication programme in England, similar to that carried out against colonies of feral monk parakeets in south-east Florida, southern California and New York City,* may become necessary, if, indeed, it is not already too late.

*Wilson Bull; 85, pp. 491–512.
Note: See also Addenda.

24. Budgerigar

(Melopsittacus undulatus)

The budgerigar* is widely distributed in suitable habitats – open country interspersed with trees and scrub – throughout Australia.

In the wild, the general ground colour of the budgerigar is a grassy-green, mottled with undulating and alternating bands of darker green, greyish-black and browny-yellow. The forehead, crown, cheeks, and under tail-coverts are a pale shade of primrose yellow; each cheek has an oblique blue patch, beneath which are three round black spots. The 4 in. (10 cm) long blue tail-feathers are narrow and acuminate. The sexes are alike except that the cock has a black cere, that of the hen being brown or cream: both measure about $7\frac{1}{2}$ in. (19 cm) in length including the tail.

The budgerigar is both gregarious and nomadic, associating in flocks of considerable size which tend to move south in late winter. In its homeland it nests, between October and December, in colonies in hollows and holes in trees, where from four to eight white eggs are laid, and sometimes two broods are reared. Both cock and hen help to rear the young, which the cock will continue to feed by himself should the hen be killed. The budgerigar's food consists mainly of seeds and berries; its voice is a continuous and melodious warble.

Budgerigars were seen in great abundance in New South Wales by Captain Charles Sturt (1795–1869) during his ill-fated explorations between 1828 and 1845. John Gould (1804–81), who is credited with

---

*The name 'budgerigar' (variously boodgerigar, budgerygah, betherrygah, boodgereegar, etc.) is derived from the aboriginal *boodgeri* meaning 'good' (to eat) and *gar* meaning a cockatoo or parakeet. The bird is also sometimes known as the Australian love-bird, the undulated grass-parakeet or the shell-parakeet. In her *Household Management* Mrs Beeton gives, under 'Australian Dishes', a recipe for 'Paraquet Pie', which doubtless refers to budgerigars.

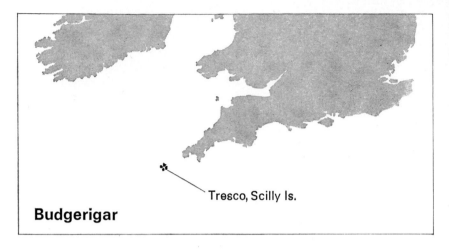

Budgerigar

Tresco, Scilly Is.

the first introduction of the budgerigar to England in about 1840
wrote:

This gentleman informed me that on the extensive plains bordering the
Murrumbidgee [river] he met with this lovely species in immense flocks,
feeding upon the seeds and berries of the low stunted bushes as they flew
to water holes in flocks up to a hundred strong.

In March 1861 the animal-dealer Charles Jamrach acquired no fewer
than six thousand 'Australian grass parakeets', which had been
consigned to London from Port Adelaide on board the clippers
Orient and *Golden Star*, which he offered for sale to the London Zoo.

Many people in Britain have, at one time or another, kept in their
gardens free-flying 'homing' budgerigars. These are allowed out of
their aviaries during the day, returning to them at night to roost.
Over the years many have failed to 'home', while others have been
deliberately given their freedom, and have become temporarily
established locally, *e.g.* at Margaretting in Essex; in Windsor Great
Park in Berkshire; and in parts of the New Forest in Hampshire.
    Research for the British Trust for Ornithology's *Atlas of Breeding
Birds in Britain and Ireland* (1976) revealed three recent nesting-sites;
at Wigginton, south of Tring in Hertfordshire, where a pair nested
in an orchard in 1971 and 1972 over half a mile (800 m) from the
aviary from which they are believed to have escaped; near Downham
Market in Norfolk where thirty birds were released in 1970–1 in

an attempt to establish a free-flying 'homing' flock: several pairs bred in 1971 and 1972, but the flock rapidly diminished and there has been no satisfactory evidence of breeding since the latter year: and at Fenstanton, south of St Ives in Huntingdonshire, where one pair nested four to five feet above ground in the stump of a decayed apple tree in a garden at Honey Hill: between a dozen and fifteen birds were seen here in 1974. On mainland Britain, however, it seems unlikely that feral budgerigars could survive in winter without artificial feeding and shelter.

In 1969, as a result of her visit to the gardens of Tresco Abbey in the Scilly Isles, Her Majesty the Queen Mother suggested to Lieutenant-Commander and Mrs Thomas Dorrien-Smith that the gardens there would be a suitable place in which to establish a colony of free-flying budgerigars.

Accordingly, an aviary, based on a design of the Duke of Bedford, was constructed during the summer in the garden below the Abbey, into which were introduced in November four pairs of free-flying budgerigars from Royal Lodge in Windsor Park. The single exit/entry hole – designed when open to prevent intrusion by sparrows – was then closed. Early in April 1970 eight nesting-boxes were placed in the aviary, and breeding commenced almost immediately. After the young had hatched, the exit/entry hole was opened for the first time. Initially, only the cocks chose to leave, always returning to roost with their families in the evening. Later in the summer the cocks were joined in the gardens by the hens and young birds, but all continued regularly to return to the aviary.

As this first brood of young budgerigars appeared to be not very strong, the colony was reinforced in the autumn of 1970 by the addition of a further six feather-clipped pairs, and the exit/entry hole was once again closed. In the following spring, after these birds had begun to rear their young, the hole was re-opened, never to be closed again. By this time the number of nesting-boxes in the aviary (which were removed during the winter to prevent over-breeding) had been increased to sixteen, most of which were occupied. The birds appeared to spend most of the time in the aviary, only leaving for flights around the gardens.

In 1972, however, it was noticed for the first time that the budgerigars were nesting outside the aviary in holes in the trunks of corduline (*Corduline australis*) and palm trees in the gardens, where they fed on the seeds of the grass *Poa annua*; the rush *Juncus bufonius*; *Arenaria leptoclados*; *Sagina* sp.; (? *apetola*), and *Coronopus* sp. As soon as the fledgelings were able to fly they were brought to the aviary by their parents who showed them how to use the exit/entry hole and how to feed from the seed-hoppers.

By 1973, when the population stood at about sixty birds, the habits of the budgerigars began to change considerably. Instead of spending most of the time within the aviary and only leaving to fly around the gardens, they became noticeably more independent: they started to live as well as to nest outside their former home (only about four nesting boxes inside the aviary being occupied) returning solely to feed or to shelter from stormy weather.

The birds living in the gardens appeared to nest continually throughout the year in natural holes in a variety of trees; nesting boxes in the aviary were no longer used, and boxes placed in trees were likewise ignored. In the spring of 1974 the trunks of palm trees opposite the aviary contained a considerable number of nests, and some seven or eight colonies of budgerigars were flying freely around the gardens. By 1975, when the population was estimated to have increased to over a hundred, the birds had been seen in small flocks on the neighbouring islands of Bryher, St Martin's, St Mary's and St

Agnes, but there is so far no evidence of breeding away from Tresco.

Following the death of Commander Dorrien-Smith his widow left the island; as a result the budgerigars are now entirely self-supporting. They have no natural predators in the Scillies, and appear to consort amicably with flocks of starlings, but it is surmised that some losses occur during a harsh winter. It will be interesting to see whether this unique British flock of these delightful little birds will become successfully established as permanent members of the Scillonian avifauna.

Note: See also Addenda.

25. Little Owl
(Athene noctua)

The range of the little owl extends through central and southern Europe as far north as the Baltic Sea, eastwards to China, and south to the North African coast of the Mediterranean.

This bird is the smallest British owl, measuring only 8½ to 9 in. (21 to 23 cm) in length. It is, moreover, the owl most frequently seen during the day, flying with a buoyant and bounding flight similar to that of a woodpecker, from which, however, it is easily distinguished by its rounded instead of pointed wings, short tail, and general 'owl-like' appearance. It is often seen in the day-time perched in the open on a branch, wall, fence-post or telegraph-wire. The little owl's plumage is greyish-brown above, barred and mottled with white. The underparts are whitish streaked with brown. The facial disk is somewhat flattened above the eye (the iris of which is a pale canary-yellow) giving the bird a rather fierce expression.

The favourite haunt of the little owl is open agricultural country-side, plentifully supplied with nesting-sites in the form of mature and pollarded trees and farm-buildings. Here, in suitable holes, the hen lays between three and five matt-white eggs in late April or early May, which hatch after twenty-eight or twenty-nine days. The normal call of the little owl is a monotonous and plaintive 'kiew-kiew', several times repeated. The alarm note has been likened to 'the bark of a toy terrier' and 'a short hysterical laugh with a catch of the breath'.*

During the 1930s, when the little owl achieved its greatest rate of expansion in England, it aroused the ire of ill-informed gamekeepers and landowners on account of its alleged eating habits. The British Trust for Ornithology commissioned Miss Alice Hibbert-Ware to carry out a detailed survey of the little owl's diet. She examined

*Stanley Morris, *Bird Song*, p. 104.

2460 food pellets* and the gizzards of twenty-eight little owls, and analysed the material from seventy-six nests, some of which were purposely taken from game-rearing parts of the country. Eighteen months of intensive investigation disclosed only two pellets each of which contained the remains of one game-bird poult. There was further evidence of a possible third game-bird fledgeling, and the remains of seven chicks of domestic fowl. It was found that 49·24 per cent of the little owl's diet consists of insects: the remains of 10217 earwigs (343 in one pellet), burying- and ground-beetles, cockchafers and weevils, were discovered as well as those of craneflies (Daddy-long-legs) whose larvae (leatherjackets) cause so much harm to farmers and gardeners. The balance of the little owl's food comprises principally small rodents, birds (4·96 per cent: mainly starlings, sparrows, blackbirds and thrushes), woodlice, earth- and slow-worms, snails and slugs, and centipedes and millipedes. Miss Hibbert-Ware's report, published in 1937,† proved conclusively that the little owl causes far more good than harm in the English countryside, and showed that Coward's fear that 'the intentional release of the Little Owl is a menace', and Professor Ritchie's stricture that

*All but the largest prey of owls is normally swallowed whole without being picked of plucked: the indigestible parts are disgorged in the form of pellets, those of the little owl measuring from 1–1½ in. (2·5–4 cm) long.
†See Bibliography.

Little Owl

main breeding range

secondary breeding range

' . . . Nature [has] also mocked us in this country in the case of the introduction of the Little Owl . . . everywhere it has betrayed its trust . . .', were totally without foundation.

Few birds in history have given rise to as much folk-lore and super-stition as owls. The ancient Greeks worshipped a goddess – Pallas Athene, daughter of Zeus – who was said sometimes to appear in the guise of an owl. The little owl was particularly abundant in the neighbourhood of Athens (hence the expression 'owls to Athens', synonymous with 'coals to Newcastle'), where it was regarded as a sacred bird. From Athens it acquired both its scientific name and also its reputation as a symbol of intelligence. A Greek coin depicting on the obverse the head of Athene and on the reverse a little owl, was known colloquially as an 'owl'. In both Europe and Asia the owl has for centuries been regarded as the bird of doom and death: in ancient Rome, and indeed among superstitious country-people in Britain until the turn of the century, an owl perching on a house was said to presage a forthcoming death. The bird even gave its name to a crime under English law known as 'owling': this was the offence of 'transporting sheep or wool out of the country, to the detriment of the staple manufacture of wool', and was usually carried out during the hours of darkness. It first appeared on the statute-books in 1336–7 during the reign of Edward III, and was not finally repealed until 1824.

The first little owls to reach England were those brought back from Rome by the eccentric but engaging squire of Walton Hall in York-shire, Charles Waterton.

Thinking that the civetta [the little owl] would be peculiarly useful to the British horticulturalist, not, by the way in his kitchen, but in his kitchen-garden, I determined to import a dozen of these birds into our own country. . . . I agreed with a bird-vendor in the market at the Pan-theon for a dozen young civettas, and having provided a commodious cage for the journey, we left the Eternal City on the 20th July 1842, [see note] for the land that gave me birth. All went well, until we reached Aix-la-Chapelle. Here, an act of rashness on my part caused a serious diminution in the family. A long journey and wet weather had tended to soil the plumage of the little owls; and I deemed it necessary

Q

that they, as well as their master, should have the benefit of a warm bath. Five of them died of cold the same night. A sixth got its thigh broke. I don't know how; and a seventh breathed its last, without any previous symptoms of indisposition, about a fortnight after we had arrived at Walton Hall. The remaining five . . . have been well taken care of for eight months. On the 10th of May 1842,* there being abundance of snails, slugs and beetles on the ground, I released them . . . at seven o'clock in the evening, the weather being sunny and warm, I opened the door of the cage; the five owls stepped out, to try their fortunes in this wicked world.

Having survived Waterton's prediliction for hot baths and his altercations with unco-operative Swiss bankers in Basle and officious Italian customs officers in Genoa (the Squire's accounts of which make amusing and entertaining reading), it is sad that the 'civettas' disappeared into the park of Walton Hall, and were not seen again.

Apart from about twenty recorded examples, which according to *The Handbook of British Birds* may have been genuine vagrants, there is no further reference to little owls in England until Lieutenant-Colonel E. G. B. Meade-Waldo turned some down in Stonewall Park, near Edenbridge in Kent, 'to rid belfries of sparrows and bats, and fields of mice':

We let out Little Owls first of all about 1874, and between then and 1880 about forty good birds went off. We knew of one nest in 1879. In 1896 and again in 1900 I 'hacked off' about twenty-five. Since then they have been comparatively abundant all throughout the district, which is roughly between Tunbridge Wells and Sevenoaks. I know of generally some forty nests in a radius of some four or five miles [6–8 km]. With us their favourite haunts are old orchards and rocks.

From Stonewall the birds spread north-west into Surrey, where the first two nests were found hear Horley in 1907. In the south, little owls soon crossed the border into Sussex, where a nest containing an egg and two fledgelings was discovered in a hollow tree on 16 June 1903. In Kent their principal early direction of expansion appears to have been to the north by way of Sevenoaks and Westerham;

---

*These dates are both as they appear in Waterton's *Essays*, so one is clearly a misprint. *The Handbook of British Birds*, Witherby and Ticehurst in *British Birds*, and Coward all give 1843 as the year in which the owls were released. Fitter gives 1841 as the year when Waterton left Rome, and 1842 as the year of the release of the birds.

they reached Dartford by the early 1870s, Swanscombe near Gravesend by 1883, and the present outskirts of London well before 1897.
In the east, little owls had penetrated as far as Shorne, between
Gravesend and Rochester, before 1893, and had reached Cuxton,
between Rochester and Maidstone, by the following year. In the south
and south-east of the county, the birds had reached Cranbrook by
1903, Bilsington, between Tenterden and Hythe, by 1906, and Boxley,
north-east of Maidstone, a year later.

In 1876 a pair of little owls were released in the park of Knepp
Castle in south-east Sussex, but both had been killed by the following
spring. In the same year the 1st Earl of Kimberley liberated six birds
in the park of Kimberley House near Wymondham in Norfolk, but
none survived for long.

Principal credit (Coward unfairly describes it as 'blame') for the
naturalization of the little owl in England belongs to that great
ornithologist Thomas Littleton Powys, 4th Baron Lilford (1833–96)
of Lilford Hall near Oundle in Northamptonshire, from whom is also
derived the bird's early alternative name, Lilford's owl.

On 25 May 1889, Lord Lilford wrote to the Revd Murray-Matthew:

I turned down about forty little owls, about the house here and over a
radius of some three or four miles [5-6 km] in the neighbourhood, early in
July last. Several were too young to feed themselves, or, rather, to find their
own food, and we recaptured more than half of those originally put out. A
very few were found dead. Several were constantly seen about; during the
summer and autumn of 1888 many disappeared entirely, but three or four
were seen, and often heard, throughout the winter. On April 23rd, 1889,
one of my keepers discovered a nest in the hollow bough of a high ash tree
in the deer-park. [This was the earliest breeding-record for Northamptonshire.] The old bird would not move, but on being gently pushed with a
stick, two eggs were visible. On May 10th two young birds about a week
old could be made out, and on the 22nd, four or five, all of different sizes.
... This is encouraging, and I shall invest largely in little owls this summer,
and adopt somewhat different treatment. Similar experiments have been
tried to my knowledge, in Hants, Sussex, Norfolk and Yorkshire, but I do
not know of a brood having been reared in a genuinely free condition in
this country, till this lot of mine.

On 20 February 1892 Lord Lilford wrote to Mr E. Cambridge Phillips:

You may be interested to hear that we have a little-owl (*Athene*) sitting on five eggs in a hollow tree not far off. I have turned out a great many of these birds during the past few years, and this is the fifth nest of which I have positive information.

Three months later, on 19 May, Lord Lilford wrote to Colonel Meade-Waldo:

A pair, if not two, of little owls have taken their young off safely at no great distance.

Two days later he wrote again:

I have not heard recently of any little-owls at a distance, and of no nests at more than two miles [3 km] from this.

On 23 June 1893, Lord Lilford wrote:

Last year we had a nest of little owls (*Athene noctua*), of which I have turned out a great many, in an ash-stump about two miles [3 km] off.

In February 1894 Lord Lilford said, in the course of an address to the Northamptonshire Field Club,

. . . for several years past I have annually set at liberty a considerable number of the little owl, properly so called (*Athene noctua*), from Holland, and several pairs of these most amusing birds have nested and reared broods in the neighbourhood of Lilford. It is remarkable that, although this species is abundant in Holland, and by no means uncommon in certain parts of France, Belgium and Germany, it has been rarely met with in a wild state in our country. I trust, however, that I have now fully succeeded in establishing it as a Northamptonshire bird, . . . they are excellent mouse-catchers, very bad neighbours to young sparrows in their nests, and therefore valuable friends to farmers and gardeners. The nest of this owl is generally placed either in a hollow tree at no great height from the ground, or in vacant spaces in the masonry of old buildings. . . . Never destroy or molest an owl of any sort. I consider all the owls as not only harmless, but most useful. . . .

Summarizing the position in *Notes on the Birds of Northamptonshire and Neighbourhood*, Lord Lilford wrote:

I have succeeded in establishing the Little Owl as a resident Northamptonshire bird. . . . I have for a considerable number of years annually purchased a number of these owls in the London market . . . we occasionally saw and frequently heard one or more of the Little Owls in the neighbourhood . . . In 1889 one of our gamekeepers, on April 23rd, found a Little Owl sitting in a hollow bough of an old ash-tree in the deer-park at Lilford; . . . he found that she was sitting upon a single egg, to which she added three, and brought off four young birds in the second week in June. One, if not two, other broods were reared in our near neighbourhood in 1889. In 1890 a nest . . . containing six eggs was found . . . on April 25th: all these eggs were hatched and the young reared to maturity. On October 15th [1890] a Little Owl was . . . discovered . . . in a rabbit-burrow in the park at Deene [seat of the Brudenells, near Kettering] . . . a bird-stuffer at Stamford had one . . . (which he called a Dutch owl) in February 1891, sent to him from Normanton, Rutland. A 'nestful' of young was discovered early in July 1891 [at] Wadenehoe Manor. . . . In 1892 I received . . . reports of Little Owls in the neighbourhood of Lilford and on April 19th a nest containing four eggs was found at Wadenhoe. . . . Mr. A[rchibald] Thorburn saw near our aviary . . . a Little Owl . . . on November 11th [1892]. In 1893 a nest containing seven young was found in the park at Lilford, on May 13th. I had authentic reports of one or two other broods at Wadenhoe and another near Lyveden. . . . I have some reason to believe that there was a brood of these owls in Wadenhoe church-tower. On December 22nd I was pleased to hear from a lady living at Stoke Doyle that several of these birds had made a settlement in some old trees in her garden. With the exception of the bird killed at Normanton and another at Elton, I have not heard of the death by human agency of any Little Owls in our district, although a few have been picked up dead from natural causes. . . . These Owls delight in taking the sun, and are active during the hours of daylight. They are infinitely useful in the destruction of voles, mice, sparrows, and insects of many kinds. . . .

In about 1890 Lord Rothschild liberated a number of little owls in Tring Park, Hertfordshire, where they soon disappeared. At about the same time, and again fifteen years later, Mr W. H. St Quintin unsuccessfully turned some down in the Park at Scampston Hall, Rillington, near Malton in the East Riding of Yorkshire. 'This is not a well-wooded district', he wrote shortly afterwards, 'and there are no rock-crevices except along the coast . . . I have given up any hope of naturalizing the birds here.' An introduction at East Grinstead in Sussex in 1900 and 1901 was rather more successful, and was followed by others in Essex (1905 and 1908) and Yorkshire (1905).

From Lilford the birds multiplied and spread with considerable rapidity, and had soon colonized the north-eastern corner of the county. From here they travelled further east to Suffolk and Norfolk by 1901, where they were first proved to have bred in 1907 and 1912 respectively; to Huntingdonshire, where they nested first near Offord Darcy in 1907; and to Cambridgeshire where they bred in the same year.

In Rutland a single bird was recorded by Lord Lilford at Normanton as early as 1891; this was followed by the discovery of owlets at Glaston, east of Uppingham, in 1895 and two dead birds near Seaton, to the south, two years later. A single bird was caught in the park at Exton Hall, north-east of Oakham, seat of Lord Gainsborough, in the same year (1897); another was killed nearby at Burley two years later, where nests were discovered in the following year (1900).

The first little owl in Nottinghamshire was caught in 1896 at Newark-on-Trent; a second was shot on the outskirts of Nottingham in 1901, but the first nest in the county was not found before 1913. In 1902 a little owl was killed near Lincoln, and a second was shot near Sleaford two years later. The species first bred in the county in 1907.

North-west of Lilford little owls were seen near the Derbyshire/Leicestershire border in 1906, nesting in the latter county three years later. A bird was heard – and the same or a second one was later caught – near Willey Park, Broseley (seat of Lord Forester) in southern Shropshire in 1899; another was killed in 1906 on the Shropshire/Staffordshire border, and another was caught near Shrewsbury in February 1908.

At Woburn Abbey in Bedfordshire the Duchess of Bedford wrote that she

knew the Little Owl since I first came to live here in 1892. The Keepers say that it existed here before that date. I should like to correct the impression which appears to be very prevalent that we have imported it ourselves. We have never imported them, and as they were here so early as 1892 I conclude that the birds here must have come from Lord Lilford's importation and not from the more recent one [by Lord Rothschild] at

Tring. There are a great many of them about the Park, and they have bred freely for many years. Owing to their habit of flying about in the daytime they are frequently seen.

In the rest of Bedfordshire little owls reached Chawston in the north-east by 1894; Turvey in the west (1897); Lidlington in the north and nearly to the Cambridgeshire border in the east (1898); and Luton in the south by the turn of the century. Nests were found at Great Barford and at Southill (respectively east and south-east of Bedford) in 1901.

In Hertfordshire little owls first bred at Ware in 1897, and by 1902 were reported by Witherby and Ticehurst as 'quite common and resident as far south as Watford'. In 1907 three fledgelings were found in a chalk-quarry near Royston, and nests were discovered near Watton-at-Stone and St Albans.

The little owl first reached Essex in 1899, when one was trapped at Harlow; the first known breeding record for the county was at Easton Park in 1910. By 1919 the bird was a common resident throughout Essex, having shown remarkable powers of colonization.

In the wooded Chiltern Hills of Buckinghamshire and Oxfordshire little owls were recorded first at Turville in 1894 and Fingest in 1896 (both north of Henley); at Thame, between Aylesbury and Oxford, in November 1901; at Bletchley (1902); and near Goring, north-west of Reading, in 1907. Further south, in Berkshire, a pair were seen in the early spring of 1907 in Windsor Forest, where they – or another pair – nested for the first time three years later.

From this period onwards the little owl's range in England increased apace. The following table gives dates of earliest proven nesting in individual counties, although in most cases little owls were present before breeding was proved:

Year	County
1909	Derbyshire/Nottinghamshire border. Hampshire (Southampton).
1910	Gloucestershire (south).
1913	Leicestershire (south).
1914	Monmouthshire; Lincolnshire.
1915	Wiltshire.
1916	Glamorganshire (earliest breeding record for Wales).

Year   County

1918   Isle of Wight; Dorset; Cardiganshire; Radnorshire.
1919   Montgomeryshire.
1920   Pembrokeshire.
1921   Cheshire; Lancashire (south).
1922   Yorkshire (East Riding); Merionethshire; Brecknockshire;
       Cornwall (north).
1923   Carmarthenshire; Lancashire (Preston); Cornwall (west).
1925   Yorkshire (south).
1930   Caernarvonshire.
1931   Denbighshire.
1933   Yorkshirc (North Riding).
1934   Durham.
1935   Northumberland (south).
1937   Yorkshire (central).

The Handbook of British Birds lists the following early isolated records of the little owl in Scotland and Ireland, which 'may have been escapes or genuine migrants':*

Date	County
February 1902	Kincardineshire.
June 1903	Co. Kildare.
November 1910	Fife.
April 1921	Roxburghshire.
August 1921	Peeblesshire.
April 1924	Berwickshire.
March/April 1925	Renfrewshire.
April 1926	Renfrewshire.

By about 1925 the main breeding range of the little owl extended over England and parts of Wales as far north as central Lancashire and Yorkshire. A few nests were reported from southern Northumberland, Durham and north-west Wales† by the 1930s; from Cumberland and Westmorland (once) by 1944; and from across the Scottish border in Berwickshire by 1958.

Between about 1940 and 1963 the little owl population appeared to be on the wane; by 1965 there was evidence of a recovery in

---

*Professor Ritchie refers to a little owl that was shot at East Grange in Fife in 1912.
†A bird was seen, but did not breed, on the Isle of Anglesey as early as 1909.

numbers, and by 1967 it was renewing its advance northwards into Cumberland, Westmorland and northern Northumberland; it now breeds throughout England and Wales and locally as far north as parts of Dumfriesshire, Kirkcudbrightshire, and Berwickshire. Recent Scottish records include one breeding site in Midlothian in 1968, and individual birds seen at Glasgow Airport on 7 September 1971 and at Kilbarchan on 14 May 1972 – both in Renfrewshire. On the latter date an owlet was found dead on a road north of Perth. On the other hand, there is little if any evidence of an increase in the population of southern and central England, where its numbers may still be declining. The early fall in population may have been caused by a succession of harsh winters, but in the 1950s was probably largely the result of the increasing use in agriculture of long-residual toxic pesticides to which predators – being at the end of the food chain – prove particularly vulnerable.

As Fitter points out, the British avifauna in the late nineteenth century had an empty niche for a largely diurnal and principally insectivorous small bird of prey; this vacancy the little owl fills today, thanks principally to the efforts of Lord Lilford and Colonel Meade-Waldo.

Note: See also Addenda.

26. Night-Heron
(Nycticorax nycticorax)

The typical race of the night-heron breeds in Europe and Asia from southern Spain and central France through Astrakhan and the southern Caucasus as far east as Indonesia, the Philippines and Japan; and in Africa from Morocco to Tunisia and Egypt in the north, south to parts of tropical and southern Africa. In the New World, where it is known as the black-crowned night-heron, it is replaced by two races: *N. n. obscurus* which inhabits southern South America and *N. n. hoactli* which extends from North America south to central South America and Hawaii.

Both on the ground and in flight the night-heron appears a compact and rather thick-set bird, whose head and retracted neck seem to merge directly into the body. The crown of the head and mantle are black with a greenish-blue gloss, the broad and rounded wings and tail are lilac-grey, and the underparts whitish-grey. Three long slender thread-like white feathers extend backwards from the nape of the neck. The bird stands about 2 ft (61 cm) high on pale yellow legs, and has a greyish bill. The young are dull brown in colour and superficially bittern-like in appearance, but are spotted and streaked with buffish white.

The night-heron is essentially a secretive and crepuscular species, frequenting wooded and reeded marshes and swamps near streams and rivers or pools of standing water. Here it remains hidden during the day, only emerging from cover at dusk to feed in the shallow water at the edge of reed-beds. Night-herons nest in colonies ('seges' or 'sieges') of considerable size, usually in bushes or trees (where they also roost) or occasionally on the ground.* The flattish and

*Intensive studies have been made of the courtship and breeding habits of the night-heron: K. Lorenz (*Proc. VIII Int. Ornith. Congress, Oxford*, pp. 207–18 – European race); R. P. Allen and R. P. Mangels ('Studies of the nesting behaviour of the black-crowned night-heron', *Proc. Linn. Soc., New York*, 50–1, pp. 1–28, 1938–9); G. K. Noble, M. Wurm and A Schmidt (*Auk*, pp. 7–40, 1938; pp. 205–24, 1942 – American races).

rather flimsy nest is built of twigs and reeds. Three or four pale greenish-blue eggs are laid between late April and early June, which hatch after some three weeks' incubation by both parents. The voice of the adult night-heron is a hoarse croak, uttered more often in flight than on the ground: the young have a curious continuous ticking call. The food consists of small fish, newts, lizards, frogs, molluscs, worms, aquatic insects and larvae, crustaceans and occasionally small mammals and birds.

The earliest, and so far only recorded, deliberate attempt to introduce the night-heron to Britain, was made by Lord Lilford in Northamptonshire. On 4 July 1868 Lilford accidentally shot a night-heron near the river Nene between Wadenhoe and Aldwincle. In 1885 he received a report from Sir Rainald Knightley that a night-heron had been observed at Fawsley House in the west of the county, and between 13 July and September of the following year a third night-heron frequented the Nene near Wadenhoe.

Encouraged by the late stay of this bird in our county, and in the hope of atoning in some measure for my offence, I turned out two young Night Herons at Lilford in the summer of 1887, but though they were seen on

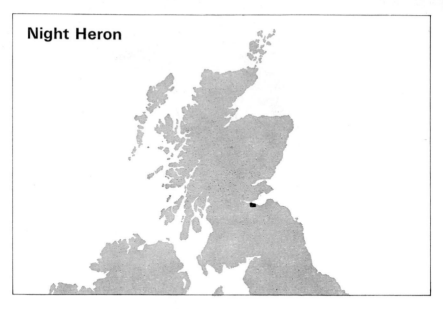

Night Heron

several occasions shortly after their liberation, I have no evidence to prove that they remained in the neighbourhood for any length of time.

Although this first and only attempt to naturalize night-herons proved unsuccessful, it is clear that Lord Lilford continued for some time to keep them in captivity. On 25 May 1889 he wrote to the Revd Murray-Matthew to tell him that . . . 'one of my night-herons laid an egg this morning on the top of a box bush, trodden to a sort of platform by a stork. Those night-herons have been here for three years, and I have great hopes of a brood.' Under the heading 'Lilford Aviary and Living Collection' Lord Lilford wrote in his daybook on 16 March 1893: 'Night-heron in aviary laid first egg of season.'

In 1936 the Royal Zoological Society of Scotland acquired three pairs of the North American race of the night-heron, *N. n. hoactli*, from the National Parks Bureau in Canada. These were kept un-pinioned in a covered aviary in the Society's gardens at Corstorphine, Edinburgh, where the first pair bred successfully two years later, and where nesting has taken place annually ever since. The colony was reinforced by an additional pair obtained from the same source in 1946.

In December 1950 a number of night-herons escaped from the aviary through a hole in the roof: five months later the entire roof was removed and the whole colony of about eighteen birds was at liberty: they soon became established outside the aviary, and started to breed in the vicinity of their former home.

When Dorward carried out his survey of the Edinburgh night-herons in 1955, he estimated their annual populations to that year as follows:

Year	Numbers
1938	9
1939	12
1940	16
1941	18
1942	15
1943	20
1944	27
1945	27
1946	27
1947	27
1948	28
1949	18 (10 going through exchange).
1950	20
1951	18 (approximately).
1952–3	20 (approximately).
1955	21

As can be seen the population increased slowly during the war years, and then remained fairly constant until 1949, before slipping back to the level of the early 1940s. Initially, the slow rate of increase was probably due to poor reproduction; the decline between 1949 and 1955 may have been also partly caused by low reproduction, coupled with predation by grey squirrels, brown rats, carrion crows, jackdaws, magpies and semi-feral domestic cats – all of which frequented the herons' breeding area – and some adult emigration. During this period night-herons were reported from outside the Society's grounds at Union Canal, Sighthill, 2 miles (3 km) to the south-west (January 1952); on the river Tyne at Haddington, East Lothian, 19 miles (30 km) east (May/June 1954); at Water of Leith, Colinton, 2½ miles (4 km) south, and on the river Almond at Cramond Brig, 2½ miles (4 km) north-west (both June 1954).

In April 1955 a party of five or six birds was observed to be flying regularly every evening to feed in the Gogar Burn, $3\frac{1}{4}$ miles (5 km) to the south-west, returning to the zoo on the following morning. It is interesting to note that although Dorward found the birds quite tame while in the zoo gardens, they became extremely wild and wary when in the Gogar Burn.

Until 1930 the night-heron was only an irregular vagrant to the British Isles, occurring at all times of the year – mainly between April and June and, to a lesser extent, October and November – principally on the south and east coasts of England – especially in Cornwall, Devon, Kent, Sussex, Norfolk and Suffolk, and as far north as Yorkshire. Some birds were observed away from the coast, but few in the north and west. Twenty-six specimens were recorded in Ireland, of which only one came from Connaught. Night-herons were only rarely reported from Wales – on the Isle of Anglesey and in Carmarthenshire, Glamorganshire and Pembrokeshire – and in Scotland from Aberdeenshire (one), Argyllshire (two or three), the Outer Hebrides (one), and the Lowlands (six). On 30 October 1936 (the year in which the species was acquired by the Edinburgh zoo) an immature night-heron, said to be of the New World race *hoactli*, was found dead at the mouth of the river Yealm near Plymouth in south Devon, after a period of strong westerly winds. Although no escapes were reported in that year from Edinburgh, this record has not been accepted by ornithologists as a valid transatlantic vagrant.[*]

In 1946 a small breeding colony of the European race of the night-heron, *N. n. nycticorax,* became established in southern Holland, where the last native birds became extinct in 1876. (An unsuccessful attempt at re-introduction was made in 1908–9). It was never determined whether this colony was formed of natural migrants from southern France or Hungary – from where immature night-herons occasionally reach Holland – or whether they were the descendants of birds from a private collection contained before the war in the neighbouring province of Zeeland. The two night-herons reported near Brookland on Romney Marsh in Kent on 11 May 1947, and on the river Frome near Wareham in Dorset on 8 November 1949, may well have come from this source.

[*]R. Moore, *Birds of Devon*, (1969). My thanks are due to Mr Robert Hudson for drawing my attention to this reference.

Of more uncertain origin are the immature night-herons recorded on the river Coquet above Warkworth in Northumberland between 15 March and late August 1953;* at Sheerness in Kent from 16 October to 11 November 1953; on Foulness Island from before 8 November 1953 to 7 January 1954; at Steeple Stone on the river Blackwater in Essex between 21 November and 24 December 1953; on the Avon below Loddiswell in south Devon from 27 March to 25 April 1954; and the adults seen at Hingham Seamere in Norfolk between 7 July and 18 August 1953; at Caton in Lancashire on 31 July 1953; and other sightings at irregular intervals in subsequent years throughout the country. Unfortunately the European and American races are too similar in appearance for clear identification in the field, but the preponderance of records in the southern half of England suggests a continental origin rather than escapees from Edinburgh.

In the wild, both in Europe and America, the breeding season of night-herons is strictly limited. The Edinburgh colony, on the other hand, has been known to nest at all times of the year except between August and October, when the adults are said to moult. The reasons for this considerably extended breeding season are probably two-fold: firstly, the birds have a regular and abundant supply of food without the necessity for constant foraging: they freely steal fresh fish put out for other birds, and kill and eat the offspring of various species in neighbouring enclosures. Secondly, the migratory urge – which has an endocrinological origin connected to the reproductive instinct – has been lost, which might also result in more continual breeding. In the Netherlands, for example, as Dorward points out, the migratory purple heron (*Ardea purpurea*) breeds from four to six weeks later than the sedentary common heron (*A. cinerea*).

Surprisingly little research has been carried out on the Edinburgh night-herons since Dorward conducted his survey in 1955. Twenty years later the population was estimated to number between forty and fifty – over double the 1955 figure. The birds flight regularly in the

*'I have little doubt', wrote the Director-Secretary of the R.Z.S.S., somewhat rashly, 'that the Night Heron seen on the River Coquet is one that has escaped from our collection.'

evening from the park to feed in burns, pools and marshes around Edinburgh – they are often, for example, seen at Musselburgh 8 miles 13 km) to the east, and beyond – returning to the zoo in the early morning. At least eight nesting sites are known to exist inside the zoo; at one time these were confined to three separate areas; the tall trees in the garden to the south of the Fellows' House; near the gate leading to the vegetable garden at the eastern end of the crane enclosure; and in the trees and bushes around the sea-lion pool. During the last few years, however, the night-herons have extended their breeding range inside the park, where they now nest in several other suitable areas, and there are reasons to suspect that one or two pairs may also be nesting outside the park.

In the last decade there have been a number of reports of night-herons from the Border counties. Because of the difficulty of differentiation in the field between the European and American races, it is impossible to tell whether these birds are migrants from the Edinburgh colony or whether they are vagrants from abroad. This unique colony of night-herons in Edinburgh, whose study has so far been so sadly neglected, surely deserves further scientific attention.

27. Canada Goose

(Branta canadensis)

Canada geese, of which there are nine or ten subspecies showing a considerable variation in size, range over the greater part of North America. The typical eastern or Atlantic race, *B. c. canadensis,* which is the one British birds most resemble, is also one of the largest; it breeds on the south-eastern side of Baffin Island, in Newfoundland, Labrador, and eastern Quebec, and on the Magdalen Islands in the Gulf of St Lawrence: in winter it migrates south to the Atlantic coast between Nova Scotia and Florida.

The Canada goose is the largest British goose, measuring over 3 ft (91 cm) in length. It is greyish brown in colour – darker above than beneath – with a lighter shade of brown on the breast, white upper- and under-tail coverts, and a black tail. The head and neck are black, with a white chin-patch which extends upwards behind the eye. The bill is black and the legs olive-green.

In Britain Canada geese frequent open grassland and marshes with lakes and ponds, especially those in parks or surrounded by mature woodland. They nest, sometimes in colonies of considerable numbers, between late March and mid-May, almost invariably near water and frequently on an island, in some well-sheltered and low-lying place. (In Yorkshire they also breed on open heather-clad moorland surrounding large man-made reservoirs.) An average of five or six off-white eggs are laid in a depression in the ground which is lined with dead leaves, grasses, reeds, down and feathers. Incubation, by the goose only, lasts for twenty-eight or twenty-nine days. A so-called 'triumph' or 'celebratory' display takes place after an intruder has been driven away – when goose and gander run together side by side with outstretched necks which they 'snake' from side to side, at the same time trumpeting loudly – which is characteristic of their court-ship display ritual. Canada geese are gregarious birds, forming flocks of up to three hundred outside the breeding season. On long flights

R

Canada Goose

- ▉ main breeding range
- ▨ secondary breeding range
- ▨ occurs but does not breed

they travel in the typical 'V'-formation common to most geese, or in long and wavy oblique lines or in single file. They feed mainly on grass, water-plants, sprouting corn and, in autumn, on gleanings from the stubble-fields. Their voice is a loud, resonant, trumpeting 'aa-honk', the second syllable of which is higher than the first.

The Canada goose has been known in domestication on the continent and in Britain for over three hundred years. Georges Louis Leclerc, Comte de Buffon (1707–88), in his monumental *Histoire Naturelle, Générale et Particulière*, states that it was breeding in considerable numbers in the park at Versailles during the reign (1643–1715) of Louis XIV.

The earliest mention of the species in Britain appears to refer to birds in the collection of Charles II (reigned 1660–85) in St James's Park in London. Here, on 9 February 1665, John Evelyn (1620–1706), saw, according to his *Diary*,

numerous flocks of severall sorts of ordinary and extraordinary wild fowle

... which for being so neere so great a Citty, and among such a concourse of souldiers and people, is a singular and diverting thing.

Francis Willughby (1635–72) and John Ray (1628–1705) saw Canada geese in the royal collection prior to 1672:

the name shows the place whence it comes. We saw and described both this and the precedent [the spur-winged goose] among the King's wildfowl in St. James's Park.*

Most of the Canada geese which have been introduced to Britain are presumed to have been of the typical form, *B. c. canadensis*: a few, however, may possibly have been of a slightly smaller and darker race, the Central or Todd's Canada goose, *B. c. interior*, from the eastern coast of Hudson Bay.† In her review of H. C. Hanson's monograph, Dr Janet Kear of the Wildfowl Trust suggested (1966) that a number of British birds share characteristics with the Giant Canada Goose (*B. c. maxima*) – which lives on the Great Plains of the central United States where, having long been thought to be extinct, it was re-discovered in 1962 – principally large size, a tendency for the white cheek patch to have a backward pointing hook at the top, and a non-migratory nature, but this has yet to be confirmed.

It has been suggested by a number of writers, *e.g.* William Thompson in his *Natural History of Ireland* (1849–56), that some of the Canada geese which have now and then appeared in winter on the western sea-coasts of the British Isles – and especially on the off-shore islands – may have been transatlantic migrants. Examples include the single bird shot in the Scilly Isles before 1863; a party of seven, of which two were shot, at the end of February 1903 on South Uist, which were described as being 'wilder even than Whitefronts and much more difficult to approach': another shot on Lundy Island in the Bristol Channel in about 1914; and others seen from time to time in the Inner and Outer Hebrides and the Orkneys and Shetlands. It is now accepted that this hypothesis is correct. Wild Canada geese from the east Canadian Arctic occur in Britain and Ireland as rare but regular transatlantic winter visitors. Their origin is known

---

Ornithologia, 1676–8.
†W. E. Clyde Todd, *Auk*, pp. 661–2, 1938. Writing in 1777 that 'Canada Geese are very plentiful in Hudson's Bay', George Forster (1754–94) – '('Hudson's Bay Birds', *Phil. Soc. Trans.*, 62, p. 414) – was doubless referring to *B. c. interior*.

because they include the smaller subspecies which are not feral in Britain; because they almost invariably occur among flocks of white-fronted and barnacle geese of Greenland origin; and because there is evidence of hybridization with the other species (barnacle, white-fronted and snow geese) found in eastern Arctic Canada and Greenland. Nearly all of these records are from Ireland (especially Co. Wexford) and Scotland (especially the Isle of Islay). Clearly, migrant Canada geese which reach Greenland in small numbers (subspecies Richardson's Canada goose (*B. c. hutchinsi*) and the Lesser Canada goose (*B. c. parvipes*) have been identified there), become attached to white-fronted and barnacle flocks and continue their migratory flight with them. There were only ten records of Canada geese in Ireland before 1900, and none between then and 1954. Since that year there have been almost annual records, mostly from the North and South Slobs in Co. Wexford; individual birds have also been seen at Downpatrick, Co. Down (1954); near Banagher (1955) and at Little Brosna (1956), Co. Offaly: Killag, Co. Wexford (1959); Bunduff (1961) and Lissadell (1970), Co. Sligo; and on Lough Kinale, Co. Longford (1963).*

The first record of an escaped Canada goose in Britain appears to be one which was shot on the Thames at Brentford in Middlesex in 1731. In 1785 Latham wrote:

This species is now pretty common in a tame state, both on the continent and in England; on the Great Canal, at Versailles, hundreds are seen, mixing with the Swans with the greatest cordiality; and the same at Chantilly. In England, likewise, they are thought a great ornament to the pieces of water in many gentlemen's seats, where they are very familiar and breed freely.

Charles Waterton's father, Thomas, kept a flock of thirteen pinioned (presumably) Canada geese at his seat, Walton Hall in Yorkshire, in the last decade of the eighteenth century. These appear to have died out shortly afterwards, for we do not hear of other Canada geese at Walton until the early 1800s, when mention is made of a flock of twenty-four on the lake. 'The fine proportions of this stately foreigner', wrote Charles Waterton, 'it's voice and flavour of its flesh are strong inducement for us all to hope that erelong it will become a naturalized bird throughout the whole of Great Britain.'

*O. J. Merne (see Bibliography). I am indebted to Mr Robert Hudson for drawing my attention to this reference.

In 1841 Waterton's gamekeeper told the 'Squire' on his return from Italy that in the preceding spring one of his Canada geese had unsuccessfully mated with a 'Bernacle' gander: in 1842 the eggs again proved addled, but in the following year, much to Waterton's surprise, they hatched successfully. The 'Squire', who had always maintained that such a union would never prove fertile, made a characteristically generous apology: 'I stand convinced by a hybrid, reprimanded by a gander and instructed by a goose.'* Charles Waterton died in 1865, and a century later Canada geese still swam and nested on the lake at Walton Hall.

Other nineteenth-century estates known to have contained Canada geese among their collections of wildfowl, include those at Lilford in Northamptonshire (Lord Lilford), Gosford House, Longniddry, East Lothian (the Earl of Wemyss), and Bicton House near Exeter, Devon.† In 1844 a pair nested at Groby Pool in Leicestershire, and William Yarrell (1784–1856) in his *History of British Birds* (1845) refers to free-flying Canada geese which were 'obtained in Cambridgeshire, Cornwall, Derbyshire, Devonshire, Hampshire, Oxfordshire and Yorkshire. Two were shot at Skerne [in the East Riding] on 29 May 1845 by Mr W. Mosey. They have also been taken in Scotland.' In 1885 a pair of Canada geese nested in Edgbaston Park on the outskirts of Birmingham, and a second pair bred at Garendon Pond in Leicestershire, where Alexander Montagu Browne, writing in *The Vertebrate Animals of Leicestershire and Rutland* (1889), described them as 'an introduced species, often found at large, especially in winter, and roaming so far afield as to give rise to the doubt if it may not soon become feral'. In 1892 two pairs bred at Rydal Water, between Grasmere and Ambleside in Westmorland.

During the present century, a number of introductions of Canada geese have been made in Britain; examples include those at Leighton Park, Montgomeryshire (*c.* 1908); at Radipole Lake, near Weymouth, Dorset (before 1932) and at West Wycombe Park, Buckinghamshire, by Sir John Dashwood, Bt, in 1933–4. It took an inordi-

---

*In 1922 a Canada gander at Holkham mated with a grey-lag goose (*Anser anser*), which remained behind when the other grey-lags flew north, and successfully hatched five eggs. The hybrid goslings – which had black rather than grey geese characteristics – joined the flock of resident Canadas, and in the spring of 1924 both they and the adults remained. Hybrids have also occurred between Canada geese and semi-domesticated Chinese geese.

†M. A. Mathew and W. S. M. D'Urban, *Birds of Devon*, 1895.

nately long time for ornithologists to accept that the Canada goose had become a part of the British avifauna. Not until 1938 did H. F. Witherby write:

Although there is no more proof that the Canada Goose (*Branta canadensis*) has travelled unaided from America to this country than that the Red-legged Partridge (*Alectoris rufa*) has crossed the Channel, they are more or less on equal footing so far as the time they have been introduced, and both have bred and flourished for a long time in a wild state. So that as the one is generally admitted to the list of British birds, it would seem that the other should be also and for this reason we propose to include the Canada Goose in the new *Handbook* [*of British Birds*].*

More recent introductions include those made in the Maidstone/ Sevenoaks/Tunbridge Wells area of Kent by the Wildfowlers' Association of Great Britain and Ireland in the 1950s; at Frampton in Gloucestershire by the Wildfowl Trust in 1953: and to Anglesey, Pembrokeshire, Staffordshire and Hyde Park in London from Stapleford Park, Leicestershire and Swinton Park, Yorkshire in 1955.

In 1953 N. G. Blurton-Jones through the British Trust for Ornithology, and from 1967 to 1969 M. A. Ogilvie of the Wildfowl Trust, conducted censuses of feral Canada geese in the British Isles. Blurton-Jones calculated the population at between 2964 and 3866 birds, and Ogilvie estimated it at 10260 or more. A summary of their findings appears in the table on the following pages.

Blurton-Jones found that the population of Canada geese in Britain was divided into a large number of isolated sub-populations. In North America all races of Canada geese have a northward moult-migration (as have a number of sub-populations in Britain), and in autumn move southwards to separate winter quarters, which British birds do not, probably because of our more equable climate and an adequate food supply.† This, however, cannot be the only reason

---

British Birds, 32, p. 119, 1938.

†The Canada geese in Perthshire, Dumfriesshire and East Lothian before the war appear to have had a regular autumn migration. There is also a regular moult migration from Yorkshire to and from the Beauly Firth, Inverness-shire and possibly Loch Rangag in Caithness (see below).

An adult, ringed in Yorkshire in 1958, was recovered in Carmarthenshire (170 miles/ 273 km south-west) in January 1963. Four birds, ringed as goslings as Yorkshire in 1962, were recovered in the Pas-de-Calais in France in January and February 1963 – the first foreign recoveries of British ringed Canada geese; these flights were doubtless a result of the severe winter.

Sub-Population	Total 1953	Remarks	Estimated Total 1967–9	Remarks
Devon (south).	—	—	220	Based on Shobrooke Park, near Exeter, where 10 goslings were released in 1949: in 1963 these had increased to 163, and were spreading. Elsewhere small numbers are in the Exe valley. In winter on the Exe estuary.
Dorset (south).	—	—	180	Poole Harbour and Crichel Lake. Breed on Brownsea Island.
Hampshire (south).	—	—	80	Needs Oar Point, at the mouth of the Beaulieu river.
Sussex and Surrey (south).	23–50	Introduced to Petworth between the wars; in the early 1920s bred at Hampden Park, Eastbourne.	290	Pulborough floods (240 in January 1968); Knapp Castle; Warnham Mill; and elsewhere.
Kent (south).	—	—	30	Dungeness.
Kent (north and central).	—	—	340	Gravel-pits and lakes around Sevenoaks; Maidstone and Tunbridge Wells: introduced by wildfowlers in the 1950s, and have flourished.

London.	—	—	130	Hyde Park (introduced in 1955 from Stapleford Park, Leicestershire, and Swinton Park, Yorkshire), and Thames near Kew.
Berkshire (south), Hampshire (north), Surrey (west), Buckinghamshire (south).	133–163	15 pairs at Englefield Green, where the whole population gathers in August and September before dispersing for the winter to e.g. Fleet Ponds.	570 (probably more)	At least 18 lakes and gravel-pits in the Aldershot/Reading/Newbury triangle; Chertsey; Windsor Park (especially Virginia Water); Ascot Place; Foliejon Park; and Wraysbury.
Wiltshire and Berkshire (north).	30–62	5–6 pairs on Buscot Lake.	50	Stourhead; Wilton Water; Broad Water; Buscot Lake; r. Kennet.
Gloucestershire.	—	—	100	Gravel-pits at Frampton-on-Severn: (41 introduced from Stapleford Park, Leicestershire, in 1953); the population is controlled by the removal of young, and endemic renal disease.
Monmouthshire.	—	—	20	Newport.
Pembrokeshire.	—	—	40	Fowborough: introduced from Stapleford and Swinton in 1955.
Worcestershire (north), Warwickshire, and Staffordshire (south and east).	84–104	Mainly at Packington Park.	380	Lakes and ponds around Birmingham, and gravel-pits near Tamworth and Burton-upon-Trent.

Sub-Population	Total 1953	Remarks	Estimated Total 1967-9	Remarks
Derbyshire.	376–437	400 winter at Kedleston Hall (Viscount Scarsdale) and Osmaston Park: they leave to breed, and are increasing: colonized Sudbury Park in 1946.	890	Kedleston (about 800); Osmaston; Allestree; Locko Park (Capt. P. J. B. Drury-Lowe); r. Trent floods (in winter) and elsewhere.
Staffordshire (west), Shropshire (east).	—	—	480	Introduced in 1955 from Stapleford and Swinton.
Shropshire (central and west).	—	—	290	Mainly around Shrewsbury and Oswestry.
Shropshire (north), Cheshire (south).	506–598	70–200 at Ellesmere, and bred at Halston and Shavington Halls. 300–400 in S. Cheshire; 35–40 pairs at Combermere.	490 (probably more)	Various waters around Ellesmere, and at Combermere; Barmere; Shavington; and elsewhere.
Montgomeryshire.	68–84	30–50 at Powis Park: introduced to Leighton Park c. 1908.	400	Severn valley near Welshpool.
Cheshire (north).	165–219	At Tabley c. 15 pairs.	550	Rostherne; Tabley; Tatton; and other meres.
Cheshire (west).	—	—	110	Aldford; Eaton Hall; r. Dee marshes and floods.

Locality			Notes
Isle of Anglesey.	—	200	Scattered: introduced in 1955 from Stapleford and Swinton.
Essex.	—	140	Hanningfield and other reservoirs and waters in N. and N.W. Minsmere.
Suffolk.	—	30	
Suffolk (north, Brecklands) and Norfolk (south).	157–225	300	Wretham Park; Mickle Mere; Livermere. c. 180 in autumn. Breckland Waters. There may be some movement to and from Holkham.
Norfolk (north).	350–500	760	Holkham Hall (Earl of Leicester) the principal British colony. 200 in 1941: 300 in January 1952. Bred at Narford Lake in 1950–2, and 40 were breeding in 1953 at Tunstead and Dilham. At Holkham, 700–1000 in 1955; 1500–2000 in 1963; 1700–2000 in 1965; considerably reduced in late 1960s.
Norfolk (Broads).	—	120	Possibly some movement to and from Holkham.
Cambridgeshire.	—	40	In the late 1950s/early 1960s bred on open waters around Cambridge, wintering near Holkham on the north Norfolk coast. By 1964 the valley of the Cam south of Cambridge had become the main wintering area. By the winter of 1968–9 up to 150 roosted at Hauxton Gravel Pits.

Sub-Population	Total 1953	Remarks	Estimated Total 1967–9	Remarks
Cambridgeshire—*continued*				Between 1969 and 1972 over 70 geese – mostly nestlings – were ringed at Hauxton, and it was found that adults returned to breed in successive years and that two of the young ringed in 1969 bred in 1972. Few birds bred at Hauxton in 1973, and in 1974 none of the four pairs which attempted to breed were successful, the goslings being taken by predators because, following the drought of 1973–4, there was insufficient water to afford them refuge. There were no large roosting flocks at Hauxton in the winter of 1969–70, and single birds remained only for short periods. Recoveries of ringed birds suggest that between August and October 1969–72 the geese dispersed south-east to the Cam Valley at least as far as Saffron Walden, and

County				
				possibly even to the valleys of the Chelmer and Pant. By 1974–5 Canada geese appeared to be wintering further south still in the Blackwater valley near Maldon in Essex. (J. D. Limentani – see Bibliography).
Northamptonshire and Leicestershire.	181–229	'Introduced long ago' to Stapleford Park (50–60 in 1920s; 50 in 1943; 60 in 1945; 80 in 1953). Also at Deene Park (Mr E. C. S. J. G. Brudenell); Blatherwycke; Burghley House (Marquess of Exeter; 30–75 between 1949 and 1951); Kingston Park (introduced about 1900; once 420 in winter; 150 between March and September; 50 resident in 1953); and around Ashby-de-la-Zouche.	180	Blatherwycke; Deene and Stapleford; and reservoirs and gravel-pits.
Lincolnshire.	–	–	340	Grimsthorpe Lake.
Nottinghamshire.	173–200	Counts for 1952–5 suggested a population of between 50 and 90 for the 'Dukeries', mainly at Thoresby and Rufford.	460	Principally Clumber; Thoresby; Worksop; and Welbeck Abbey (Duke of Portland).
Yorkshire (south-east).	–	–	40	Hornsea Mere, near Hull. (There is an isolated colony of about 80 birds at Castle Howard, 12 miles (19 km) north-east of York; Mr G. A. G. Howard).

Sub-Population	Total 1953	Remarks	Estimated Total 1967–9	Remarks
Yorkshire (south).	107–127	Mainly around Wakefield and Barnsley; 5–10 pairs bred at Bretton Park; visited Wentworth Woodhouse.	200	Around Barnsley.
Yorkshire (central).	302–328*	At Swinton Park (Earl of Swinton: 9 pairs) since *c.* 1900; in hard winters they travel south to Gouthwaite Reservoir; increasing at Ilston, Leighton and Roundhill reservoirs; 5 pairs at Ripley Park; 3 pairs at Studley Park; 1 pair at Swinsty in 1952; 1 pair at Lindley in 1953; large numbers at Harewood (Earl of Harewood) in July and August.	1310	Many lakes and reservoirs between Leeds and Masham. From Yorkshire, Nottinghamshire and Derbyshire a moult migration, involving over 700 birds *per annum* by 1974, takes place to the Beauly Firth, Inverness-shire, and to Loch Rangag in Caithness.
Lancashire. (Liverpool).	115–156	Mainly at Knowsley Park (Earl of Derby) which the birds leave in autumn to winter in Liverpool parks, returning to Knowsley in early March to breed. Bred on ten other lakes near Liverpool.	170	Liverpool area.

*See note on page 271

Lancashire (north) and Westmorland (south).	—	—	140	Winter in the Lune Valley south of Kirkby Lonsdale, and breed between there and Sedbergh.
Northumberland	—	—	20	Colt Crag Reservoir, and nearby.
Scotland.	147–264*	80–180 in Dumfriesshire: at Kinmount since 1934; in 1953 17 or 18 pairs. 5–7 pairs at Lochmaben depart in autumn to better feeding-grounds on the Solway Firth. About 15–30 nest annually at Loch Leven, Kinross. Mid- and East Lothian birds feed in autumn in Aberlady Bay. 50 pairs nested up to 1939. Between 4 and 20 in Perthshire. 1 or 2 pairs on Abercairney Loch; 1 pair on Ochtertyre Loch; a few on Buchanty Loch migrate in autumn to Tay estuary. 28–70 in Morayshire: 20 on the Beauly Firth in summer leave in September. In June 1947 a pair was seen on the east side of Mull, and a pair with 2 goslings nested on	100	About 50 at Kinmount, Dumfriesshire; 30–50 on Colonsay; a few pairs breeding in Perthshire and Renfrewshire. Small parties of birds sometimes seen elsewhere, *e.g.* on the Tweed, are or include moult-migrants *en route* to or from the Beauly Firth and Loch Rangag. Since 1953 there has been a further marked decline in the Scottish sub-population.

*Ogilvie gives incorrect totals of 330–98 for central Yorkshire, and 119–94 for Scotland.

Sub-Population	Total 1953	Remarks	Estimated Total 1967–9	Remarks
Scotland—*continued*		Colonsay. According to Dr John Berry: 'Within the last fifteen years [to 1953] the number of full-winged Canada geese breeding in Scotland appears to have declined by about 90%. . . . Before the war there were probably at least a couple of thousand, and quite possibly several thousand . . . chiefly composed of flocks of 20–30 in the neighbourhood of lochs in private estates . . . retained in hard weather by artificial feeding.		
Northern Ireland	47–120	Introduced from Baronscourt to Caledon, Co. Tyrone, in 1935: 1 pair on Lough Neagh: a few at Ward Park, Co. Down. (In southern Ireland, a few were kept on St Ann's Estate, Clontarf, near Dublin until 1914: between 1912 and 1920 up to 20 were in the Dublin Zoo, which were free-flying but	70	Strangford Lough, Co. Down. In 1970 Merne (see Bibliography) found Canada geese in the following collections in Ireland: Co. Cork, the Lough (about 55): Co. Wexford, Burrow Rosslare (4); North Slob (2): Co. Dublin, St Stephen's Green, Dublin (2): Co. Down,

established no colonies. Occasional birds have been shot, mostly in the east, during the last 40 years. Sub-population nil in 1953.)

Strangford Lough – Castle Ward, Killyleagh, Ringdufferin, Newtownards – (over 110); Tollymore, Newcastle, Castlewellan (3 – previously more); Kircublin (2): Co. Antrim, Ballymoney (8): Co. Tyrone, Caledon (3 – expected to increase); Omagh (2 – expected to increase): Co. Fermanagh, Ely Lodge, Enniskillen (32); Irvinestown (6). Total: 229.

2964*–3866

10 260

*Blurton-Jones (whose error is repeated by Ogilvie) gives the total as 2954.

S

why these sub-populations do not inter-connect and do not increase in range. Nor is it caused by a lack of pressure of numbers or the absence of suitable neighbouring habitats; there is evidence that some pairs fail to breed through shortage of space, and a number of suitable waters outside sub-population areas are not colonized. The isolation of these sub-populations and the failure to colonize outside waters, suggests that the movements within a sub-population such as the summer assemblies and spring dispersals which take place on a number of British waters, may well correspond to similar pre- and post-migratory movements in North America. It is interesting to note that Canada geese introduced to Sweden have developed a definite migration pattern to separate winter quarters in West and East Germany, perhaps due to the severer weather in these more northern latitudes.

In 1964 R. H. Dennis established that a moult-migration, then involving over three hundred adult and immature birds, takes place annually between Yorkshire and the Beauly Firth in Inverness-shire, 260 miles (418 km) to the north-west: many birds are known also to make the return journey to Yorkshire. This moult-migration has been studied by A. F. G. Walker, who found that the northward flight, possibly by way of the Firth of Forth and the Great Glen, appears to take place before the middle of June. The moult is probably completed at Beauly by late July, and the return southward migration, possibly on routes slightly east of south and along the east coast, starts in the third week of August and reaches a peak in late August and early September. It has been suggested that the destruction of eggs (see below), which results in many frustrated breeders who have no link to their breeding-grounds, may have a direct bearing on this moult-migration, which is also perhaps based on ancient adaptation linked to heredity. It is known that a number of immature birds do not make this northward flight to moult, but remain behind in northern England.

The moulting flock at Beauly has grown from 250 in 1968 to 300 (1969); 232 (1970); 378 (1971); 520 (1972); 630 (1973), and 704 in 1974. In 1973 600 were caught for ringing, of which 214 were discovered to have been previously handled in Yorkshire, and 18 in Derbyshire (Kedleston) and Nottinghamshire (the Dukeries). The inclusion of birds from the north-midland counties indicates a continuing development of the moult-migration pattern, and partially explains the increasing numbers at Beauly. In August 1973 nine moulting Canada geese were observed on Loch Rangag in Caithness, where twenty-six birds were reported in 1974, suggesting the likelihood of a further moulting site.* It seems clear that much work remains to be done in this field before the full reasons for this partial moult-migration become apparent.

During the 1950s, Canada geese came increasingly into conflict with agricultural interests due to their predilection for new-sown grass-leys and spring and winter wheat. Adult Canada geese in Britain have few enemies other than man, though goslings sometimes fall prey to pike, otters, mink and mute swans. Effective population control is achieved by limited egg-destruction (which, if it is not correctly timed will cause the goose to lay a second clutch, and which is contrary to the Protection of Birds Act 1954), and limited shooting which gives immediate relief from damage. Unfortunately, most British Canada geese are too tame to rate very highly as 'sporting' birds; shooting takes place, therefore, more as a form of pest-control.

Another method of control which has been used is to capture the birds when they are flightless at the time of the summer moult, and transfer them to previously uninhabited waters. Between 1953 and 1957 The Wildfowl Trust, and in the late 1950s the Wildfowlers' Association of Great Britain and Ireland, between them caught and transferred some 1400 or more birds. Most of these settled down and successfully established new colonies, although some, surprisingly, 'homed' from distances of over a hundred miles (160 km). This deliberate policy of introducing Canada geese into previously un-colonized areas in order to reduce in size various sub-populations, was one of the primary reasons for the enormous (three-fold) increase in the overall population of the species between 1953 and 1969 (see Table). The new colonies flourished, and the old sub-populations rapidly regenerated themselves unless further reductions were made by egg-destruction or shooting. As a means of control, therefore, transplanting was a failure. Although the total population of Canada geese has almost certainly not increased in the 1970s at the rate it did in the middle and late 1950s during the transplanting programme, and in the 1960s, it is still apparently rising. Winter shooting would seem to be the most effective method of limiting the population of this fine bird, whose introduction to Britain is yet another example of the dangers of transplanting an alien species outside its normal range.

*Scottish Birds, 8(4), p. 230, 1974, 8(8), p. 417, 1975.

28. Egyptian Goose

(Alopochen aegyptiacus)

The Egyptian goose* ranges over the greater part of Africa south of the Sahara, except in areas of waterless desert and thick forest; it is becoming scarce in Egypt (where it was formerly found throughout the Nile Valley), and is now extinct in Palestine and Syria.

The plumage of the Egyptian goose, which is subject to a considerable amount of individual variation, is most distinctive. The principal ground-colour is brown or brownish grey above, shading to greyish fawn beneath; this is also the colour of the head and neck – the latter being marked with darkish-brown streaks. The bill and legs are pale pink. White wing-coverts, black primary and pea-green secondary wing-feathers combine to produce a most striking appearance when the bird is in flight. The most noticeable features, however, are the chocolate-brown patches encircling each eye, and one on the lower part of the breast. The sexes are alike, and each measures a little over 2 ft (61 cm) in length.

In Britain the Egyptian goose frequents much the same kind of habitat as that favoured by the Canada goose. It nests on the bare ground, under bushes, on rock ledges, or in the disused nests of other birds in trees. Some six to eight eggs are laid between late March and May, which hatch after twenty-eight to thirty days' incubation by the goose. The food consists mainly of grasses, but other vegetable matter and small aquatic animals are also eaten. Egyptian geese swim with ease, but are more terrestrial and more quarrelsome than most other geese. The normal call is a harsh bark, frequently repeated: during courtship the gander produces a strange rasping puff.

*The term 'goose' is somewhat misleading; the Egyptian goose is not in fact a true goose – tribe *Anserini* – but is assigned to the tribe *Tadornini* – the shelducks and sheldgeese.

The Egyptian goose was first domesticated by the ancient Egyptians by whom it was regarded, like the ibis, as a sacred bird. The Greeks and Romans carried on the tradition of domestication, and the bird was commonly kept in captivity in Europe in the seventeenth century, and probably even earlier. Willughby and Ray in their *Ornithologia* (1676–8) describe correctly (p. 360) 'The Gambo-Goose or Spur Wing'd Goose', which they saw in St James's Park. Their illustration (plate LXXI), however, clearly shows an Egyptian goose with dark eye and breast patches. There is no indication of the provenance of the specimen used by their artist, but the implication is that the species may have existed in England in the seventeenth century.* In 1785 it was described by Latham as ' . . . common at the Cape of Good Hope, from whence numbers have been brought to England: and are now not uncommon in gentleman's ponds in many parts of this Kingdom'.

In January 1795 a stray Egyptian goose was shot at Thatcham, west of Newbury, in Berkshire, and in 1803 or 1804 a further six were killed near Buscot – having possibly strayed from the lakes of

*My thanks are due to Mr Robert Hudson for drawing my attention to this reference.

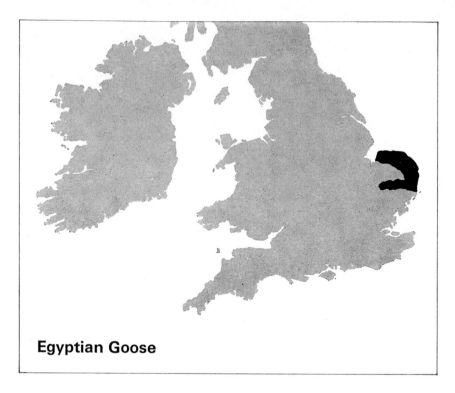

Egyptian Goose

Buscot House, near the Berkshire/Wiltshire/Oxfordshire/Gloucester-shire border. In December 1808 two were killed at Harleston on the Norfolk/east Suffolk border – the earliest record of escaped birds for East Anglia. In his *Diary*, published by Sir Ralph Payne-Gallwey in 1893, that noted wildfowler Colonel Peter Hawker (1786–1853), wrote that he 'killed 2 [Egyptian geese] in Norfolk and 3 at Longparish in Hampshire in the winter of 1823, and the next year again during some tremendous gales from the west, a flock of about 80 appeared near the same place when 2 more were killed'. In April 1830 a flock of five Egyptian geese was seen on the 'Fern' Islands (Farne Islands), in Northumberland; until 1880 the species was a regular winter visitor in small numbers on the nearby Fenham Flats, where it was referred to by the local wildfowlers as the 'Spanish goose'.* Other references during the first half of the nineteenth century to (presumably) escaped Egyptian geese, include three shot in Fin Glen, Campsie, 9 miles (14 km) north of Glasgow, in November 1832; two shot in Dorset in 1836; a party of nine seen on the Isle of

*R. Perry, *A Naturalist in Lindisfarne*, 1946.

Man in September 1838; a small number observed on an unspecified stretch of the river Tweed in February 1839; four shot on the Severn near Bridgwater in Somerset in February 1840; one out of a flock of four shot near Leicester on 4 March 1843; five seen on Romney Marsh in Kent in July 1846; one seen at Blenheim and another at Shelswell in Oxfordshire in December 1847; one shot at Shoreham in Sussex on 5 January 1848, and another at Shermanbury six days later; two killed at Ormsby Broad in Norfolk in March of the same year, and one on Derwent Water on 2 May 1849. Small parties of Egyptian geese were also occasionally reported from Ireland.

By the mid-nineteenth century a considerable number of free-flying and self-perpetuating colonies of Egyptian geese were established on estates in southern and eastern England, as well as some in other parts of Britain. The largest flocks were probably those maintained at Blickling Hall; Gunton Park; Holkham Hall (the Earl of Leicester), and Kimberley Park (the Earl of Kimberley) in Norfolk; at Bicton House and Crediton (respectively south-east and north-west of Exeter) in Devon; at Woburn Abbey by the Duke of Bedford, and at Gosford House, East Lothian, in Scotland. From these and other estates Egyptian geese frequently strayed to various parts of the country. In their *Birds of Devon* (1895) M. A. Mathew and W. S. M. D'Urban wrote: 'The Egyptian and Canada Goose and the Summer Duck, being so frequently kept on ponds in a semi-domesticated state, breeding freely and wandering at will over the country, are on the same footing as the Mute Swan and Pheasant. . . .'

In the early years of the present century a pair of Egyptian geese probably from Woburn, bred near London for several years. A single bird appeared at Hamper Mill in the Colne Valley between Rickmansworth and Watford in Hertfordshire, in the spring of 1934; two years later it was joined by a second bird, and after failing to breed in 1936 and 1937 the pair successfully hatched three goslings in the following year. They, or possibly another pair, were observed on a sewage-farm at Reading in Berkshire, about 25 miles (40 km) to the south-west, in May 1935, and on Connaught Water in Epping Forest, Essex, some 20 miles (32 km) due east, in the spring of the following year. A single bird, perhaps one of the same pair or of their offspring, lived on Ruislip Reservoir in Middlesex, only 3 miles (5 km) south of Hamper Mill, from 1941 to 1942. In 1957 a pair which had taken up residence on the ponds in the Royal Botanic

Gardens, Kew, became frightened and moved to the Thames, where Richard Fitter saw them at Strand-on-the-Green, Chiswick, in May.

Between 1954 and 1956 one or two pairs of Egyptian geese bred on Hillington Lake, 8 miles (13 km) north-east of King's Lynn, in Norfolk; others were seen on Stradsett Lake, 8 miles (13 km) south of King's Lynn, and up to fifteen at Beeston Hall, 18 miles (29 km) south-east of King's Lynn. A single pair nested on the lake in Gunton Park, 16 miles (25 km) due north of Norwich. In 1958 two pairs bred successfully at Fustyweed gravel-pits and two or three pairs on Salhouse Broad, 8 miles (13 km) north-east of Norwich. By 1959 the largest full-winged flocks of Egyptian geese were those at Holkham and Beeston, with smaller numbers at Woodbastwick near Salhouse, and at Woburn.

The Egyptian goose is today principally found in north Norfolk in the Holkham/Beeston area and in the valley of the river Bure. Between 1971, when the species was admitted to the British and Irish List, and 1975 colonies or individuals were reported from Barton, Beeston Hall and Beeston St Lawrence, Blickling, Cley, Didlington, Felbrigg, Gunton, Hillington, Holkham, Houghton, Horeton, Lenwade, Lexham, Lyng, Narborough, Narford, Salhouse, Sparham, Stradsett, Swanton Morley, West Acre, West Newton, and Wroxham. There has also been some spread to the Breckland area of Suffolk. In September 1973 a total of ninety-eight Egyptian geese were counted at Holkham, and in February 1974 141 were seen on the lake at Beeston Hall; these two counts may have included an element of duplication, as there is believed to be some intercommunication between these and other centres. In the spring of 1972 a pair was present on Fritton Lake in Suffolk, between Great Yarmouth and Lowestoft, and others were seen at Sotterley, south-west of Lowestoft,* while in 1974 a pair bred in the Leiston area, between Saxmundham and the coast. On 25 June 1974 a pair with three goslings were seen on Effingham Ponds in Surrey, where the parents may have been recently introduced.

The study of the Egyptian goose has, like that of the night-heron, been sadly neglected by ornithologists, most of whom have regarded it as not worthy of scientific attention. Much, therefore, remains to be discovered about the bird's habits and ecology. The failure of the species to spread and multiply in Britain to the extent that the

---

*Trans. Suff. Nat. Hist. Soc., (16) 3, p. 135, July 1973.

mandarin duck and Canada goose have done suggests that ecological conditions for it here are verging on marginal. What its limiting factors are have yet to be discovered, but it may be relevant that drier climates prevail in its African homelands.

In 1963 G. L. Atkinson-Willes estimated the free-flying Norfolk population of Egyptian geese at between 300 and 400 birds, of which about 200 were in Holkham Park. Recent counts suggest that the present total is about the same as in 1963, but there has been some expansion of range, small numbers now being found in from ten to fifteen centres in the Broads area and scattered along the valley of the river Bure.

Note: See also Addenda.

29. Mandarin Duck
(Aix galericulata)

The mandarin duck* is a native of eastern Asia, where it is found mainly in China and Japan, inhabiting well-established forests intersected by rivers, streams and lakes, seldom far from the coast. Here mandarins breed from about 40°N on the Chinese mainland, principally in the Tung Ling (Eastern Tombs) Forest (formerly the Imperial Hunting Grounds) north of Peking, to approximately 51°N on the island of Sakhalin and about 55°N on the river Uda in far eastern Russia. Between these extremes mandarins breed in northern Korea; in the Kirin Forest – another important centre – north of Antung in Manchuria, and throughout the watershed of the Amur river; on the Japanese islands of Kyushu (at Isahaya east of Nagasaki), Honshu (mainly on Lake Ashi near Toyko at about 36°N and at between 2000 and 3000 ft (609 and 914 m) around Mt Fuji), and on Hokkaido. Mandarins winter on the Chinese mainland from around 23°N in Kwantung north of Canton, to approximately 38°N in Kiangsi north of Shanghai – their principal winter quarters – and on the islands of Taiwan and Honshu, and in Assam in eastern India.

The mandarin duck – known in China as *Yüan Yang* and in Japan as *Oshidori* – reputedly received its name, by analogy, from seventeenth-century English merchants out of their sense of its superiority over others of its kind, implied in the title 'Mandarin' – in Chinese *Kwan* or *Kwūn* – a public official, counsellor or minister of state.†
During the period of the Ch'ing Dynasty (1644–1912), the 'mandarin' language was the Chinese spoken in official and legal circles, and

*For much of the early history of the mandarin duck I am indebted to Christopher Savage's excellent monograph (see Bibliography).
†The mandarin orange (*Citrus nobilis*) may have acquired its name for the same reason.

[FEMALE]

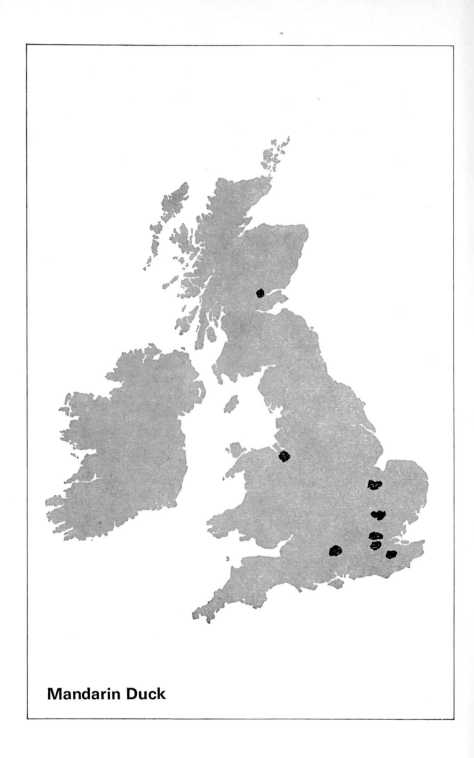

Mandarin Duck

the mandarin drake was the emblem worn by the seventh grade of civilian mandarin officials.

Since around 200 B.C. the mandarin duck has been revered by the Buddhists in China and Japan for its beauty and behaviour and, being monogamous, has for long been regarded as a symbol and example to mankind of connubial bliss, fidelity and mutual affection; this is shown by the many appearances it makes in Chinese and Japanese literature and art from the sixth century B.C. onwards. The mandarin was for hundreds of years a protected bird in China and Japan, which it was unlucky, if not illegal, to destroy.

The earliest references to the mandarin in literature occur in the writings of the disciples of Confucius (551–478 B.C.). Savage quotes an example translated by Sir Arthur Waley:*

> Mandarin ducks were in flight,
> We netted them, snared them.
> Long life to our Lord,
> Well may blessings and rewards be his!
>
> There are mandarin ducks on the dam,
> Folding their left wings.
> Long life to our Lord,
> Well may blessings for ever be his!

The mention of the folding of the left wing portends 'blessing heaped upon blessing'.

A legend, known to the Japanese as Buddha Ohmu Kyó, speaks of a pair of mandarins foretelling the advent of the Lord. As a child Siddartha Gautama (Buddha, 'The Great Enlightened One', born c. 507 B.C.) reputedly possessed a pair of mandarins as pets, which so impressed him by their behaviour and display of mutual affection that in later years he made many references to them in his preaching. Another legend relates that Amida Buddha sometimes assumed the form of a mandarin in order to lend emphasis to his teaching.

Savage quotes three examples of references to the mandarin in the literature of the T'ang Dynasty (620–900 A.D.), translated by Soame

*Chinese Poems, George Allen and Unwin, 1946.

Jenyns;* in *The Beautiful Women* by Tu Fu occur the following lines:

> My husband holds me in light esteem,
> But his new mistress seems as beautiful as jade.
> Even the morning glory has its passing hour.
> The mandarin duck and drake do not roost apart,
> But wrapt in his new favourite's smiles,
> How can he hear his old love's sighs?

From *The Song of Never-Ending Grief* by Po Chu-I come the lines:

> The mandarin duck and drake tiles glitter coldly in the hoar frost,
> The Emperor is cold beneath the kingfisher quilt,
> For who is there to share it with him?

Referring to this verse, Savage points out that if a Chinese wishes to refer to a pair of anything, he associates it with the mandarin duck; thus, for example, there is a mandarin duck double pillow, a mandarin duck double bed-cover and a mandarin duck sword with a pair of blades emerging from a single handle.

The following lines are taken from *A Song of Chaste Women* by Mêng Chiao:

> The Wu-t'ung trees grow old together
> The mandarin duck and drake pair for life,
> Even so the chaste woman prides herself on
>   following her husband to the tomb,
> And throws away her life.

On the same theme, a Japanese legend tells of a falconer named Sanjō who killed a mandarin drake, whose mate, however, escaped. The same night he dreamed that a beautiful but sad and accusing woman appeared before him. The next day the escaped duck flew up to him and tore her breast open at his feet with her bill. Sanjō, the legend concludes, was so full of repentance that he became a monk.

'A picture is a voiceless poem' say the Chinese, meaning that paintings are not supposed to represent facts but rather to suggest a

---

*Poems of the T'ang Dynasty, Lowe and Brydone, 1940.

poetic idea; this is certainly true in most Chinese paintings of mandarins, which tend to emphasize the theme of love and beauty at the expense of anatomical accuracy.

The earliest representations of mandarins in Chinese painting date from the period of the Sung Dynasty (960–c. 1279 A.D.); an early example is a fine pen-and-ink drawing of a pair of mandarins beneath a rock and a sunflower, by Tsao Chung (Chao Chung) – known to the Japanese at Chō-shō: this drawing bears the seal of the collection of Emperor Kao Tsung, and his inscription 'Marvellous pen of Tsao Chung'. In the library of Cambridge University is a long scroll dating from the Ming Dynasty (1368–1644) painted in about 1430 by Li-y-ho showing, in part, a pair of mandarins swimming beneath a peony tree (*Paeonia moutan*), the emblem of spring, love, and feminine charm.

In the Department of Oriental Antiquities at the British Museum are two fine Japanese paintings of mandarins; the earlier one, painted on silk by Naorubu Sōhyei (1519–92 – Kano School) shows a waterfall plunging through a cloud into a mist-enshrouded stream which tumbles through a rocky gorge overhung by maple trees; a female mandarin in the water gazes up at her mate who is standing on a rock beneath a peony. The later picture, depicting a pair of mandarins swimming together under snow-laden plum-blossom – an emblem of winter – was painted in the seventeenth century by Murata Sohaku. The composition of this picture is remarkably similar to that of Li-y-ho. A painting from the nineteenth century – Okyo School – of mandarins on a grassy bank under a tree, is contained in the Soame Jenyns collection.

The beginning of the eighteenth century saw the introduction to Japan from Europe of the art of print-making, and a number of prints of mandarins may be seen in the Victoria and Albert Museum: one, by Koriusai (c. 1733), shows a pair swimming amidst reeds with snow-clad mountains in the background; on another – a fan-print by Kitao Shigemasa (1738–1819) – a mandarin duck in a stream gazes up at her mate who stands on a flat rock beneath a peony – a somewhat similar composition to the silk painting by Naorubu Sōhyei; on a third – an illustration from the *Sketchbook of Birds and Animals* by Fukuzensai Kagen (1741–86) – a pair of mandarins, the female more accurately depicted than in the two previous examples, are shown swimming together beneath a peony branch. Hiroshige (1796–1858) produced many prints of mandarins, the most frequently reproduced of which shows a pair displaying to each other under a 'fuyo', a variety of rose-mallow.

As in paintings so in pottery and porcelain, the earliest representations of mandarins date from the Sung Dynasty: the Victoria and Albert and British Museums possess a number of fine examples of translucent Ting ware of this period decorated with mandarins. Also in the Victoria and Albert museum are a fine late-eighteenth century bowl and plate, each showing a pair of mandarins in full plumage beneath a peony branch. A number of pieces of Chinese embroidery from the Kien Lung period, mostly showing mandarins swimming among reeds with peonies overhead, may be seen in the same collection.

Mandarins are also well represented in the field of Japanese and Chinese sculpture, appearing, frequently in pairs, in a variety of materials such as rose-quartz, ivory, jade, lacquer, gold and bronze. A magnificent bronze of a mandarin drake entwined with snakes, from Shi-chai-shan, Yünnan, in south-west China, of the western Han Dynasty and dating from the second or first century B.C., was shown at the *Genius of China* exhibition at the Royal Academy in 1973–4. It is interesting to note that the site where this superb piece was excavated in 1955 – one of the finest examples known of the mandarin in Chinese art – is several hundred miles away from the present-day range of the mandarin in China.

The mandarin drake possesses an amazing and striking plumage which makes him, without doubt, one of the most beautiful ducks – indeed one of the most beautiful birds – in the world. He is unique in appearance, with a purple-green crested head;* a broad white eye-stripe which extends from the lores back along the side of the head through the crest; a chestnut ruff; a glossy purple breast with four alternate black and white vertical stripes on either side; buff flanks; white underparts; a greenish-brown back, and a pair of magnificent golden-orange 'fans' or 'sails', which are formed of enormously enlarged and expanded secondary wing-feathers; his bill is vermilion and his legs a buffish-orange.

The head, neck and crest of the duck are a soft lilac-grey, with a small greyish ruff; the chin and lores are white, and a stripe of the same colour runs back from the eye into the crest at the nape of the

*Galericulata means 'wearing a little wig'.

neck; the back is brown with a greenish sheen; the breast and sides are brownish grey; the wings and tail vary from pale brown to dark sepia, with two of the secondary wing-feathers having a narrow diagonal white fleck edged in black; the bill is greyish brown and the legs brownish yellow.

In eclipse plumage the drake closely resembles his mate except that the eye stripe inclines to be more distinct, and the sides usually have more brown and less grey about them. The diagonal white fleck on the secondaries which is always present in the duck but never in the drake is, however, the sole definite distinguishing mark between the sexes in eclipse plumage. Mandarins measure between 16 and 17 in. (40 and 43 cm) in length – a little larger than a teal – and weigh from $1\frac{1}{4}$ to $1\frac{1}{2}$ lb (560–672 gm).

The earliest mention of the mandarin duck in Europe appears in 1599 in the writings of the Italian naturalist Ulissi Aldrovandi (1522–1605), where he describes a bird called *Querquedula indica** which was portrayed in a painting brought to Rome by Japanese envoys. The mandarin was first introduced to Britain shortly before 1745, in which year a drawing of *La Sarcelle de la Chine* ('The Chinese Teal'), as it was called, in the gardens of Sir Matthew Decker, Bt (1679–1749), on Richmond Green in Surrey, was made by George Edwards (1693–1773) in his *Natural History of Birds*. (Sir Matthew, sometime M.P. for Bishop's Castle in Shropshire and in 1729 appointed High Sheriff for the county of Surrey, was also an influential Director of the East India Company, and imported many species of exotic flora and fauna to the grounds of his home in Richmond.) Edwards also quotes from *History of Japan* (1727) by Engelbrecht Kaempfer (1651–1716), one of the earliest Europeans to visit that country;

Of ducks there are several different kinds, and as tame as the geese. One kind particularly I cannot forbear mentioning, because of the surprising beauty of its male, call'd 'Kinmodsui' which is so great, that being shewed its picture in colours, I could hardly believe my own eyes, till I saw the Bird itself being a very common one.

The *Encyclopaedia Britannica* (1777) refers to 'The *galericulata*, or Chinese teal of Edwards, which has a hanging crest. . . . The English in China give it the name of "mandarin" duck'. In 1783 Buffon

*Lat: *Querquedula*: a teal.

T

included an illustration of a mandarin drake standing on a rock in his *Histoire Naturelle* ... : seven years later John Latham (1740–1837) first described the female mandarin – '*Femina feminae sponsae similis*'; the female of *Aix sponsa*, the wood duck, is indeed superficially very similar to the female of its close relation, the mandarin.

In 1830 two pairs of mandarins were purchased for the considerable sum of £70 by the London zoo, where they bred for the first time in England four years later; here a male was drawn from life by John Gould (1804–81) for his *Birds of Asia.*

The first recorded specimen of an escaped mandarin in Britain was shot near the Thames at Cookham, Berkshire, in May 1866.

At the beginning of the present century the 11th Duke of Bedford introduced some mandarins to his collection of waterfowl at Woburn Park, Bedfordshire, where they have flourished ever since: it was probably one of the duke's mandarins which spent the autumn of 1908 in Regent's Park, London. By the outbreak of the First World War the mandarins at Woburn numbered over three hundred, but because of the difficulty of feeding the birds in war-time their numbers shrank by half; a similar decline in numbers occurred between 1939 and 1945. The success of the ducks at Woburn is due almost entirely to the ideal surroundings; the many old trees in the park and profusion of small ponds and streams, many encircled by rhododendron bushes, offer the birds excellent nesting-sites and resting-places; the woods, too, are prolific in oak, chestnut and beech-trees which provide the acorns, chestnuts and beechmast that form the mandarin's staple autumnal diet in Britain. Although the local conditions are so good, there unfortunately seems little chance that many birds will spread from Woburn, as the surrounding countryside is not well suited to their habits; in 1973, however, a pair succeeded in rearing five ducklings from a clutch of ten eggs at Old Linslade in Bedfordshire, a few miles south of Woburn, and mandarins were reported to be present throughout the year on the neighbouring Eversholt Lake.

In the years immediately preceding the First World War, Sir Richard Graham, Bt, of Netherby in Cumberland on the Border Esk, obtained some mandarins from the Wormald and McLean game-farm in East Anglia (whither they had been imported direct from the market at Canton) where the birds were artificially reared.

Many of these birds failed to breed because, it is believed, the Chinese caponized them to protect their trade. A few, however, did manage to breed successfully at Netherby and along the Esk for a number of years but did not spread into Scotland and elswhere as had been hoped, and by about 1920 the colony was extinct.

In 1918 Viscount Grey of Fallodon added some mandarins bred by Wormald and McLean to his bird sanctuary in Northumberland, where he already had a large and famous collection of waterfowl. They thrived and spread widely throughout the surrounding country-side, and it was believed that they would eventually become firmly established. Unfortunately, after Lord Grey's death in 1933 the sanctuary and collection only lasted a short time, and the mandarins soon disappeared completely.

From 1910 to 1935 Lieutenant-Colonel E. G. B. Meade-Waldo kept a flock of free-flying mandarins at his home at Stonewall Park near Edenbridge in Kent, but for some reason, probably the un-suitability of the habitat, the birds failed to establish themselves. A pair which bred at Seal near Sevenoaks, 8 miles (13 km) north-east of Chiddingstone, in 1935, and a further two pairs seen there three years later, presumably came from this source.

The next collection of waterfowl to contain mandarins was that of Mr Alfred Ezra at Foxwarren Park, near Cobham in Surrey, who in 1928* was presented with six pairs by M. Jean Delacour.† They bred in the following year and soon spread outside the sanctuary, eventually helping to form a separate colony which has proved most successful, and from which, in autumn and winter, mandarins regularly disperse south-west through Surrey as far as Haslemere; northwards to southern Buckinghamshire (where in 1955 a pair bred at Dropmore) and south-western Middlesex, and as far east and west as Staines and Reading respectively.

In January 1930 Ezra, with the help of J. Spedan Lewis and W. H. St Quintin, made an attempt to establish the mandarin as a free-flying duck in the parks of London. Ninety-nine full-winged birds imported from China were set free in the grounds of Bucking-ham Palace and Hampton Court and in Regent's, the Central and Greenwich Parks. None of these birds remained for very long; two

*In January of the same year a duck and two drakes were seen on the lake in Regent's Park in London, and on 18 November Dr Carmichael Low saw a single bird on Staines Reservoirs.

†My thanks are due to Mr David Tomlinson for drawing my attention to the source of Mr Ezra's birds.

ringed specimens were traced to Hungary and one to Sweden.* A further fifteen (pinioned) pairs were released in Regent's Park in the following year; a year later only one or two remained, and it was clear that the experiment had failed.

The mandarins at Virginia Water on the Surrey/Berkshire border, which now form one of the two largest colonies in the country, are presumed to be the descendants of Ezra's original stock at Foxwarren that found their way up the Bourne (where they were first seen by Derek Goodwin in 1929 or 1930), a small stream which flows from Virginia Water into the Thames at Chertsey; they were probably supplemented by migrants from the earlier experiments at Hampton Court and in inner London. By the early or middle 1930s they were well established, and have increased steadily every since. In addition, they have spread throughout Windsor Great Park and Windsor Forest in which they have found for themselves a perfect habitat; oak, chestnut and beech-trees abound, and there are any number of small ponds and streams which mandarins love so well; there is a wealth of rhododendron undergrowth to provide the necessary shelter and seclusion, and nesting-sites are in profusion; vermin which prey on birds and their nests, such as foxes and grey squirrels, are kept to a minimum by the park gamekeepers, and almost the only enemies the mandarins have to fear are unenlightened people who shoot them as they fly to and from the park and many, alas, have been killed in this way. In Windsor Park they frequent Great Meadow Pond, Johnson's Pond and Obelisk Pond among other waters, and since the war have been seen at Titness Park near Sunninghill, where there are sometimes over a hundred birds in winter; Bagshot Park (where they first bred in 1950); Sunninghill Park; Ascot Place; Foliejon Park, Winkfield; and Englemere Lodge, Ascot. The Virginia Water birds have formed probably the most successful and certainly the most truly wild colony of mandarins in Britain, and it is to them that we most look for the future expansion of the species in this country.

In 1935 Ronald and Noel Stevens added some free-flying mandarins to their collection of waterfowl at Walcot Hall in Shropshire:

*Another remarkable example of the mandarin duck's powers of flight was provided by a pair which had been reared and ringed at Ekeberg, Norway, in 1962. After being frightened by an explosion, they left Ekeberg at midday on 8 November, and were shot at Seaton Burn, Northumberland, at 5 p.m. on the following day, having travelled a distance of 560 miles in 29 hours, at an average speed of 19·3 m.p.h. (*British Birds*, 57, p. 585, 1964).

at first they refused to breed because hordes of jackdaws appropriated all the suitable nesting-sites. Once the jackdaws were disposed of the mandarins began to breed freely, and by the outbreak of war numbered about a hundred birds. The house was requisitioned during the war, when the mandarins dispersed as far as north Wales and Lancashire, where they were first seen in 1944. After the war they returned to form a small but apparently stable colony.

Since the war, other free-flying colonies of mandarins have been started at – among other places – Tillingbourne Manor, Surrey; Leckford, Hampshire; Bassmead, Huntingdonshire; Eaton Hall, Cheshire; Apethorpe Lake (Lord Brassey) and Milton Park (Major Hugh Peacock) – where flocks of up to fifty birds were counted in 1973 – in the Soke of Peterborough; the Zoological Gardens in Regent's Park in London, and on the river Tay in Perthshire. This last colony is of particular interest being, as it is, so far removed from any other established centre of feral birds.

The mandarins on the Tay originate from a collection started in 1946–7 by the late Mr J. Christie Laidlay at Holmwood on the outskirts of Perth. In 1962, and in successive years, half a dozen mandarin ducklings were feather-clipped, to keep them at Holmwood for their first breeding season. When Mr Laidlay died in April 1964 the flock consisted of between six and ten birds. The collection was continued by Mr Laidlay's widow, who in February of the following year flushed a party of seventeen mandarins from bushes on the banks of the Tay. In November 1973 the flock had increased to between thirty and forty birds, and had reached about double that figure by the spring of 1975. The majority breed in some of the fifty or so nesting-boxes in the trees in the garden (several have been rescued from chimneys in their search for nesting-sites), the most popular boxes being those placed in cherry trees in a vermin-proof pen by the river. Shortly before the eggs are due to hatch, the remaining occupied boxes are transferred inside the pen, which in 1975 contained some forty birds. The young hatched there are feather-clipped and are identified by ringing – different coloured rings being used each year: ducklings hatched in the wild usually return with their parents to Holmwood by October to feed with the pinioned birds. At least three pairs are known to have nested successfully away from Holmwood: at Bonhard, Murrayshall, and at the Earl of Mansfield's seat, Scone Palace, some 2 miles (3 km) to the north. Mandarins from Holmwood have been shot at Methven

Castle, 6 miles (10 km) to the west (one bird has been seen hanging in a game-dealer's shop in Perth) and have been reported from Glencarse on the north bank of the Tay, 5 miles (8 km) to the east, and as far as Dunbog on the south bank, east of Newburgh in Fife, some 12 miles (19 km) from home. Flocks of between twenty and thirty mandarins may be seen flighting up and down the Tay, or floating in on the tide at sunset.

In recent years there has been a marked expansion of range of the Windsor Park/Virginia Water population into other parts of Surrey and Berkshire as well as into Buckinghamshire and Middlesex. By 1969, when a survey of the district was carried out, between 110 and 120 mandarins had colonized parts of the middle reaches of the river Mole and suitable neighbouring habitats. Two pairs were at Mickleham; at least eight pairs lived between Leatherhead and Stoke D'Abernon (where there were ten pairs); four pairs were in Cobham Park and ten pairs were on the Mole in the Cobham area; between ten and fifteen pairs were to be found between Cobham and Esher, and two pairs were on Claremont Lake and Esher Common. Mandarins were observed in the Thursley district, between Godalming and Haslemere, and up to three were reported on Vann Lake, between Dorking and Horsham, during most of the year. The largest single concentration was a flock of forty-two which were seen feeding on acorns in October at Stoke D'Abernon. Nesting, with broods of up to ten ducklings, was reported at Bookham Common; Little Heath Pond; Black Pond on Esher Common; Norbury Park, Stoke D'Abernon (two families); Woodlands Park, Leatherhead, and possibly at Brockham Green.

In Berkshire, an adult was flushed from a nest-hole in a tree in Swallowfield Park, south of Reading, in 1971, and mandarins bred at Bray, near Maidenhead in 1971 and 1972, and at Widbrook, near Cookham in 1973, when a single pair also nested at Wraysbury in Buckinghamshire. Throughout 1969 a mandarin drake was present in Bushy Park, Middlesex, where in August it was joined by a duck. Two ducks and a drake, and later three ducks, were seen in the same year at Potters Bar – the first record for this district. In 1972 two pairs of mandarins bred at Wrotham Park near Barnet. All these birds are believed to have come from the collection started in about 1930 by Mr J. O. D'Eath at Monken Hadley in Hertfordshire. Some fifteen years ago Mr D'Eath allowed a number of mandarins to fly free, and these have now become well established in the neighbourhood. In most years about a dozen birds are ringed and set at liberty to reinforce those already in the wild: in 1975 these numbered between twenty and thirty pairs, some of which return each year to nest in natural and artificial sites.

In about 1972–3 a small number of feral mandarins became established at Leonardslee, near Horsham in west Sussex, the home of Sir Giles Loder, Bt, where in 1976 several broods – one comprising eight ducklings – were reared successfully.

There are, in addition, numerous isolated records of feral mandarins from many other parts of the country. In 1959, for example, a full-winged drake bred with a pinioned duck at Margaretting in south Essex. In the 1950s and 1960s mandarins were reported from Brown Candover, Lymington and Sowley Pond, Hampshire. In 1971–2 mandarins were seen at Ledsham, Aldford, Poulton, Huntington and on Rostherne Mere in Cheshire, which may well have come from the thriving colony nearby at Eaton Hall. In 1972 a pair nested at Alverstone near Sandown on the Isle of Wight; in the following year seven were seen at Newchurch (which may have been the offspring of this pair) and breeding again occurred. A mandarin drake mated to a mallard duck which nested at Walton Dam near Chesterfield in Derbyshire in 1973, were less successful: the first clutch of at least three eggs was sucked by a magpie; the second, of five eggs, was broken, but hardly surprisingly the eggs proved infertile. In the spring of 1974, several pinioned birds escaped from a collection near Loch Lomond; a fully-winged duck was seen on the loch on 13 December, as a result of which feral breeding was suspected.

The haunts of mandarins in Britain are much the same as those in the Far East – rivers, streams and lakes surrounded by mature deciduous woodland, especially those with a wealth of rhododendron undergrowth. The drakes emerge from eclipse into full plumage by late September or early October, and soon start displaying by raising and fluttering their 'fans' or 'sails', by expanding their crests to expose their yellowish-white cheeks, and by alternately dipping their heads forward and jerking them backwards. The majority are paired by the end of December, and in mid-March the pairs disperse in search of suitable nesting-sites; these consist of cavities or holes in trees – ancient oaks appear to be the most favoured, though in areas of marginal habitat pollarded willows are sometimes used – which vary from a few inches to up to several feet in depth, and are normally

between 10 and 50 ft (3 and 15 m) from the ground. In late April
or early May between nine and twelve glossy buffish-white eggs are
laid in the unlined nest, which are incubated by the duck only for
from twenty-eight to thirty days; the ducklings, which are fully
fledged by early autumn, normally drop to the ground and take to
the water within twenty-four hours of hatching. The summer moult
of the adults is complete by the end of June, and is followed by a
short flightless period of less than a fortnight: full plumage is re-
assumed by the end of September or the beginning of October.

The wing-area/weight ratio of mandarins is one of the highest of all
the ducks and, coupled with their unusually long tails, gives them
great aerial manoeuvrability, enabling them either to make a normal
landing on water or to plunge gannet-like with half-closed wings to
within a few feet of the surface before checking their descent.
Although their long legs enable mandarins to walk smoothly and
easily on land, in the water they swim somewhat jerkily in the
manner of moorhens. They are strong, though not particularly
swift, fliers seldom ascending to higher than tree-top level.

On the ground or in the water both sexes normally utter little
more than low twittering sounds, apart from their alarm notes, that
of the drake being a short 'uib', that of the duck a plaintive 'ack'.
During courtship the drake utters a booming 'uab' and in flight
produces a shrill piping whistle.

Mandarins feed, mostly at night, on insects in spring, and largely
on 'pink-toothbrush' water-weed (*Polygonum amphibium*) and frogs
in summer: in autumn nuts – especially acorns, chestnuts and
beechmast – are their staple diet, supplemented by gleanings from
the stubble fields.

The future of the mandarin in the Far East is in some doubt; after
the war it appeared to be rapidly becoming extinct due almost
entirely to the disastrous deforestation of its two main breeding-
grounds, the Tung Ling and Kirin Forests. The deforestation was
carried out by the Chinese Republic following the policy of the
Soviet Union in making Manchuria economically and agriculturally
self-supporting with heavy industrial areas and collective farms.
Predatory animals of the cat tribe are numerous in China, and birds
of prey, especially the black-eared kite (*Milvus migrans lineatus*), also
take heavy toll of the mandarin population. On the credit side is the

fact that the Chinese do not willingly kill mandarins, as for 2500 years they have looked upon the little ducks with respect and admiration as a symbol of love, conjugal fidelity and true happiness. The Chinese are not great flesh-eaters in any case, and the mandarin is a rather dirty eater in the East, consuming fish-spawn, water-snails, worms, frogs, and occasionally small mice and voles, in addition to the nuts and corn which it prefers in Britain.

In 1963 Cheng Tso-hsin wrote:

Usually it [the mandarin] lives together in pairs, and during the winter large flocks of about 100 mandarin ducks have been seen on small islands in rivers, river banks and bays, though mostly on lakes within villages and upland areas. . . . The mandarin duck is a valuable bird well known by the people. In the parks they are tame, and placed there for people to enjoy. Live birds are also shipped to foreign countries, though it is not wise to capture them in large amounts.*

It may be, therefore, that the outlook in China is not as gloomy as it appeared to be thirty years ago. The mandarin's close relation, the wood duck, now flourishing in America but only recently almost exterminated because of deforestation, especially in the eastern and mid-western states, has successfully adapted its breeding habits to the problems presented by man, and now nests in hollow gate-posts and similar artificial sites.

The future of the mandarin in this country is largely up to us; climatically, the British Isles are probably ideal for mandarins, as they find here approximately the same mean annual temperature as in China, but without the need for extensive migration; mandarins, as representatives of the tribe *Cairinini* – Perching Ducks – which spend much of their time in trees, fill an ecological blank among British waterfowl, and for this reason alone we should be glad to have them. Unfortunately, breeding-grounds in Britain are rather limited, as coniferous forests are quite unsuitable, and only certain deciduous trees are favoured; but unless mandarins are re-establishing themselves in the wild in China, it may well be that in the years to come this country will be their principal home. It is even possible that there may already be more free-flying feral mandarins in Britain than are left in the whole of China, although up-to-date information

China's Economic Fauna: Birds, Peiping, 1963. [English language translation by Joint Publications Research Service, Washington, U.S.A., 1964]. I am grateful to Mr Robert Hudson for drawing my attention to this reference.

from the People's Republic on the present position is meagre and lacking in detail. This lays a heavy responsibility upon us to ensure the firm establishment of the species in Britain. Provided that mandarins continue to have access to suitable breeding-grounds and are not slaughtered indiscriminately, there is every possibility that they will succeed. If they cannot find suitable breeding-sites they will not nest at all. Lack of such sites was the cause of the near extinction of the wood duck in America, and is the main cause of the mandarin's decline in China.

It is difficult to estimate accurately the present population of mandarins – which were admitted to the British and Irish List as recently as 1971 – but the total is probably in the region of about 1000 birds (B. Campbell and I. J. Ferguson-Lees in *A Field Guide to Birds' Nests* (1972) put the British breeding population at 'probably over 250 pairs'), and at present may be slightly increasing. Quite apart from the various conservation arguments put forward in favour of the establishment of the mandarin in Britain, we should feel privileged to welcome to this country what is arguably the most charming, beautiful and attractive duck in the world.

Note: See also Addenda.

30. Wood Duck or Carolina Duck

(Aix sponsa)

The wood duck or Carolina duck (sometimes also known as the summer duck) is a native of the eastern half of the United States and Canada, ranging from Nova Scotia and Lake Winnipeg in the north as far south, in winter, as Florida and Texas. An entirely separate population in British Columbia and California extends as far south as the San Joaquin Valley.

There are only two species in the genus *Aix* – the wood duck here described, and the mandarin duck (*A. galericulata*): their appearance and behaviour are so similar that, notwithstanding that they come from entirely different parts of the world, they are usually regarded as congeneric.

In breeding plumage the drake wood duck is almost as spectacular in appearance as his mandarin counterpart. His metallic-green head has a long purple occipital crest bounded by two narrow white lines, with a white throat and chin. The mantle is a dark greenish brown, and the wings are a metallic blue, green and black. The long and broad tail-feathers are a dark glossy green, the elongated coverts being black and dark brown. Triangular white spots mark the chestnut upper-breast, which passes into white on the lower-breast and underside. Broad black and white bars on the sides enclose buff-coloured flanks streaked with black, with a maroon-tinged rump. The iris is orangey red, the bill yellow, red, white and black, and the shortish legs yellowy brown. In eclipse plumage the drake wood duck resembles his mate, who herself is similar to the female of the mandarin duck except that the white eye-line is missing. The habits and habitat of the wood duck, its feeding, flight and nesting, are remarkably similar to those of the mandarin – ten to fourteen ivory coloured eggs hatching after twenty-eight to thirty-two days of incubation.

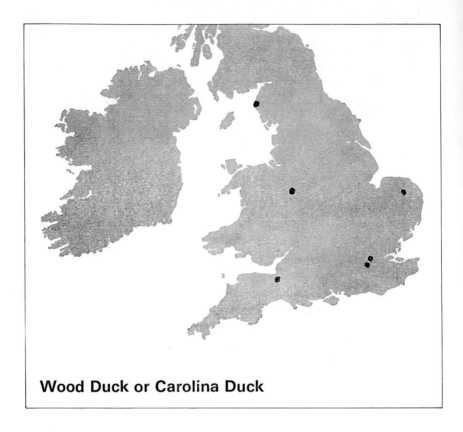

Wood Duck or Carolina Duck

Although the wood duck is now flourishing in North America, due largely to its adaptation to nesting in such artificial sites as holes in gate- and fence-posts, not long ago it was almost exterminated as a result of widespread timber-felling operations.

'It's tree-loving habit', wrote W. Dutcher around the turn of the century,

is one of the causes of its decrease. The increase of population in the country, and the consequent clearing of the land for agricultural purposes, the ruthless destruction of the forests, and the draining of swamp-lands, have lessened the number of breeding sites. This applies particularly to the eastern and middle-western section of the country. In many localities where the wood-duck was known to breed until within a few years, it is not now found, owing to the fact that every tree suitable for nesting has been cut down. This cause of decrease is largely due to the habits of the species and in some degree to the unwise practice of deforestation, which, unhappily, is so common in these days.

Various theories have been suggested to account for the success of the mandarin duck and the comparative failure so far of the closely related wood duck to establish itself as a feral breeding bird in Britain. Probably the principal reason is the higher duckling survival rate of the mandarin, in which the fledgeling period lasts for from six to eight weeks compared to the ten weeks of the wood duck, whose young are therefore vulnerable for a considerably longer period to ground and water predators.

Mathew and D'Urban writing in *The Birds of Devon* at the end of the nineteenth century, describe the 'Summer Duck' as 'breeding freely and wandering at will over the country': Bicton House near Exeter possessed at the time one of the most comprehensive collections of wildfowl in England, and the flock of fourteen wood duck seen near Plymouth in 1873 and the single bird shot at Slapton Ley south of Dartmouth (respectively some 40 and 30 miles (64 and 48 km) south-west of Bicton) may well have come from this source. The Duke of Bedford at Woburn, Viscount Grey of Fallodon in Northumberland and Mr Alfred Ezra at Foxwarren Park in Surrey, are among those who have in the past kept flocks of unpinioned wood duck – usually in the company of mandarins – on their waters.

In the late 1960s a pair of wood duck were acquired by the owner of an estate near some ponds at Puttenham Common between Guildford and Farnham in Surrey. The birds, which were unpinioned, bred in succeeding years, and free-flying adults, presumably from this source, were reported in the Cutt Mill/Puttenham district in increasing numbers from 1968 onwards. In June 1969 a duck was seen with two ducklings, and in the following month the same, or possibly a different duck was present with four young. By 1972 it was estimated that there were upwards of ten pairs in the neighbourhood; the birds in the collection remained unpinioned and frequently visited nearby waters where it is thought that they probably bred.

In 1969 a family party of five wood duck was seen on Virginia Water in Windsor Great Park; others were reported, sometimes in the company of mandarins, on a number of lakes in the Virginia Water/Sunningdale area: the source of these birds is unknown. Between 20 and 29 May 1973 a pair successfully hatched and reared four ducklings on Virginia Water.

In September 1970 a pair of wood duck with six well-grown ducklings was seen on a sewage-farm near Guildford, having presumably nested in the wild somewhere along the valley of the river Wey. The source of these birds, too, is unknown: apart from a pair which bred at Cobham a few years after the war, the only previous breeding-season record of a wood duck in the district is of a single bird in 1967.

Between 1964 and 1970 small numbers were occasionally seen on Vann Lake between Dorking and Horsham, where an unsuccessful attempt to breed was made in the spring of 1968. A number of well-grown young were reported later in the year, but it seems unlikely that they resulted from breeding in the wild.

In the winter of 1972 up to fifty free-flying wood duck were counted at the Tropical Bird Gardens at Rode in Somerset. At least one pair with young has been seen nearby on the river Frome, and it is suspected that others may also be nesting in the wild.*

In 1973 wood duck bred on the estuary of the river Duddon near Broughton in Cumberland; two possible sources have been suggested as the origin of these birds; they may have escaped either from a private collection at Drigg, about 12 miles (19 km) to the north-west or from a reserve of the Wildfowlers Association of Great Britain and

*D. H. S. Risdon, 'Report on the Tropical Bird Gardens at Rode, 1972', *Avicult. Mag.*, 79, pp. 52–5, 1973.

Ireland at Millom near the mouth of the estuary. Fourteen miles (22 km) to the north-east an attempt is being made to introduce wood duck to Grizedale Forest between Coniston Water and Windermere in north Lancashire.

In the winter of 1972–3 some free-flying wood duck were observed near East Dereham in Norfolk, and it is understood that breeding has taken place in the north-west of the county near the Wash.

In 1974 a wood duck's nest was found at Cranbury Park near Winchester in Hampshire, a few miles away from a bird-farm at Leckford.

From 1972 to 1975 inclusive three or four adult wood duck with ducklings were observed on Bishops Offley Mill Pool, 4 miles (6 km) west of Eccleshall in Staffordshire: as no nests were found nearby, it is suspected that the birds are breeding on a private estate upstream of the pool.

In 1972 a report of the British Ornithologists' Union Records Committee contained the following statement by Dr J. T. R. Sharrock:

Wood Duck (*Aix sponsa*). Nine records of successful breeding by feral birds are known during 1968–71, at six sites in Surrey and one in Norfolk. The Records Committee considers that the evidence suggests that the populations may not yet be self-supporting and recommends retention of the species in Category 'D'. The situation will be reviewed in 1977.*

*Category 'D' is an appendix to the British List for species worth recording but not yet given formal admission to that list.

31. Ruddy Duck

(Oxyura jamaicensis)

The North American ruddy duck breeds from central British
Columbia, the Great Slave Lake in the North-West Territories and
northern Manitoba, south to central California, New Mexico,
northern Texas, Iowa and Illinois. It winters from southern British
Columbia, New Mexico and southern Illinois east to the Chesapeake
Bay, and south to southern California, the Gulf of Mexico, the
Carolinas, and Costa Rica. Ruddy ducks resident in the West Indies
have been separated from the North American birds under the name
O.j. rubida, but the distinction is of doubtful validity.

In breeding plumage the drake ruddy duck has a ruddy-coloured
mantle with a black crown and neck tinged with a purplish sheen.
The cheeks, chin, and the sides of the head below the eye are white.*
The breast and flanks are finely barred with brown, and the underparts and undertail-coverts are silvery white. The bill is a distinctive
pale blue. The duck has a generally light-brown mantle and neck –
sometimes inclining to brownish red in summer – barred with
darker-brown streaks. Her head has a blackish cap and a white chin,
with a dark-brown line surmounted by a whitish line spotted with
brown running below the eye. The chin and underparts, as in the
drake, are a silvery white, and the bill is olive-green. In eclipse
plumage the drake resembles the duck, apart from a brownish-black
crown and the retention of the distinguishing white cheek-patch.
Both sexes measure about 15½ in. (39 cm) in length. The food consists
of waterweed, insect larvae and the seeds of aquatic plants, which are
strained from the mud while diving in water up to nine or ten feet
(3 m) in depth. The legs of the ruddy duck are positioned well back
on its body, like those of a grebe, causing it to be somewhat ungainly
on land. In the water, however, it swims buoyantly and dives with
ease, and is also able to submerge slowly without diving. Like a grebe

*Races *andina* and *ferruginea* show progressively smaller white head patches.

it can swim with only its head and neck showing above the surface. Ruddy ducks prefer still to moving water – even outside the breeding season – and Hudson refers to only two records on tidal waters in Britain; a party of five off Hilbre Island in the Dee estuary in September 1959, and a single bird at Chittening on the Severn estuary in December 1974. The call-note is a clear 'chuck-chuck'.

The display of the drake ruddy duck is one of the most unusual of all the *Anatinae*. Ruddy ducks belong to the tribe *Oxyurini* ('stiff-tails'), and in courtship the tail of the drake is erected at right-angles to the body and is spread out in a fan-shape. At the same time the breast is puffed out, by means of air-sacs beneath the feathers, in resemblance to a pouter pigeon; an inconspicuous crest is raised on the head, and the bill is drummed rapidly on the expanded breast to produce a popping sound.

Ruddy ducks build a platform-like nest in patches of thick cover such as rushes, reed-beds or other aquatic vegetation in or near shallow water, usually about 8 in. (20 cm) above water-level (but sometimes actually floating), in which an average of six to eight eggs are laid from mid-May to early June which hatch after twenty to twenty-five days' incubation. This is one of the latest breeding seasons of all British waterfowl. The earliest recorded feral brood was seen in Leicestershire in April 1975 after an exceptionally mild winter: at the other end of the season, a duckling was seen in 1965 on Chew Valley reservoir as late as 31 October.

In the 1930s a pair of ruddy ducks in the collection of Mr Noel

U

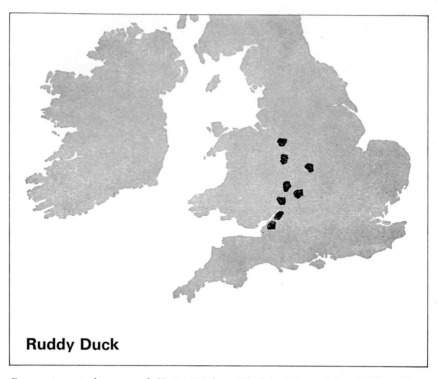

Ruddy Duck

Stevens nested successfully at Walcot Hall in Shropshire. This collection became dispersed during the last war.*

The ruddy ducks now living and breeding in considerable numbers in several south-western and west Midland counties are the descendants of some which escaped from the Wildfowl Trust's reserve at Slimbridge in Gloucestershire.† In 1948 the Trust obtained three pairs from the United States, which bred at Slimbridge in the following year. In general, however, the birds proved difficult to rear artificially; four ducklings were successfully reared naturally in 1952; five in 1954; twelve (from three pairs) in 1955; seventeen in 1956; about forty (from twelve pairs) in 1957; about forty (from ten pairs) in 1958; and between ten and forty in subsequent years. The Wildfowl Trust has for some time made a point of pinioning the young of exotic species, but as the stock at Slimbridge increased this became increasingly difficult. The first full-winged young (two birds) escaped from Slimbridge in the winter of 1952–3. From 1956 most of the

*Avic. Mag., pp. 104–5, 1938.
†For information on the spread of the ruddy duck in Britain I am indebted to Robert Hudson's paper in *British Birds* (see Bibliography).

ruddy ducks at Slimbridge were permitted to rear their own young in the Rushey Pen, and more ducklings were thus able to escape pinioning. As only some thirty birds in all were counted on 31 October 1957, it became clear that – since some had died or had been transferred to other collections, such as the Trust's reserve at Peakirk in the Soke of Peterborough – up to twenty had departed. It is believed that up to 1973 about seventy ruddy ducks flew away from Slimbridge; the majority departed before and during the severe winter of 1962–3, when all the full-winged birds left and most of the pinioned ones died.

The earliest reports of ruddy ducks in the wild date from 1954; in April of that year a single drake was reported from Hingham, Norfolk, 14 miles (22 km) south-west of Norwich, and in June a second was seen at Carsebreck Loch (Carsebreck Curling Pond), 17 miles (27 km) south-west of Perth. These birds are presumed to be those which escaped from Slimbridge – respectively 155 miles (249 km) and 317 miles (510 km) away – in the winter of 1952–3. In August 1954 a party of five 'stifftails' (three drakes and two ducks) was reported on Aqualate Mere in Staffordshire, but it is uncertain whether or not these birds were *jamaicensis*. In November 1957 a single young male ruddy duck was reported from Villice Bay on the Chew Valley reservoir south of Bristol in northern Somerset. During the ensuing winter the number of ruddy ducks at Chew increased to four, and the same number – which may well have been the same birds – were reported from the neighbouring Blagdon reservoir. Single individuals were also observed from time to time at the nearby reservoir of Barrow Gurney. These birds moulted into adult plumage in the early spring of 1958, and in subsequent years were joined by several other immature birds.

In the early days of their freedom, the escaped ruddy ducks at Chew and Blagdon took to the air with some frequency, although their flights were inclined to be at low levels and of short duration. After landing, they sometimes rose half out of the water and flapped their wings at great speed for up to half a minute. As the birds became established and their confidence increased, they preferred to escape from potential danger by diving beneath the surface of the water and taking refuge in nearby reed-beds, from which they seldom emerged during the breeding season. In 1958 and 1959 these drakes were seen to be displaying – albeit in a restrained manner – to each other and, on at least one occasion, to a female mallard and to a mallard drake in eclipse plumage. It was not until December 1959 that the first female ruddy ducks, presumably further escapees from Slimbridge, appeared at Chew Valley reservoir, where one pair bred in 1960 and a brood of ducklings was seen in May.

Although it was at one time believed that the English feral population arose solely from escaped birds and their progeny, it is now known that there was one deliberate release, when three or four juvenile females were captured at Slimbridge and transported to Chew as potential mates for the drakes already there. The exact date is not recorded but it was some time during the second half of 1961. This release thus took place after feral breeding had already commenced at Chew; the fact that the Chew breeding numbers remained at one or two pairs throughout the 1960s is further evidence that this release had little or no influence on the establishment of a feral breeding population.

At the same time ruddy ducks were also establishing themselves in the west Midlands, where birds first appeared in Staffordshire in September 1959, and were proved to be breeding in 1961 at Gailey and at Belvide. They multiplied so rapidly – many suitable nesting sites being available – that they soon outnumbered the whole breeding population of the lower Severn. These west Midland colonies, which are currently breeding in six or seven counties, are almost certainly derived from escapees from Slimbridge and their progeny.

The table opposite, compiled from Hudson and relevant County Bird Reports, gives particulars of the principal records and spread of the ruddy duck in England from 1957 to 1976.

As the table shows, the increase in population and range of the ruddy duck in England during the past nineteen years – apart from a temporary setback following the severe winter of 1962–3 – has been more rapid and spectacular, over a comparable period of time, than that of any other species of feral British waterfowl. The principal reasons are probably the availability of suitable habitats; the existence of a vacant niche for a freshwater bottom-feeding duck; lack of shooting pressure; mild climatic conditions; and, until the early 1960s, some possible reinforcements in the shape of escapees from Slimbridge.

The isolated colony in Leicestershire appears to be unique in that the birds both winter and breed on their natal water. Elsewhere there is a distinct tendency for regular seasonal dispersals; the birds leave their smaller breeding waters in the autumn and congregate in

First Record	Breeding	Population	Remarks
Somerset 1957, Villice Bay Chew Valley Reservoir (4 drakes).	First British feral breeding record: 1960, Chew Reservoir. 1–2 pairs to 1971: 4 in 1972: 6 in 1973: about 10 in 1974: 10 drakes present in April 1975. Sole Somerset breeding site. 5–6 breeding pairs in 1975.	November 1962: 7 December 1963: 15 October 1966: 24 February 1969: 42 January 1971: 55 January 1973: 68 December 1973: 96 December 1974: 100–110 November 1975: 120 (of which 111 were at Blagdon). 12 January 1976: 154 8–9 February 1976: 130 (during cold weather).	Numbers divided between Chew – which remains the only breeding site – and Blagdon Reservoirs, with some interchanging. Main congregations November/December, disperse March/April. Also seen at Barrow Gurney (1958); Cheddar (1963); Durleigh (1969); Sutton Bingham (1969). Many must be winter visitors from elsewhere, perhaps the west Midlands.
Gloucestershire Full-winged juveniles began leaving Slimbridge in significant numbers in 1957.	1963, Frampton. 1 pair 1969, 1970, 1971. 1975 breeding population: 1 pair (usually unsuccessful due to predation by pike).	14 (maximum) on 22 November 1969.	Seen at Frampton in May and June 1973 and 1974. (There are no longer any full-winged birds at Slimbridge.)
Worcestershire 1961, Bittell Reservoir.	1971, Westwood Park and Upton Warren; 1974 1 pair at Westwood, 2 pairs at Upton: 1975 breeding population, 3–4 pairs, (1 pair at Westwood, 2–3 pairs at Upton).	—	Seen almost annually in summer from 1970 at Pirton Pool, near Pershore, but not proved to breed.

First Record	Breeding	Population	Remarks
Herefordshire January 1963, Hereford.	One pair present on Eywood Pool, Titley, in July 1975. Probably now breeding.	—	—
Warwickshire 1962, Alvecote Pools.	In 1974 probably at Middleton Hall Pools where juveniles were seen in September, and possibly at Alvecote. Birds again present at both in 1975. Breeding population probably 2 pairs.	6 (maximum) at Alvecote in November 1972.	Has occurred annually since 1971, mainly at Alvecote, where a pair was present in May and June 1974, but breeding was not proved. Seen 14 August 1963 and in 1973 in Edgbaston Park, at Draycote in 1972, and a pair at Packington Park in the summer of 1974. 4 birds at Alvecote Pools from April to December: up to 5 (including some juveniles) at Middleton between July and November 1974.
Staffordshire 1959, Belvide Reservoir (early September). 27 September 3 males and 2 females at Gailey. (In August 1954 five 'stifftails' (*? jamaicensis*) were seen on	1961, Gailey and Belvide; bred again at Gailey in 1962: believed to breed most years at Belvide (3 pairs 1962 and 1974 – other years 1–2 pairs). Bred at Copmere near Eccleshall in 1968, and 1–2 pairs perhaps since. Bred White Sitch near Weston-under-Lizard 1971–2. 4 displaying pairs on Copmere and 5 pairs on Aqualate Mere, where breeding is suspected but unproven, in April	December 1962: 13 Winter 1965/6: 15–20 September 1967: 20 November 1969: 35–40 December 1972: 70–80 November 1974: 110 October 1975: *c.* 190 birds at Belvide. February 1976: 221 at Belvide and Blithfield reservoirs,	Flocks largest between October and November, dispersing in March/April. Movement between two principal centres, Belvide and Blithfield. Flocks form at Belvide during September–November, and by December many move to Blithfield where they remain until the spring. Also seen in 1970 at Cannock reservoir and Betley and in 1973 at Aqualate and Knighton.

		highest number ever recorded in the county.	
Shropshire 1962: four localities.	1965–6 Crosemere, where breeding has probably taken place annually since. At Ossmere from 1971 – currently 3 pairs: Hawk Lake (1–2 pairs in 1974): ? Cloverley Pool and Shavington (1–2 pairs at each in July 1974) and ? Colemere, where 7 birds were seen in September 1974. Possibly nesting sites at Fenmere (Berth Pool and Birchgrove Pool) near Baschurch; Tittenley (near Shavington); Norton Mere (near Tong); Allscott Pools (near Telford); and Marten Pool (near Chirbury). In 1975 2 pairs at Cloverley Pool: 3 pairs at Shavington; 2 pairs at Fenmere – all probably breeding. 1975 breeding population 12–15 pairs, of which 4 pairs were at Crosemere.	17 at Crosemere in November 1969. 4 pairs at Crosemere: 2 pairs at Ossmere: 6 drakes, 5 ducks, 6 ducklings at Hawk Lake, 1975.	Seen with increasing frequency – mainly in spring and summer – since 1969 throughout the county, especially from Ellesmere and Market Drayton north to the Cheshire border. Most birds probably winter on Staffordshire reservoirs.
Cheshire 1959 (25 September), Hilbre Island (flock of 5 – an exceptional coastal record).	? prior to 1971 at Barmere and Quoisley Mere: Barmere, Cholmondely and ? Quoisley in 1972; Redesmere and Capesthorne (both near Macclesfield) in 1973, and 3 pairs at Redesmere in 1974. 2–3	13 on Barmere in January and up to 35 in October 1974.	Smallest numbers in winter indicates some dispersal. Non-breeding birds seen on the Sandbach Flashes in 1972, and annually since then at Rostherne and Tatton Meres, both near

First Record	Breeding	Population	Remarks
Cheshire—*contd.*	displaying pairs on Rode Pool (near Alsager) in 1973, 1974 and 1975, but breeding not confirmed. 1 pair bred at Oakmere near Delamere in 1975. 3 males (presumed breeding) at Quoisley in June 1975. 1975 breeding population, 12–15 pairs.		Knutsford, at the northern point of the bird's 'regular' British range. Reported on Great Budworth Mere in 1974, but with no proof of breeding. Seen at Marbury, near Malpas, Combermere, Doddington Park and Budworth, 1975.
Derbyshire January 1973, Drakelow near Burton-on-Trent.	One pair probably bred Osmaston Park (near Ashbourne) 1975. One drake seen in July: a duck with half-grown young seen in September.	—	—
Leicestershire 1961.	1973, 2 pairs at Swithland reservoir: 2 pairs in 1974 (3 ducklings in May and further 6 in August); 1975, brood seen on 20 April. 3 pairs displaying in the spring, and 2 broods seen in April and June at Groby Pool. 1975 breeding population 2–3 pairs.	11 on Swithland on 9 February 1975 and the same number (possibly the same birds) at Groby in December.	Resident on Swithland and nearby Groby Pool.
Hertfordshire 1960, Tring.	1965–8, 1 pair on Tring reservoirs.	—	Irregular visitor since 1969: Hertfordshire is the only county where breeding has so far failed to result in colonization. The birds at Tring are believed to be some which Mr J. O. D'Eath allowed to

considerable numbers on larger reservoirs with natural margins and shallow bays (outside the breeding season extensive reed-beds are not a necessity) such as Belvide, Blithfield, Chew and Blagdon in Staffordshire and Somerset. Few ruddy ducks are known to breed in the lower Severn valley, their principal breeding area being further north in the west Midlands: it is thus apparent that large numbers of winter visitors to waters in Somerset must come from the west Midlands, presumably arriving and departing along the valley of the Severn. The Staffordshire winter flocks reach a peak in late October or early November (at least a month earlier than those in Somerset), and there may thus be some onward movement – albeit on a small scale – from the former to the latter, although the majority of the Somerset winter visitors almost certainly arrive direct from their breeding waters.

Away from their main breeding and wintering areas, Robert Hudson records that ruddy ducks have been seen in Essex; Nottinghamshire; Wiltshire; Northamptonshire (at Naseby Reservoir in June 1975); Huntingdonshire; Cambridgeshire (possibly some being escapees from the Wildfowl Trust's reserve at Peakirk); Lincolnshire (1968); Lancashire (1959, 1968, 1972 and 1975); on Island Barn Reservoir (five in June 1973) and elsewhere in Greater London; Hertfordshire (where they bred from 1965 to 1968); Buckinghamshire; mid-Glamorgan (September 1972); south Glamorgan (January 1973); Flintshire (January 1970); and Montgomeryshire (where a non-breeding pair summered in 1974). The most northerly English records are from north Yorkshire, where single birds were seen at Fairburn Ings in 1968 and 1969 and at Gouthwaite reservoir and Masham in 1974. The extreme northerly British record is that of a drake on Unst – the most northern of the Shetland Isles – in May 1974, over 500 miles (804 km) from the nearest feral breeding site. The source of these wanderers remains unknown.

Robert Hudson estimated the 1974 post-breeding population of feral ruddy ducks, which were admitted to the British and Irish List three years previously, at a minimum of 250 birds, including forty-five to fifty breeding pairs together with their progeny. By 1975 there were believed to be a total of between fifty and sixty nesting pairs in a total of eight or nine counties, and a post-breeding population of between 300 and 350 birds, of which about 120 were in Somerset, about 190 in Staffordshire, and between ten and fifteen in each of Shropshire, Cheshire and Leicestershire. During the decade between

1965 and 1975 the rate of increase (based on estimates of pairs and counts of winter flocks) has averaged at least 25 per cent per annum.

Determined efforts are made by the Wildfowl Trust to pinion the young of all exotic species, and since 1969 not more than one or two ruddy ducks are thought to have escaped each year. From 1973 the birds were once again being reared by the Trust artificially, thus still further reducing the likelihood of any young remaining full-winged. There exists, however, the problem of the occasional ruddy duck egg being laid in the nest of another species, and consequently being overlooked. It will be interesting to see what further progress the ruddy duck is able to make in its colonization of Britain; there is an abundance of so far unexploited but apparently suitable habitat for the species in Yorkshire, East Anglia, the east Midlands and the Home Counties, which may well be colonized as waters in the west Midlands become fully populated.

Note: See also Addenda.

32. Capercaillie

(Tetrao urogallus)

The home of the capercaillie is the coniferous forests of northern and central Europe; here it ranges from Scandinavia and northern Finland and Russia south to the Italian Alps, and eastward to the Balkans and Carpathians.

The cock capercaillie is one of the largest British birds, measuring from 33 to 36 in. (84 to 91 cm) in length. His general colour is a dark slaty grey broken by fine wavy lines: the wing-coverts are dark brown, the throat and sides of the head and neck black, and the breast a dark glossy blue-green. The tail is black with some white markings, and the upper tail-coverts are tipped with white. A patch of bright red skin stretches above the eye. The hen capercaillie is considerably smaller than the cock, measuring between 23 and 26 in. (58 and 66 cm) long. Her plumage is mottled and barred with rufous buff, black and greyish white, the neck and throat being a chestnut buff.

Capercaillies are strong fliers, and on the ground run or walk like other game-birds. They are equally at home perching and roosting in trees or at ground level, and are seldom seen other than singly or in small parties. In spring the cocks assemble at special display grounds known as 'leks'; here they engage in mock battles, leaping up and down fanning out their tails, drooping and noisily flapping their wings, the while uttering their strange 'song'. The nest consists of a hollow scraped in the ground by the hen, usually at the foot of a tree, in which between five and eight yellowish eggs with reddish-brown blotches are laid from mid-April onwards. Incubation – by the hen only – varies from twenty-six to twenty-nine days. The voice of the cock capercaillie is one of the most extraordinary of all birds; commencing with a resonant rattling sound, which is followed without a break by the noise of a cork being drawn from a bottle, it concludes with the impression of water being squirted from a

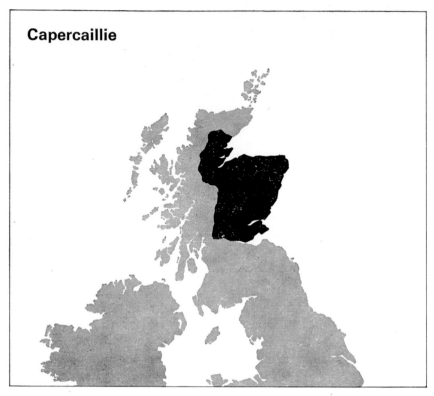

Capercaillie

siphon, followed by the sound of muffled knife-grinding. From October to April capercaillies live almost entirely on the buds and young shoots of conifers, favouring particularly Scots pine and larch. In spring and summer they eat buds and flowers of various plants – especially heather – as well as many small insects and fruits.

The capercaillie is unique among vertebrates in representing the only successful re-introduction to Britain by man of a species in the British fauna.* The bird was originally indigenous here, but became extinct in the second half of the eighteenth century: it was successfully re-introduced during the nineteenth century, and is now firmly re-established in suitable habitats in the Highlands of Scotland.

*The avocet (*Recurvirostra avosetta*) which was extinct as a breeding bird between 1842 and 1947; the bittern (*Botaurus stellaris*) which did not breed here between 1868 and 1911; the black-tailed godwit (*Limosa limosa*), (1847–1952); the osprey (*Pandion haliaetus*), (1903–54), and Savi's warbler (*Locustella luscinioides*), (1855–1960) – all re-established themselves naturally as British breeding species.

[FEMALE]

The capercaillie* is first referred to in the literature of Britain in 1527 by Hector Boece (1465?–1536).† In his translation, entitled *Chronicle of Scotland* (1528–9), of Boece's work, John Bellenden (fl. 1533–87) writes:

Many other fowls are in Scotland which are seen in no other parts of the world, as the capercailye, a foul more [larger] than a raven, which lives only on the barkis of trees.

He goes on to say that in the summer of 1529 James V went 'to Atholl to the huntis. . . . The Earl of Atholl maid great and gorgeous provisioun for him in all things pertaining to ane prince . . . with fleshis . . . black-cock, muir-foull and capercaillies.' In his *De Origine Moribus et rebus gestis Scotorum* (1578) John Lesley (1527–96), Bishop of Ross, wrote:

In Ross and Lochaber and in other places among the hills and knowes which are not lacking in fir trees, sits a certain very rare foul called the Capercalye, to name, with the vulgar people, the horse of the forrest. . . .

In a letter written to Lord Tullibardine on 14 March 1617 James VI suggested that his cousin the noble Earl should 'now and then send to us by way of present . . . the known commoditie yee have to provide, capercaillies and termigantis . . . the raritie of these foules will both make their estimation the more pretious. . . .' (*Old Statistical Account of Scotland*; xx. 1617). In the following year John Taylor, the 'Water Poet' in his *Visit to the Brea [Brae] of Marr* mentions the presence, among other varieties of game, of 'partridge, moorecoots, heathcocks, caperkellies and termegants', provided for him by his host 'the goode Lord Erskine'. An old 'Scots Act' of Parliament (Jas. VI. xxx. 1621) restricted trading in 'powties, partrikes, moorefoulles, blackcoks, grayhennis, termigantis, quailzies, caperkalzes etc.'. Sir Robert Gordon in his *History of the Earldom of Sutherland* (1630), described the forests of that county as having 'great store of partridges, pluivers, capercaleys, blackwaks, murefowls, heth-hens . . .' On

---

*Variously capercalze, capercailzie, capercally, caperkally, capercaili, capercaleg, caperkellie, etc. The name is thought to be derived either from the Gaelic *capull* which means a 'great horse', and *coille* meaning 'a wood', or according to the Revd Dr T. M'Lauchlan, from *cabher* (*coille*) which means 'an old man' (of the wood). The species is also sometimes referred to as the 'wood-grouse', 'cock-of-the-wood', or 'mountain cock'.

†*Scotorum Historiae*, 1527.

3 February 1651 Mr J. Dickson of Perth wrote to the Laird of Glenorquhy to tell him that 'I went and shew your Capercailzie to the King in his bedchamber, who accepted it weel as a raretie, for he had never seen any of them before'. In a pamphlet published in 1678 and entitled *A Description of Angus . . .* Robert Edward, the Minister of Murroes, Montrose, wrote: 'The mountains and heaths [of that county] abound with partridge and grouse [in original Latin *capricalis*, *i.e.* capercaillie], and plover'. Sir Robert Sibbald (1641– 1722), the Scottish physician and antiquary, in his *Scotia Illustrata* (1684) both describes and figures the cock and the hen of 'Capricalea'.

In *Letters from a Gentleman in the North of Scotland* (1754) Edward Burt (d. 1755) wrote:

Of the eatable parts of the feathered kind peculiar to the mountains is, First, the Cobber-kely, which is sometimes called a wild turkey, but it is not like it other than in size. This is very seldom to be met with, being an inhabitant of very high and unfrequented hills, and is therefore esteemed a great rarity for the table.

A considerable amount of doubt and controversy has existed as to when the indigenous capercaillie finally died out in Britain. *The Handbook of British Birds* (1941) states that it 'became extinct in Scotland and Ireland about 1760, and England* perhaps a century previously'. Vesey-Fitzgerald (1946) considered that the last British birds were killed, in Ireland, in about 1760.

When Lachlan Shaw (1692–1777) described the capercaillie as 'becoming rare' in his *History of the Province of Moray* (1775), it was by that time almost certainly extinct in that northern county. A number of authorities give Glenmoriston/Strathglass as the area where the last indigenous capercaillie was to be found. John Latham (1740–1837) wrote in 1783: 'the last bird of this kind found in Scotland was in the Chicholm's [Chisholm's] great forest of Strathglass'. *The Old Statistical Account of Scotland* (1798) states that 'the last caper-coille was seen in Glenmoriston and in . . . Strathglass,

*The probable remains of capercaillies have been discovered in deposits in Somerset, Yorkshire and in Co. Durham. N. Ticehurst has suggested that 'cock-of-the-wood' is the literal translation of *gall. de bosco* in the Bursar's Roll of the Monastery of Durham for 1–7 December 1348. ('Some Birds of the Fourteenth century,' *British Birds*, 17, pp. 29–35, 1923.)

about forty years ago'. Elsewhere it refers to the last bird being killed 'about fifty or sixty years ago in the woods of Strathglass', *i.e.* between about 1740 and 1760. In 1947 Darling wrote: 'the last capercaillie of the old stock disappeared in about 1771 from the Glenmoriston area', and Edlin (1952) stated that the species was exterminated in the same district in about 1770. Pennant (1772) and Harting (1880) wrote that the last native bird was seen in Chisholme Park, Inverness-shire. Sir William Jardine in the *National Library of Ornithology* (1834) suggested that the last indigenous capercaillie was killed between about 1774 and 1784, and Prichard was of the opinion that 'in the year 1780 at latest (and probably considerably earlier) no caper "display'd his breast of varying green" in woods Irish or Scottish'.

It was left to Dr Ian Pennie to discover a reference to what was probably the last native capercaillie in Scotland. Writing in 1950-1 he quotes from Angus (1886) the inscription on the back of a watercolour drawing at Balmoral Castle in Aberdeenshire: 'Two coileach-coille, capercailzie, shot on the occasion of a marriage rejoicing in 1785, and which Sir Robert [Gordon] is confident are the last of the native birds heard of in Scotland. Pennant saw the bird at Inverness in 1772.' Fitter (1959) concurs with this date. Pennie goes on to say that Angus found the memory of the ancient race lingering in Braemar as late as 1870. The woods of Ballochbuie contain the remnants of trees of the ancient Caledonian forest, and it is not unreasonable to suppose that one or two capercaillies may have survived until as late as 1785.* It is probable that the indigenous Scottish capercaillie formed a separate geographical race, but this remains uncertain as the only known surviving specimen of the native stock is a cock in the Hancock Museum, Newcastle-upon-Tyne.

The story of the decline and fall of the capercaillie in Ireland is much the same as that in Scotland. The earliest written reference to the bird is probably that made by Giraldus Cambrensis in his *Topographia Hibernica* (1184), where he writes that *Pavones sylvestres hic abundant.* John de Trevisa translates (1387) *pavonibus* as 'pekokes' in Ranulf Higden's *Polychronicon*, but it is generally thought that the species

---

*A number of even later records – *e.g.* one shot near Fort William by a Mr Henderson in 1807 and another killed at Borrowstowness by a Captain Stanton in 1811 – undoubtedly arise from unrecorded introductions.

referred to was the capercaillie. In their *Ornithologia* (1676–8) Willughby and Ray wrote: 'this bird is found on high mountains beyond the seas, and, as we are told, in Ireland (where they call it 'Cock-of-the-wood' or of the Mountain), but nowhere in England'. In his *Ogygia, seu rerum Hibernicarum chronologia* (1685) Roderic O'Flaherty (1629–1718) mentioned that in parts of Connaught the capercaillie was known as the 'cock-of-the-wood'. In an Irish Statute (II Anne, cap. 7) it is stated that 'the species of cock of the wood (a fowl peculiar to this Kingdom) is in danger of being lost'. In his *History of Cork* (1749) Smith wrote that the capercaillie 'is now found rarely in Ireland since our woods have been destroyed'. John Rutty in his *Natural History of Dublin* (1772) wrote: 'one of these [birds] was seen in Co. Leitrim about the year 1710. But they have entirely disappeared of late, by reason of the destruction of our woods'. According to Pennant, a small number of capercaillies remained around Thomastown in Co. Tipperary in 1760, and John Scouler (1804–71) stated that 'it remained in the County of Cork until as late as 1750'. James Fennell wrote in *The Field Naturalist* (1834) that the capercaillie had been extinct in Ireland for nearly seventy years, and in Scotland for fifty.

The demise of the indigenous capercaillie was brought about almost entirely by the deforestation of its woodland habitat in the seventeenth and early eighteenth centuries; at first this was done in order to clear the country of wolves,* and later to provide timber. The re-afforestation programmes of the eighteenth century – one of the earliest of which was that initiated by the Earl of Haddington in 1705 – came too late to prevent the extinction of the capercaillie – an end which was hastened by over-shooting.

A number of early, but unsuccessful, attempts were made to re-introduce the capercaillie to Scotland, Ireland and, surprisingly, England. In *British Animals* (1828) Fleming mentions that 'recent attempts have been made to recruit our forests from Norway, where the species is still common'. Edward Buxton in Blaine's *Encyclopaedia of Rural Sports* refers to the first of these attempts when he describes

---

*In England the wolf (*Canis lupus*) became extinct during the reign of Henry VII (1485–1509); in Scotland the last wolf was killed in Morayshire in 1743: the last wolves in Ireland died out in Co. Wicklow, and possibly Co. Down, in the 1760s.

how in about 1823 the Welsh-born naturalist Llewellyn Lloyd sent a pair to his cousin and fellow-Quaker, Mr Thomas Fowell Buxton* of Northrepps and Cromer Halls in Norfolk where, hardly surprisingly, considering the unsuitability of the countryside, they failed to establish themselves. Between 1827 and 1831 attempts were made by the Earl of Fife to introduce capercaillies from Sweden to his Mar Lodge estate at Braemar, Aberdeenshire: one hen died shortly after arriving in Montrose Bay, while another laid twenty-four eggs, all of which proved to be addled. In the late 1830s five hens and a cock were introduced to the grounds of Dunkeld House in Perthshire by the Duchess of Atholl: after the death of the cock the surviving hens were sent to the Marquess of Breadalbane at Taymouth Castle in the same county. In 1842 unsuccessful attempts were made to introduce capercaillies to Taplow Court in Buckinghamshire, and to Knowsley Hall in Lancashire by Lord Derby. Three years later Mr David Carnegie failed to establish two cocks and four hens from Sweden at Stronvar in Perthshire. Other unsuccessful introductions were made at Eslington, Northumberland by Lord Ravensworth, between 1872 and 1877; near Hebden Bridge in the West Riding of Yorkshire in 1877; at Glengariff, Co. Cork by Lord Bantry with three pairs sent from Sweden by Lloyd, and at Markree, Co. Sligo, by Colonel Edward H. Cooper in 1879. In a letter from Sweden in December 1846 (published in the *Sporting Review* in October 1847) Lloyd wrote: 'It is to be hoped that the capercaillie with which I some years ago supplied his lordship [Lord Bantry] have succeeded.'

In the *Natural History of Ireland* (1850), Thompson recalls that he was told by Mr G. Jackson, Lord Bantry's head-gamekeeper, that

Lord Bantry received three brace of capercaili from Mr Lloyd about seven years ago [*i.e.* about 1842]. They arrived safely, and were to all appearances doing well for the first six months, when one of them was observed to mope about and appear quite solitary. In a few days it died. I do not know of any others that have been introduced to Ireland.

Isolated individuals – presumably from unrecorded introductions – were seen in Buckinghamshire in 1855, in the Forest of Dean in 1880, and at Herstmonceux in Sussex in 1924–5.

In 1967 and 1968 unsuccessful attempts were made to introduce

---

*In July 1841 Buxton received a baronetcy for his services in the campaign to abolish slavery.

capercaillies to Forestry Commission property at Grizedale –
between Coniston Water and Windermere in the Furness District of
Lancashire – where experiments are in progress concerning the inter-
relationship between forestry, wildlife and human recreational
interests. On 18 May 1970 fifty-two capercaillie eggs were brought
from Scotland to Grizedale, where they were hatched under bantams.
On 24 August 1971 twenty cocks and fifteen hens were released in an
enclosure in Grizedale Forest, where they were later given complete
liberty. No breeding was observed in 1971, and although in the
following year lekking calls were heard, there were still no apparent
signs of nesting. As Grizedale covers over 8000 acres (3237 hectares),
however, unobserved breeding may possibly have taken place. The
Furness district has an annual rainfall of some 70 in. (178 cm), which
provides a far from ideal climate for both capercaillies and for one
of their favourite foods, wood-ants. Although colonies of these
insects have also been transplated to Grizedale, it seems unlikely that
this latest attempt to re-introduce capercaillies to England will
succeed.

The first successful attempt to re-introduce the capercaillie to
Britain was that made in 1837–8 at Taymouth Castle, Perthshire,
home of Lord Breadalbane. In his *Gamebirds and Wildfowl of
Sweden and Norway* (1867), Mr Llewellyn Lloyd describes how this
came about:

It is fortunate for the sporting world that the Capercali, after the lapse
of more than a century, is once more included in the British Fauna, and I
feel proud in having been a contributor in a small degree to so desirable an
event. . . .

For a long while no one would move in the matter, but at length in the
autumn of 1836 the late Sir Thomas Fowell Buxton, then recently returned
from Taymouth Castle, where he had been much struck with the great
capabilities of the woods for the naturalisation of the Capercali, took up
the affair in good earnest. . . . 'Influenced by the desire, in which I am sure
you will concur,' so he wrote to me, 'to introduce these noble birds into
Scotland, coupled with that of making Lord Breadalbane some return for
his recent kindness to me, I request you to procure for his Lordship, at
whatever cost, the requisite number.' He at the same time placed his head
keeper at my disposal – no slight sacrifice for a Norfolk game-preserver.
It was, indeed, an onerous commission, as prior to this time it had been a
matter of difficulty to procure even a brace of living Capercali in Sweden;

but by distributing placards throughout the country offering ample rewards, and by instructing the peasants how to knot their snares so as not
to kill the birds, my object was at length gained and within a few months
of the Baronet's letter, twenty-nine Capercali [13 cocks and 16 hens],
followed up shortly afterwards by twenty more, were on their way from
Sweden to Taymouth Castle, and with the exception of a single one killed
by accident, all reached their destination in safety [on 24 June].

The arrival of this magnificent collection in Scotland created quite a
sensation; everyone was delighted that matters thus far had gone well. . . .
In September 1837, not very long after the arrival of the twenty-nine, he
[Buxton] wrote me as follows: 'I have just returned from Taymouth,
where I have been reminded of you very frequently by the Capercali.
I saw eighteen of them in excellent health and plumage a few days ago;
the other ten, six hens and four cocks, were turned out, and there is
reason to hope that they are going well – so that, thanks to your energy
in collecting them, Larry's [Laurence Banvill, Buxton's Irish gamekeeper]
care in bringing them over, and Lord Breadalbane's anxiety for their
welfare, our experiment is likely, I trust, to succeed; and Scotland to
be re-stocked with this noble bird. They are greatly admired by everyone,
and very deep interest is felt about them. . . . Nothing can surpass the woods
into which they are to be turned out, and the protection they will receive,
and as Lord Breadalbane's territory is so large, I hope that they will not
be disposed to leave such excellent quarters.'

In the summer of 1838 a further sixteen hens arrived at Taymouth
from Sweden. In September of the following year, James Guthrie,*
head-keeper to Lord Breadalbane, estimated that there were between
sixty and seventy capercaillies at large on the estate.

'I have great pleasure in informing you', Lord Breadalbane wrote to
Lloyd in a letter dated 1842, 'that the Capercali have thriven most
excellently. The experiment of putting the eggs under the Grey Hen
was attended with perfect success, and there are now a goodly
number of these birds hereabouts.' In *The Memoirs of Sir Thomas
Fowell Buxton, Bt* (1848) his son, Charles, wrote:

When the Queen visited Lord Breadalbane in 1842, he kindly permitted
my brother and myself to shoot the first of these birds that had been
killed in Scotland for a hundred years, in preparation for Her Majesty's
dinner. They were so extremely wild, that it took the whole day to get six

---

*Guthrie appears to have been a peculiarly schizoid character; although lavishing
much care and even affection on the game-birds under his care, he was capable of acts
of considerable and premeditated cruelty against any 'vermin' which threatened the
welfare of his charges.

shots. We could just see them vanishing from the tops of the tall larches, while we were still a great distance from them. . . .*

Llewellyn Lloyd concludes:

It is very satisfactory to add that the Capercali have subsequently flourished in the Highlands in an extraordinary manner. Less than four years ago, indeed [1862], Lord Breadalbane himself told me he imagined there were fully one thousand of these birds on the Taymouth property. His head keeper, moreover, in a letter to a friend, estimated them at double that number. . . . Sir Alexander Campbell, a near relative of Lord Breadalbane told me that the Capercaili were then as common about Taymouth Castle as the Black Cock . . . they had spread from Taymouth over all the more wooded parts of the Highlands as far as Aberdeen.

By the time that Harvie-Brown published his scholarly history and survey of the capercaillie in Scotland, the birds had spread far beyond Taymouth throughout the length of the Tay valley as far as Dunkeld, (reaching Tummel Bridge in 1844; Blair Atholl in 1845; the junction with the river Isla after 1847; and Blair Drummond in 1860) as well as to many other suitable areas in the county: they were, and indeed probably still are, most numerous in the Black Wood of Rannoch between Tummel Bridge and Bridge of Gaur: up to sixty birds have been seen here in the centre of a single capercaillie-drive in winter, and the same number have been counted on a 'lek' ground in the early spring. When Dr Pennie conducted his comprehensive survey in 1950–1, he found capercaillies widely distributed throughout the wooded parts of the county much as they had been in 1879, although there had been some changes in individual habitats mainly due to deforestation. Between 1879 and 1950 Dr Pennie found that some districts had been abandoned, but none where any suitable habitat remained. The birds were tending to concentrate in mature woods and were, under pressure from deforestation, colonizing less mature woodland than formerly. Dr Pennie estimated that the total population was about the same as in Harvie-Brown's time, and it is likely that the position is similar today.

In the neighbouring county of Angus, where capercaillies are at

*Another report states that the first of the re-introduced capercaillies was shot in 1839 and was exhibited for sale in a game-dealer's shop in Prince's Street, Edinburgh.

present widely but thinly distributed in suitable woodland, the first bird – doubtless from Perthshire – appeared at Lindertis near Kirriemuir in 1856. Introductions of capercaillies were made by the Earl of Airlie at Cortachy Castle in 1862 and at Stracathro three years later; the birds had reached Glamis by 1863 and Brechin by 1870, and were soon firmly established throughout the county.

In 1870 capercaillie eggs from Crieff in Perthshire were introduced to Drumtochty in Kincardineshire: the hatched birds spread rapidly, and reached Inglismaldie near Stracathro by 1878. In this year the first imported capercaillie in the county was shot at Fetteresso; another was killed at Fasque, three more at Scolty Head near Banchory, and one at Inchmarlo on the opposite side of the Dee. The status of the capercaillie in Kincardineshire today is much the same as in Angus.

An unsuccessful attempt to introduce capercaillies to Aberdeenshire was made at Inverernan, Strathdon, between 1870 and 1873. The birds eventually spread into the county along the valleys of the Dee and Don, the first one being shot at Kincorth in 1879. The natural spread of capercaillies was supplemented by eggs hatched at Balmoral in 1885, and in 1896–7 the first wild birds' nests were found at Hazlehead.

In Banffshire a number of capercaillies were kept in an aviary at Duff House in 1851, where, however, they soon died. In the spring of 1886 a clutch of eggs was placed in a grey hen's nest at Altmore near Keith, and on 13 October following a hen was shot in Glenbarry. The first arrivals in the county probably came from Arndilly near Craigellachie over the border in Morayshire.

The Earl of Fife made an early but unsuccessful attempt to introduce capercaillies to Morayshire, when in about 1852 he imported some birds from Norway to Lochnabo near Elgin. Further unsuccessful attempts followed: at Castle Grant where eggs were put under grey hens in 1860; at Lochnabo again in 1878, and by the Earl of Moray in 1883. In the following year a number of birds and a nest containing ten eggs were seen at Darnaway: in 1890 fifteen capercaillies were seen in one day, and two years later nine nests were found. In 1884 capercaillies were introduced to Clunas Woods, near Ferness, Nairn. Four years later a clutch of eggs was placed in the nest of a grey hen at Gordon Castle, where in 1897 the Duke of Richmond introduced a number of adult birds which had multiplied to about thirty by 1907. In 1898–9 two cocks and three hens travelled south, presumably along the Spey Valley, from Gordon Castle to Arndilly. The first bird was killed in Drum Wood at Carron in 1903, and in 1907 no fewer than twelve clutches of eggs hatched successfully at Craigellachie. In 1950 Dr Pennie found capercaillies well distributed in most of the larger coniferous woods of Moray and Nairn, where they were especially plentiful in the coastal regions of the upper valley of the Spey and in Darnaway Forest. Recent reports suggest that the position is little changed today, apart from some local population contractions due to losses of suitable habitat.

On a number of occasions prior to 1843 capercaillie eggs were despatched from Perthshire to Lord Lovat's estate at Beaufort Castle, Beauly, Inverness-shire. In 1860 the Duke of Atholl sent some birds to the Earl of Seafield's estates at Abernethy and Kinveachy, where, however, they all soon died. An introduction by Lord Tweedmouth to Guisachan (Cougie) in 1868 was no more successful; a total of thirty-nine birds were killed in their pens by polecats, and although a number of eggs subsequently hatched and the young reared, they did not flourish. An introduction to Invereshie near Kincraig in 1873 was similarly unproductive. The earliest record of a wild capercaillie in Inverness-shire was that of a cock shot at Dalwhinnie on 27 November 1881. In 1892 Mr W. Dalziel Mackenzie of Farr introduced some capercaillies to Strathnairn, south-east of Inverness, where they failed to establish themselves. Two years later a further twelve, imported from Norway, also died. In November and December 1895, thirty-one Norwegian capercaillies were introduced, and these were reinforced with further supplies over the next five years. The area chosen was a 35 acre (14 hectare) mature plantation of Scots pines and larches, with birch and bracken undergrowth and a plentiful supply of berry-bearing shrubs, interspersed with patches of open ground. In the spring of 1895 some capercaillie eggs were sent from Kinfauns in Perthshire to Mr W. MacNicol of Dorbrack Lodge, who placed them in the nests of grey hens on the Abernethy estate of a neighbour, Mr D. Collie. At about the same time a supply of eggs was received at Guisachan from Taymouth Castle. In 1950 Dr Pennie found capercaillies widely spread throughout Inverness-shire, being especially numerous around Moniak; Loch Dochfour by Loch Ness; at Farr in Strathnairn, and in Glen Kyllachy, Strathdearn; in the valley of the Spey between Abernethy and Rothiemurchus; in the Queen's Forest of Glenmore, and around Loch Laggan. Information received from local landowners, who report that between sixty and seventy capercaillies are sometimes shot during a day's driving, suggests that the position has not materially altered in the past twenty-five years, except where timber-felling has caused some reduction in range and concentration of numbers.

Two capercaillies shot at Foulis, 6 miles (10 km) north of Dingwall, in 1896, and at Fairburn, west of Muir-of-Ord, in 1898, suggest so far unrecorded introductions to Ross-shire. In a letter to Dr Pennie in 1949, Mr Hugh Frazer wrote: 'capercaillie were introduced to the Brahan Estate some sixty years ago . . .' An unsuccessful introduction made to Rosehaugh on the Black Isle in Easter Ross in 1888 was followed by a successful transplant of birds from Foulis Castle in 1910–11. In about 1930 capercaillie eggs were successfully hatched at Coulin, Kinlochewe, from where four or five birds flew to Achnashellach where they appear to have vanished. Capercaillies are still to be found in suitably wooded districts of Ross-shire – a considerable number being reported at Tain on the Dornoch Firth in 1971. A single hen was seen on the Forestry Commission's property at Rumster in Caithness, some 50 miles (80 km) or so to the north-east, in September 1972.

An unsuccessful attempt was made to introduce capercaillies to Skibo Castle near Bonar Bridge in Sutherland in 1870. In 1950 Dr Pennie reported them as 'breeding but scarce' in Rosehall Wood near Lairg; 'increasing' at Inveroykell, and present in 'fair numbers' at Skibo. The current status of capercaillies in Sutherland is uncertain.

In Fife capercaillies were introduced with success by Admiral Lord Keith to Tulliallan in 1864 (where they were still present though no longer breeding in 1950), and at Lathirsk near Falkland ten years later. A solitary hen was reported in February 1971 at Tentsmuir (near the coast between Tayport and St Andrews), where a cock was observed in January and November in the following year. The first capercaillie was shot in Stirlingshire at Stenhousemuir in 1856. In 1863 a nest was discovered at Dunmore opposite Tulliallan, but the eggs were apparently addled. The species first bred successfully at Torwood in 1868. In May 1971 a pair was seen at Cambusbarron, south of Stirling.

A hen capercaillie was shot in 1867 at Ross Priory in Dunbartonshire, where by 1950 Dr Pennie reported that they were 'breeding in several places around Loch Lomond', but were 'not numerous and decreasing'. A pair nested successfully at Torrinch near Loch

Lomond in 1972 – the first breeding record for the area for ten years.

Although an unsuccessful attempt was made to hatch capercaillie eggs in Blackmount Forest by Bridge of Orchy prior to 1867, adult birds were seen at Ardgour on the far side of Loch Linnhe, on Loch Leven-side, and at Camusnagaul by Fort William, between 1868 and 1870. Five years later they were established at Ardkinglas at the head of Loch Fyne.

As early as 1843 six hens and one cock capercaillie were sent from Taymouth Castle to Brodick Castle on the Isle of Arran: these were followed three years later by a further eight hens and two cocks from Sweden. By 1855 there were about forty capercaillies at Brodick, which had increased to seventy ten years later, but were extinct by 1910. Attempts to colonize the island of Bute in 1922 and Islay in the later 1920s all failed.

Between 1860 and 1870 six pairs of capercaillies were introduced to Douglas in Lanarkshire, where they soon died. In 1868 a solitary cock was seen and shot at Auchengray, Airdrie. From 1902–4 Mr H. B. Marshall placed some capercaillie eggs under domestic fowl and wild grey hens at Rachan, Broughton, Peeblesshire, without success. After an abortive attempt in 1929, Mr F. R. S. Balfour succeeded in establishing capercaillies at Dawyck Park, Stobo, Peeblesshire, in the following year when a pair were despatched from Deeside in August, supplemented in November by a cock and three hens from Finland. These were followed in October 1931 by a further hen and two cocks. A small breeding colony resulted, which increased up to the outbreak of war in 1939, when timber-felling caused it to decline.

A hen capercaillie was seen at Dalmeny in West Lothian (where it or a different hen was killed six years later) in 1871, and a cock at Kettlestone opposite Tulliallan in the following year. Between 1876 and 1906 six capercaillies were seen in Midlothian, and a wandering cock was observed near Selkirk in 1936. The first bird in Renfrewshire appeared near Paisley in 1872, followed by others in 1891 and at Ardgowan in 1896. In Ayrshire capercaillie eggs from Taymouth were successfully hatched at Glenapp in 1841–2, where the birds survived until 1848. Eggs from Brodick Castle were hatched by grey hens at Sanquhar in Dumfriesshire in 1865, but the birds only survived for four years. In 1869 a cock capercaillie was shot at Auchencairn, Kirkcudbrightshire, and another was killed at Newton Stewart, Wigtownshire. Isolated records of capercaillies include three seen at Tynron in Dumfriesshire in November 1905, and one shot at Tarbolton Moss in Ayrshire in the following month.

The maximum spread of re-introduced capercaillies in Scotland was reached by the outbreak of the First World War, when they were breeding from Golspie in Sutherland in the north to Stirling in the south-west, and from Cowal, Argyllshire in the west to Buchan, Aberdeenshire in the east. The general direction of advance from the original introduction at Taymouth was north-eastwards and south-westwards along the river valleys. For some unexplained reason it is normal for the hens to precede the cocks by several years in colonizing new areas: this has led to a number of blackcock/capercaillie hybrids,* which are often the first indication that capercaillies have moved into a new district. Although less common than formerly, these hybrids still occur, and were also known among the original indigenous stock. Pheasant/capercaillie hybrids are also sometimes found, but are rare.

During the last sixty years or so there has been little advance of capercaillies into fresh territory; two world wars resulted in large-scale timber-felling programmes, which in some places were responsible for the near-extinction of local populations. Capercaillies appear to need a fairly dry climate in which to breed; this belief is borne out by the failure of the bird to establish itself in western Scotland and in Denmark (where a number of introductions, between about 1870 and 1924 and later, all failed), and its comparative success in eastern and central Scotland. One reason for the capercaillie's lack of ability to colonize new areas of Scotland is the tendency, forced on it by the felling of mature woodlands with dry heather and bilberry undergrowth, to colonize younger plantations, where the damp brushwood and lack of food is responsible for the deaths of many fledgelings. Capercaillie eggs and chicks fall prey to jays, magpies and hooded crows, and the adults have in the past been persecuted by the Forestry Commission for eating pine shoots, by farmers for gleaning corn, and by landlords for sport. Today, fortunately, more enlightened views prevail, and capercaillies are firmly established once more in suitable habitats, principally in the valleys of the Tay, Dee and Spey and their tributaries, and in Morayshire, Banffshire and parts of Argyllshire.

*Originally dignified with a separate scientific name, these birds are known in Scandinavia as *Rackel-fogel*.
Note: See also Addenda.

33. Pheasant
(Phasianus colchicus)

The pheasant is a southern Palaearctic and north-eastern Oriental species; here it ranges from south-eastern Russia and northern Asia Minor as far north as the Aral Sea, Mongolia and Manchuria, east to Korea, Japan and Taiwan, and south to Burma and south-western China. Highly prized for sporting purposes and for its decorative qualities, the pheasant has been introduced successfully to many parts of Europe and North America, as well as to Hawaii and New Zealand.

The plumage of the cock pheasant is, perhaps, the most striking and spectacular of all British birds. The outstanding feature is the reddish-buff lanceolate tail, barred with black, which may measure up to 20 in. (51 cm) in length. The head and neck are a dark metallic green glossed with purple, usually with some trace of a white collar. The eyes are surrounded by scarlet wattles, and the back of the head is crowned with ear-like erectile tufts of feathers. The body is the colour of burnished copper, boldly marked on the breast and sides with blackish-purple crescents, and on the back with black-edged buff horseshoes. The hen is a uniform sandy brown with blackish markings. A cock pheasant measures between 30 and 35 in. (76 to 89 cm) long including the tail, and a hen from 20 to 25 in. (51 to 63 cm), with a tail some 8–10 in. (20 to 25 cm) in length.

The plumage of cock pheasants is subject to a greater degree of individual variation than that of any other British wild bird, with the possible exception of the male ruff (*Philomachus pugnax*) during the breeding season. The common pheasant (*Ph. colchicus*), which has brownish wing coverts and no white collar, was the form first intro-duced to England: it came originally from the banks of the river Phasis (Rioni) in Colchis (Kutais) in Georgia and Armenia on the south-eastern borders of the Black Sea. In England it was followed, after a period of several hundred years, by the Chinese ring-necked pheasant (*Ph. c. torquatus*), which is a native of eastern China. This form has a greenish or blue-grey rump with rusty-orange patches on the flanks, greyish-blue wing-coverts, and an incomplete white collar.

A number of other forms have been introduced to Britain from time to time: these include the Prince of Wales's pheasant (*Ph. c. principalis*) from north-western Afghanistan and southern Turkestan, which has a rufous rump, narrow bars on the tail, whitish wing-coverts and no white collar: the Mongolian pheasant (*Ph. c. mongolicus*) from Kirgizskaya and Chinese Turkestan (a rufous rump with a deep greenish-purple gloss, whitish wing-coverts, and a broad white collar): *Ph. c. satscheuensis* from Western Kansu (a greenish grey-blue rump with rusty-orange side patches, a broadly barred tail, greyish-blue wing coverts, with an incomplete white collar): and Pallas's pheasant (*Ph. c. pallasi*) from Ussuriland and Manchuria, which has a complete and broad white collar. The green-plumaged Japanese pheasant (*Ph. versicolor*) has also been introduced on a number of occasions: from it is almost certainly derived the melanistic mutant to which the name '*tenebrosus*' has been applied, in which the cock has a peacock-like plumage closely resembling *versicolor*, while the colouring of the hen is somewhat similar to that of the red grouse, with green-tinged shoulders. A buff or cream-coloured mutant of *Ph. colchicus* has been called, for no apparent reason, the Bohemian pheasant. All these forms have freely interbred (many of the present-day stock showing some trace of the white neck ring), and it is now virtually certain that no pure *colchicus* birds remain in Britain.

The pheasant is extremely catholic in its choice of habitat, favouring particularly partly cultivated, partly wooded country, with areas of thick undergrowth and dense plantations well supplied with water. It flourishes both on the light and sandy soil of heath- and common-land, and also in damp and marshy districts overgrown with reed- and sedge-beds. Pheasants are, in general, both fast runners and strong fliers, though their flights are seldom of long duration. Breeding begins in early April, when an average of between eight and fifteen brownish-olive eggs are laid in a hollow scraped in the ground by the hen, lined with grass and dead leaves. Incubation, normally

[FEMALE]

only by the hen, lasts for between about twenty-three and twenty-five days. The pheasant's diet is very varied, comprising mainly leaves, fruits, seeds, nuts, roots, insects, earthworms and slugs. The crowing of the cock pheasant is a sound well known to country-people, and is used both as an alarm-note and also when displaying to the hen.

The common pheasant was, according to legend, first introduced to Europe by Jason and the Argonauts, when in about 1300 B.C. they brought it back with them from Colchis to Greece on board the *Argo*, on their return from searching for the Golden Fleece. From Greece pheasants were introduced to Italy, though the exact date of their arrival in that country is unknown. Marcus Terrentius Varro (116–27 B.C.) makes no mention of the pheasant in the volume of his *Rerum Rusticarum* treating with *villatica pastio* (feeding domestic stock within the villa precincts), compiled in 54 B.C. In his *Naturalis Historia,* published in 77 A.D., Pliny the Elder (23–79 A.D.) describes Colchis as the original home of the pheasant. The Roman epicure and author M. Gavius Apicius,* writing in the first century A.D., gives recipes for making pheasant (and rabbit) rissoles (*russeola*). Rutilius Taurus Aemilianus Palladius, in his *De Re Rustica* compiled in about 350 A.D., largely from information culled from Lucius Junius Moderatus Columella's similarly named work of the first century A.D., gives a precise procedure for husbanding pheasants:

In the rearing of pheasants this point should be kept in mind, that only young birds should be used for breeding, that is, birds of the preceding year; for old birds cease to be prolific. Pheasants mate in March or April. One cock will do for two hens, for the pheasant is not such a tyrant in love as are some other birds. The hen has one brood in the year, and the clutch is about 20 eggs. It is best to incubate these under ordinary barndoor fowls, and each foster-mother should cover fifteen eggs. The young birds hatch on the 30th day. For the first 15 days they should be fed on barley-meal steamed and cooled off, and on the meal a little wine should be sprinkled. Later you will give them kibbled [coarsely ground] wheat, locusts and ants' eggs [pupae].

For fattening, Palladius continues:

The bird should be shut up and fed with wheat-meal made up into the

*See page 95.

smallest possible lumps. One peck will suffice for thirty days, or if you prefer barley-meal, one and a half pecks will fatten the bird in the same time. But you must be sure to spray the meal lightly with oil, and to place it into the mouth so that it does not get under the root of the tongue, because that will kill the bird. And you must be absolutely sure that the birds have digested what you have given them before you feed them with new food, for food lying heavy on a bird's stomach will be the end of it.

For many years it was generally believed by zoologists that the pheasant was first introduced to Britain by the Romans. In 1933 however, Dr P. R. Lowe announced that bones discovered in a Romano-British midden at Silchester near Reading in Berkshire, which had previously been thought to be those of pheasants, were in fact those of ordinary domestic fowl. 'No authentic fossil Pheasant bone', writes Lowe, 'unquestionably contemporaneous with the Roman occupation has been described. I have so far been unable to discover any material in museums which would point to the existence of the Pheasant in the British Isles contemporaneous with the Romans.'

The earliest documentary evidence of pheasants in Britain occurs in a manuscript* (in the British Museum) of about 1177 entitled *De inventione Sanctae Crucis nostrae in Monte Acuto, et de ductione ejusdem apud Waltham.* This contains details of rations specified by the Earl of the East-Angles and West-Saxons (later King Harold II) for six or seven members of the canons' household at the monastery of Waltham Abbey in Essex, between Michaelmas (29 September) 1058 and Ash Wednesday (the first day of Lent) 1059:

Erant autem tales pitantiae [rations or commons] *unicuique canonico: a festo Sancti Michaelis usque ad caput jejunii* [Ash Wednesday] *aut xii merulae* [blackbirds], *aut ii agauseae* [? magpies] *aut ii perdices* [partridges], *aut unus phasianus* [pheasant], *reliquis temporibus aut ancae* [geese] *aut gallinae* [domestic fowl].

This document shows that one pheasant was assumed to be the equivalent of a dozen blackbirds or two magpies or a brace of partridges. It also proves that pheasants were certainly known in Britain, if only in captivity, before the time of the Norman Conquest,

*Edited by Professor W. Stubbs and published in 1861 under the title of *The Foundation of Waltham Abbey.*

(having perhaps been introduced in the early years of the reign (1042–66) of the Saxon king, Edward the Confessor), and evidence exists that they started to become naturalized shortly afterwards. Sir William Dugdale (1605–86) in his *Monasticon Anglicanum*, published between 1655 and 1673, quotes from a charter dated 1098, during the reign of William II, in which the Bishop of Rochester in Kent assigned to the monks of that city sixteen pheasants, thirty geese and three hundred fowl from four separate manors. Two years later, under Henry I, a licence was granted to the Abbot of Malmesbury to kill hares and pheasants. Thomas à Becket is said to have eaten pheasant on the day of his murder, 29 December 1170. In 1249 the Sheriff of Kent was commanded to produce twenty-four pheasants for a feast for Henry III. Fifty years later, in the time of Edward I, the Bursar's Roll for the Monastery of Durham lists '*In xxvj perdicibus et uno fesaund empt. de G. P'trikur, vs*' – the pheasant costing four pence.

From the twelfth century pheasants began to appear with increasing frequency in English literature. *Sir Orfeo* (a metrical romance of about 1320 in which the story of Orpheus and Eurydice is re-told) shows that by that date the pheasant was becoming fairly plentiful in the countryside:

> Of Game they fonde grete haunt,
> Fesaunt, heron and cormerant.

T. Percy's *Reliquies of Ancient English Poetry* (a collection of ballads, historical songs and rhyming romances of around 1325, published from the *Percy Folio* in 1765) first distinguishes between the sexes, when it speaks of: 'partrich, fesaunt hen, and fesant cocke'.

The Vision of Piers Plowman of about 1377 makes an early, if negative, reference to the pheasant as an item of food: 'He fedde hem with no venysoun ne fesauntes ybake.' When George Neville (*c.* 1432–76) was enthroned as Archbishop of York in 1464, no less than two hundred 'fessauntes' were eaten at the celebratory feast. Nicholas Upton (?1400–57) in his *De Animalibus et de Avibus*... (written before 1446) describes at length how pheasants were reared and fattened for the table.* Alexander Barclay (?1475–1552) describes, in his *Eclogues*, published in 1515, 'The crane, the fesant, the pecocke, and

---

*He also states in the same work that the pheasant was first brought to Europe from the East by 'Palladius ancorista'.

curlewe' as suitable birds for a feast or banquet.* Eighteen years later
Sir Thomas Elyot (?1490–1546) in *The Castel of Helthe* was of the
opinion that 'Fesaunt excedeth all fowles in swetenesse and holsom-
nesse'. In the domestic accounts for 1512 of Henry, 6th Earl of
Northumberland, pheasants are shown to cost 'xiid' each. Provision
was made in the Privy Purse accounts of Henry VIII in 1532, for
the payment of a salary to a French priest who acted as 'fesaunt
breder' at the royal palace at Eltham in Kent.† In the household
accounts of the Kytsons of Hengrave in Suffolk for 1607, an allowance is made for providing wheat to feed partridges, quail and
pheasants.

Exactly when the pheasant became fully naturalized in the English
countryside as a feral breeding bird remains uncertain. It was
clearly well enough established by the late fifteenth century to
warrant legal protection from the crown. A couplet from a ballad of
this period on the *Battle of Otterburn* –

> The Fawkon and the Fesaunt both
> Amonge the holtes on hee.

– quoted by Fitter, suggests that, by this time, the bird had suc-
ceded in penetrating at least as far north as Northumberland.

Fynes Moryson, who was in Ireland between 1599 and 1603,
wrote that in that country there was 'such plenty of pheasants as I
have known sixty served up at one feast, and abound much more
with rails, but partridges are somewhat scarce'. Fitter refers to a
sixteenth century poem entitled *Asking for a Pheasant* by Gruffydd
Hiraethog (discovered by Iorwerth Peate of the Welsh Folk Museum),
which states that 'there never has been one [pheasant] in Merionethshire in wood or brushwood, but now they fill everywhere in the
region of Llanddwye'. In the late 1580s pheasants from Ireland were
introduced to Haroldston West in St Bride's Bay, Pembrokeshire.

The earliest reference to pheasants in Scotland appears to have
been made by Bishop Lesley who in 1578 recorded that '. . . the foul

---

*The approximately contemporaneous *Boke of Kervinge*, printed by Wynkyn de
Worde, gives 'alay' as the correct term for carving a pheasant.
†The *London Gazette* for 1725 (No. 6360/2) refers to 'His Majesty's Pheasantry in
Bushy Park'.

2B

called the storke, the fasiane, the turtle dove, the feldifare, the nichtin-
gale, with other nations are common, but are scarce with us': he
describes the partridge as being in size 'sumthing les than the
fasiane'. An old Scots Act dated 8 June 1594 forbids the killing 'at
any time hereafter . . . [of] deir, harts, phesanis, foulls, partricks, or
any other wyld foule whatever' in or near the royal policies.* It was,
however, well over another two hundred years before the pheasant
penetrated as far north as the Scottish Highlands. Mowbray, in
his *Domestic Poultry* (1830), wrote: 'In 1826 a solitary cock-pheasant
made his appearance as far north as a valley of the Grampians,
being the first that had been seen in that northern region.' Fitter
gives some dates of early introductions of pheasants to the Highlands,
e.g. to Skibo in Sutherland in 1841; to Monymusk, Aberdeenshire,
shortly before 1842, and to Shieldaig in Ross-shire in about 1860.

The earliest reference to the Chinese ring-necked pheasant (named
torquatus by J. F. Gmelin in 1789) in Britain is that made by Thomas
Pennant writing in 1768:†

Mr. Brooks the bird-merchant in Holborn, shewed us a variety of the
common pheasant which he thought came from China, the male of which
had a white ring round its neck, the other colors resembled those of the
common species, but were more brilliant.

*See Richard Gray's *Birds of the West of Scotland*, and Hugh Gladstone's *Birds
mentioned in the Acts of the Parliament of Scotland, 1124–1707*, 1926.
†George Edwards in his *Natural History of Birds* (1743) wrote: 'I have three Sorts
of Chinese Cock Pheasants, and the Hens of two of them.' Although he does not specify
which breeds he is referring to, they were probably such 'ornamental' ones as the golden
pheasant (*Chrysolophus pictus*) and the silver pheasant (*Gennaeus nycthemerus*), both
of which first arrived in Britain at about this time. (See pages 342 and 354.)

(The white ring encircling the neck of a pheasant on a sixteenth century tapestry in the Victoria and Albert Museum was almost certainly applied during late eighteenth or early nineteenth century repair-work, by a craftsman who had previously seen a specimen of *torquatus* in England.)

In 1783 Latham wrote:

The ring-necked pheasant . . . a fine variety of this bird is now not uncommon in our aviaries.

Four years later he continued:

I have scarce a doubt but that these birds will hereafter become fully as plentiful in this Kingdom as the Common Pheasant. It is well known that several noblemen and gentlemen have turned out many pairs into their neighbouring woods, for the purpose of breeding.

In 1828 the same author stated:

This beautiful variety . . . is now at large in England. These were, it is said, first introduced by the late Duke of Northumberland [died in 1786] by the name of Barbary Pheasants [the name given also by Montagu in his *Dictionary*, 1802], and many were bred and turned out at large, at His Grace's seat at Alnwick. Lord Carnarvon did the same at Highclere in Berkshire, and the late Duchess Dowager of Portland [died in 1785] at Bulstrode, Buckinghamshire, besides many private gentlemen, by which means the breed is daily becoming more common; it is true that these mix and breed with the Common Sort, and that in such produce the ring on the neck is less bright, and sometimes incomplete, but which of the two will ultimately preponderate, in respect to plumage, can hardly be conjectured.

The earliest colour reproduction of the ring-necked pheasant appears to be that contained in *Portraits of Rare and Curious Birds . . . from the Menagery of Osterley Park* (1794) by W. Haynes.

The Japanese pheasant (*Ph. versicolor*) is first referred to in Britain by John Gould in his *Birds of Asia* (1857):

About the year 1840, living examples were brought from Japan to Amsterdam, and of those a male and a female were purchased by the late

[13th] Earl of Derby [died in 1851] at a very high price: unfortunately the female died before reaching the menagerie at Knowsley, leaving the Noble Earl in possession of the male only. No other example having been brought to England, it is from this single male and a female of the common species that all the Green Pheasants, now becoming so numerous in the British Islands, have sprung. The produce of the first cross was, of course, a half-breed; the old male being placed again with these half-breeds, the result was a three-quarter race; and these breeding again with the old bird, the produce became as nearly pure as possible. On the dispersion of the late Earl of Derby's living collection, the old cock and the purest portion of his progeny were purchased by Prince Demidoff, and . . . sent to Italy. John Henry Gurney, Esq. of Norwich, and other gentlemen, became the possessors of the less pure stock. Some of Mr. Gurney's birds were turned out in the woods at Easton [six miles (10 km) west of Norwich] . . . thus giving rise to the Norfolk varieties.

In 1864 some common pheasant/Japanese hybrids were bred in Sussex by the Acclimatisation Society:

The results obtained by A. C. McLean, Esq. of Haremere Hill, Hurst Green, Sussex . . . by crossing the Common Pheasant with the Japanese

Pheasant are highly satisfactory. The bird produced by this cross is very beautiful in plumage, and Mr. McLean has succeeded in breeding several hundreds of them. . . .

In the same year 'The Society acquired by purchase three fine specimens, one male and two females, of a cross between Versicolor and Torquatus Pheasants; these birds are doing well, and justify the expectation that they will increase and multiply.'*

During the remainder of the nineteenth century and into the early twentieth further supplies of Japanese pheasants were imported into Britain: in England they were turned down in Bedfordshire, Cheshire, Cornwall, Cumberland, Durham, Gloucestershire, Herefordshire, Kent, Norfolk, Northamptonshire and Northumberland, and in Scotland on the Island of Bute and in Dumfriesshire. In 1932, however, Seth-Smith wrote; 'for the last twenty years or more not more than some half-a-dozen pairs would appear to have reached Europe'.

Although classifiers and systematists for many years debated the subject, it is now generally accepted that the Japanese pheasant was the originator of the melanistic mutant pheasant, previously described. Specimens bearing a close resemblance to this dark-coloured variety first appeared in England in 1867 (within a period of some thirty years from the date of the first introduction of the Japanese pheasant), having apparently failed to develop in either the common pheasant population during the past eight hundred years or so, or in the common pheasant/ring-necked crosses of the previous hundred years. These melanics first began to appear in any numbers in the British countryside in the 1920s. In the winter of 1925–6 the Hon. M. U. (Marquis) Hachisuka bought one in the market in Cambridge, and subsequently named it *tenebrosus*. J. L. Peters, however, in his *Check-List of Birds of the World* (Vol. 2, 1934) wrote: 'Hachisuka has named *Phasianus colchicus* mut. *tenebrosus*, an aberration now not uncommon among half-wild pheasants of mixed blood in England. The practice of designating "mutations" or other aberrations by trinominal names is indefensible.' In *Birds of the Palearctic Fauna: Non-Passeriformes* (1965), C. Vaurie does not even refer to '*tenebrosus*', since it is a name not accepted in taxonomy. (The reader who

*Acclimatisation Society 4th Annual Report, pp. 19, 21, 1864.

wishes to do so will find the controversy concerning the origin of *'tenebrosus'* fully discussed by Rothschild,* Lowe, Seth-Smith, Parker, Eltringham, Delacour and Fitter.)

As mentioned above, other varieties of *Ph. colchicus* have, from time to time, been introduced to Britain. According to Seth-Smith, 'A small number of Mongolian Pheasants (*Ph. c. mongolicus*) were introduced about twenty years ago, and also a few Prince of Wales's Pheasants (*Ph. c. principalis*) and small numbers of certain other races, or their eggs, may have reached this country.' *Ph. c. mongolicus* was first introduced, on the island of Bute, in about 1898, where it flourished for a number of years, but had died out by 1927. It was subsequently introduced in Dumfriesshire in 1910; in small numbers in Kent in 1912, where it was still to be seen around Sevenoaks thirty years later; in Norfolk prior to 1930; and at Woburn Abbey in Bedfordshire by the Duke of Bedford in the early years of the present century; it was still to be seen unpinioned in the woods near Whipsnade in 1947. *Ph. c. principalis* was liberated on Bute and in Kent and Norfolk at about the same time as *mongolicus*. Pallas's Pheasant (*Ph. c. pallasi*) was introduced to Norfolk before 1930, and *satschuensis* was turned out on two separate estates in Kent, in 1942. All these races have freely bred in the wild with nominate *colchicus* and *torquatus* – which had themselves previously interbred with each other and with *versicolor* and so-called *tenebrosus* – so that, as previously stated, the wild pheasants of the British countryside are a mongrel amalgam of these, and possibly other, varieties.

---

*Correspondence in *The Times* in May 1975 revealed that Lord Rothschild was responsible for an attempt to introduce pheasants to Hyde Park in the early 1920s, where 'they bred successfully for several years'. Since he is known to have been particularly interested in so-called *tenebrosus*, these birds may well have been of that variety.

Note: See also Addenda.

34. Golden Pheasant

(Chrysolophus pictus)

Two species of 'ruffed' pheasants are naturalized in small to moderate numbers in Britain; these are the golden pheasant and Lady Amherst's pheasant, both of which were admitted to the British and Irish List only as recently as 1971.

The home of the golden pheasant is the mountainous districts of central China. The cock, who is a most gorgeously plumaged bird with a very long brown tail, measures about 2 ft (61 cm) in length. He has a bright scarlet breast and underparts, and green and blue wings. The rump is gold and the crown of the head bears a crest of golden feathers. The back of the head and neck are covered with a gold and black hood or ruff, which can be expanded and extended forward in aggression or display. The hen is a lighter shade of brown than the hen of the common pheasant, and has a longer and more broadly barred tail.

The first golden pheasant in captivity in Britain was in the collection of Eleazar Albin (fl. 1713 59), who describes it in his *History of Birds* (1731–8).* In 1819 the bird was referred to as 'The golden pheasant of China, the most beautiful of this genus'. The earliest mention of a specimen in the wild refers to a single bird seen in Norfolk in 1845, where it was at first mistaken for a common pheasant/red-legged partridge hybrid. According to the 4th Annual Report of the Acclimatisation Society, 'From China were shipped, late in the year 1864, three silver and three golden pheasants, out of which only two arrived alive on the 31st December last. . . .' J. A. Harvie-Brown and T. E. Buckley in A *Vertebrate Fauna of Argyll and the Inner Hebrides* (1892) record that golden pheasants were introduced to a number of estates in the western Highlands, 'perhaps

---

*G. C. Bompas in his *Life of Frank Buckland* (1888) suggests 1725 as the year of the first introduction of the golden pheasant to Britain.

[FEMALE]

to none more successfully than to Gigha Island [off the west coast of Kintyre], around the mansion-house [near Ardminish]'.

In 1905 Sir Herbert Maxwell, Bt, wrote:

About ten years ago the Duke of Bedford turned some [hybrid golden × Lady Amherst's pheasants] out in the woods of Cairnsmore, near Newton-Stewart, where they have become perfectly acclimatized and have bred freely. From Cairnsmore they have spread to neighbouring properties. . . . [Some eventually reverting to pure golden characteristics.] Three years ago . . . I turned down [at Monreith, Wigtownshire] two cocks and four or five hens, which have bred each season since.

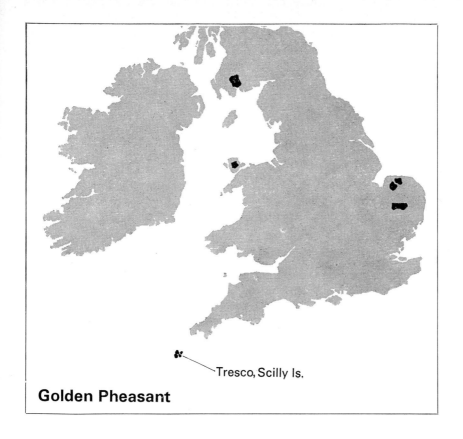

Tresco, Scilly Is.

Golden Pheasant

During the 1890s a number of golden pheasants were introduced to
Tortworth in Gloucestershire; they soon became established in the
surrounding beechwoods, where they bred successfully for a number
of years. J. M. McWilliam in *The Birds of the Island of Bute* (1927)
records that in that year they were interbreeding with common
pheasants on Bute; Fitter mentions that golden pheasants were
regularly shot around Sevenoaks in Kent until 1942. Ten years later
Edlin reported that they were living ferally and under protection at
Penninghame north-west of Newton Stewart in Galloway, and at
Hinton Admiral near Bournemouth in Hampshire, to which they had
flown from nearby Hurn Court after a forest fire. In 1925 the second
Lord Montagu of Beaulieu introduced some golden and Lady
Amherst's pheasants to the Beaulieu Manor Woods near South-
ampton where they soon began to interbreed: a fine hybrid is con-
tained in the natural history collection in Palace House, Beaulieu.

During the present century golden pheasants have been introduced to a number of estates in East Anglia, notably at Elveden Hall near Thetford in Suffolk by the late Earl of Iveagh. The Breckland area of west Norfolk and west Suffolk – especially between Thetford and Brandon and in Thetford Chase – and Galloway, on the Kirkcudbrightshire/Wigtownshire border, where they continue to breed in moderate but apparently increasing numbers, remain today their present strongholds in Britain. In the former area in 1974 they were reported from Santon Downham, Swaffham Heath, St Helen's Well, Shadwell, West Tofts, Roudham, West Harling and East Wretham.* In the same year in the latter area they were observed at Penninghame; on the coast at Creetown; and in Kirroughtree Forest, to the east of Newton Stewart, which held the largest population – estimated at about 250 – living in stands of fifteen- to thirty-year-old Scots pines and larches. A new breeding area for the species was reported in 1974 when a few pairs were discovered to be nesting in Cardrona Forest, Peeblesshire.† Smaller numbers are at liberty elsewhere – for instance on Tresco in the Scilly Isles (where two pairs have been released and chicks were seen in 1975); on the Isle of Anglesey; and in the Sandringham/Wolferton area of Norfolk. In 1974 golden pheasants were reported at Holkham on the north Norfolk coast. In 1972 a feral cock mated with an introduced hen at Sandy in Bedfordshire; although young were reared successfully, they all disappeared later that summer. Two years later golden pheasants were reported in Charle Wood and Maulden Wood south of Bedford, where the nest of a tree sparrow (*Passer montanus*) was found to be almost entirely composed of golden pheasant feathers. Between 1968 and 1971 a pair of golden pheasants frequented the Kingley Vale Nature Reserve on the South Downs in Hampshire, 4 miles (6 km) north-west of Chichester,‡ and a single bird was reported in August 1974, from Murrayshall, Scone in Perthshire. Golden pheasants are today bred all over the country in small gardens as well as on large estates and game farms. They are often turned down in entirely unsuitable habitats – open farmland without copse or covert to provide shelter – from which they soon become dispersed. Unfortunately they interbreed freely with Lady Amherst's pheasants, so their blood is frequently no longer pure.

*Norfolk Bird Report for 1974.
†Scottish Bird Report for 1974.
‡R. and W. Williamson, *British Birds*, 66, pp. 12–23, 1963.
Note: See also Addenda.

35. Lady Amherst's Pheasant

(Chrysolophus amherstiae)

The second species of 'ruffed' pheasant at large in Britain is Lady Amherst's pheasant, which is a native of the mountains of south-western China and Upper Burma: it is named after Sarah, first wife of William Pitt Amherst, 1st Earl Amherst of Arracan (1773–1857), who in 1823 was appointed Governor-General and Viceroy of India. The beautifully attired cock of this species – which is arguably the most magnificent of the male pheasants – is approximately the same size as that of the golden pheasant, but has a longer tail boldly barred with black and white. The crown of the head and throat are dark green, the rump is gold and red, the wings are blue and green, and the breast and underside are white. The ruff or hood, which is scalloped with black and white, can be displayed in the same manner as that of the golden pheasant. The hen is similar to the hen of the golden pheasant, but has greyish-blue legs and a greenish-blue rim around the eye.

Lady Amherst's pheasant was first introduced to England in 1828, although it apparently did not breed here until 1871. Subsequent introductions were made at Mount Stuart on the Isle of Bute, at Cairnsmore near Newton Stewart, and at Woburn in Bedfordshire in the 1890s; to the Beaulieu Manor Woods in Hampshire in 1925; in Richmond Park in Surrey in 1928–9, when a number were reared from eggs, and again in 1931–2, when twenty-four full-winged adults were turned down; in Whipsnade Park in Bedfordshire in the 1930s; and in 1950 at Elveden in Suffolk by Lord Iveagh. In the late 1950s P. H. Carne told Fitter that Lady Amherst's pheasants were 'natura-lized within an 8 or 10 miles (13 to 16 km) radius of Exbury [south of Beaulieu], in the south-east corner of the [New] Forest'. At least eleven were present in this area in 1973.

Today, their main centre is in the east Midlands, where feral birds breed in limited numbers in the woods of south Bedfordshire, *e.g.* in

[FEMALE]

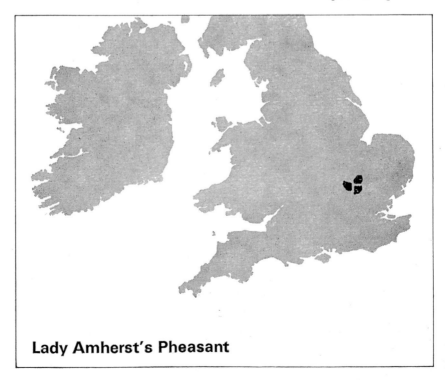

Lady Amherst's Pheasant

Charle Wood, Flitwick Plantation and Eversholt (near Woburn), Old Warden (near Biggleswade), Maulden Wood and Breakheart Hill (both near Ampthill), and a dozen pairs in 1972 in Luton Hoo Park, near Luton, as well as in neighbouring parts of Buckinghamshire (Brickhill and Mentmore Park) and Hertfordshire (Ashridge). In 1973 Lady Amherst's pheasants bred successfully at Guist, $5\frac{1}{2}$ miles (9 km) south-east of Fakenham, and at Quidenham, 10 miles (16 km) east of Thetford, both in Norfolk. Their feral populations are much smaller and more localized than are those of the golden pheasant, with which – and to a lesser degree with the common pheasant – they freely interbreed, so hybrids all too commonly occur.

Note: See also Addenda.

36. Reeves's Pheasant
(Syrmaticus reevesi)

Reeves's pheasant (which like Reeves's muntjac is named after John Reeves)* is a native of the mountains of central and northern China. The cock Reeves's pheasant is the largest of the pheasants discussed here, measuring 3 ft (91 cm) in length with a 6 ft (182 cm) long tail. His plumage is generally tawny in colour, the back and rump being predominantly mustard-yellow and the underside a rich chestnut-brown. The crown of the head is white, surrounded by a black band, and the throat and neck are also white.

The first Reeves's pheasant to reach Europe was one presented to the Zoological Society in Regent's Park by John Reeves in 1831. Three years later it was announced that 'A second male specimen of the Reeves's Pheasant . . . has also been sent to the Menagerie by John Reeves, Esq.'† The species first bred in the Society's gardens in 1867. In its Annual Report for 1864, the Acclimatisation Society announced that 'a fine male pheasant with seven-eighths of the characters of Reeves Pheasant, is placed with three females of the Common Pheasant'.

In 1870 Lord Tweedmouth liberated a pair of Reeves's pheasants, which he had received direct from Peking, at Guisachan (Cougie) at the western end of Strathglass, Inverness-shire. After a further four cocks had been introduced the birds bred freely – over twenty being reared in the first year – but were inclined to wander – some flying as far as Brahan and Contin in Strathconan, some twenty miles (32 km) to the north. In 1892 a dozen cocks were counted, and by the middle 1890s the species was said to be fully naturalized on this estate, and at the nearby property of Balmacaan in Glen Urquhart.

*See page 145.
†*Proc. Zool. Soc.*, II (34), 1834.

[FEMALE]

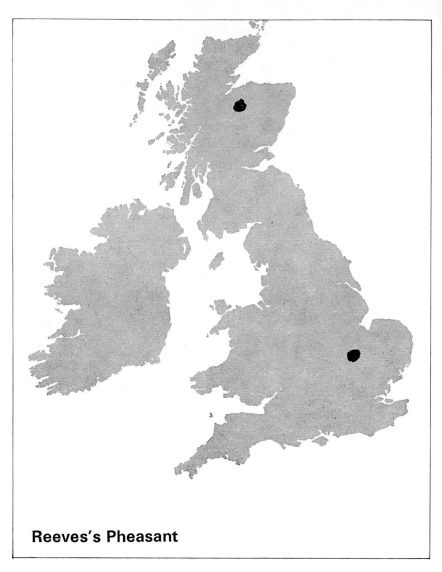

Reeves's Pheasant

In 1894 Newton's *Dictionary of Birds* announced that 'Other species of Pheasant have been introduced to the coverts of England . . . *P. reevesi* from China. . . .' These introductions included those made by Lord Lilford at Lilford Park, Northamptonshire; by the Duke of Bedford at Woburn Abbey, Bedfordshire; at Tortworth in Gloucestershire and at Bedgebury in Kent. Further attempts at naturalization were made at about the same time in Scotland; at Duff House and Pitcroy in Aberdeenshire; at Tulliallan in Fife by a daughter of Admiral Lord Keith; at Mount Stuart on the Isle of Bute by the Marquess of Bute, and in Kirkcudbrightshire. In Ireland, Colonel Edward H. Cooper unsuccessfully attempted to naturalize Reeves's pheasants at Markree Castle, Co. Sligo. Of these centres of introduction, it is believed that only in Woburn Park are there any Reeves's pheasants living ferally today. On 1 June 1972 a single cock was seen in the village of Heath and Reach, two miles (3 km) south of Woburn. This is the only recent indication that Woburn birds occasionally leave the park.

In 1950 a small number of birds were introduced at Elveden Hall in Suffolk by the late Earl of Iveagh: how these fared is uncertain.

In 1970 some Reeves's pheasants were released on Kinveachy Forest near Boat of Garten in Inverness-shire; between September and December of that year two cocks were reported from Kincraig, a few miles to the south. In the following year three cocks were seen at Kinveachy on 7 April, and a single cock was observed near Doune in Perthshire on 24 September. In December 1972 five or six cocks and two hens were seen on Kinveachy, where they had almost certainly bred in the wild, and in 1974 they were again reported to be present and presumably breeding.* Unfortunately, cock Reeves's pheasants are rather pugnacious, and also interbreed freely with common pheasants; they are, therefore, not always welcomed by landlords of sporting estates, and their blood tends to become diluted with that of other species.

*

A number of other so-called 'ornamental' pheasants have been introduced to Britain from time to time with varying degrees of success. Only one species, however, has ever shown any sign of becoming naturalized. This is the SILVER or KALIJ PHEASANT (*Gennaeus nycthemerus*), which ranges over the mountain-forests of north-eastern Burma, central Thailand, southern China, Vietnam, Laos and south-western Cambodia.

The upper-parts of the cock silver pheasant are white, marked with vermiculated black lines. The underparts and throat are bluish purple. The head is crowned with a long black crest, and on each cheek there are bright red wattles. The feathers of the tail are long and curved – the central pair being almost white.

Scottish Bird Reports, 3(7), p. 128, 1971; p. 347, 1972; 8(4), p. 234, 1973 and 1974.

2C

[FEMALE]

According to Bompas and other authorities, silver pheasants were first introduced to England in 1740, and soon became naturalized in a number of districts. A small stock was subsequently turned down in Rutland in the 1880s. In the following decade they were introduced to Woburn in Bedfordshire and Cairnsmore in Galloway by the Duke of Bedford, and at Mount Stuart (where they had died out by the 1920s) by Lord Bute. In Richmond Park in Surrey where a number were introduced in 1928–9, they apparently did not survive for long. In 1905 Sir Herbert Maxwell, Bt, wrote that the silver pheasant 'takes as naturally to British woodlands as the other [common pheasant]; it used to be pretty to see them flying from side to side of the wooded banks of Ayr in the park at Auchencruive [2½ miles (4 km) north-east of Ayr]'.

Unfortunately, silver pheasants interbreed freely with other pheasants, and, in addition, are persistent egg-stealers, which tends to make them unwelcome on game preserves. Those few at large in and around Woburn today are all hybrids.*

*Elliot's pheasant (*Syrmatieus ellioti*), a native of the mountains of south-eastern China which was first seen in Britain in the 1880s, has been acclimatized but not naturalized in the woods outside Woburn Park.

Note: For Reeves's pheasant see also Addenda.

37. Red-Legged or French Partridge

(Alectoris rufa)

As its name implies, the principal home of the red-legged or French partridge is France, south of the *départements* of Loire and Jura, from where its range extends into the Swiss lowlands, north and central Italy, Corsica and the Balearic Islands.

The red-legged partridge is a larger bird than the native British grey or common partridge (*Perdix perdix*), measuring between $13\frac{1}{2}$ and 14 in. (about 34 cm) in length. The upper-parts are a uniform medium brown, shading to grey on the crown of the head. The throat and cheeks are white, edged with a black band which passes upwards through the eyes. Beneath the throat is a gorget patch of black streaks and spots. The breast and flanks are a soft lavender blue-grey, the flanks being barred with black, white, and chestnut stripes. The belly and underside of the tail are rufous buff, and the legs and beak are red. Both sexes are similar in appearance.

Red-legged partridges prefer to live on light sandy and chalky soils, such as those found on heath- and down-land. To escape from danger they tend to run rather than to fly, and when in the air are more inclined to scatter than to remain in coveys. In late April or in May between ten and fifteen yellowish-white eggs, sparingly spotted with ochreous-red, are laid in a scrape in the ground scantily lined with grass and dead leaves. The incubation period is from twenty-three to twenty-four days. The display-note of the cock red-legged partridge is a harsh 'chucker-chucker', which has been likened to a labouring engine getting up steam. The food consists principally of leaves of clover and grasses, corn, seeds, insects and spiders.

Although the common or grey partridge is a native British bird, the red-legged partridge is an alien species which was first introduced to England some three hundred years ago.

Sir Thomas Browne (1605–82) made the first, though negative, reference to the red-legged partridge in England when, in about 1667, he wrote, 'Though there be here [Norfolk] very great store of partridges, yet the French red-legged partridge is not to be met with.' In their *Ornithologia,* published between 1676 and 1678, Francis Willughby and John Ray state: 'We have been informed that the red-legged partridge (*Perdix ruffa*) is found in the isles of Jersey and Guernsey. . . . This kind is a stranger to England; howbeit, they say it is found in the isles of Jersey and Guernsey, which are subject to our King.' Fifty years later Ephraim Chambers (d. 1740) in his *Cyclopaedia* (1728) confirmed that the position remained unchanged: 'The red-legged partridge is not found in England, but is sometimes shot in the islands of Guernsey and Jersey.' The earliest reference to the original introduction of the red-legged partridge to England occurs in *Rural Sports* (1801), written by William Barker Daniel (?1753–1833):

So far back as the time of Charles II [reigned 1660–85] several pairs of these red-legged partridges were turned out about Windsor to obtain a stock; but they are supposed to have mostly perished, although some of them, or their descendants were seen for a few years afterwards.

This statement was enlarged upon in the correspondence column of *The Field* on 19 March 1921:

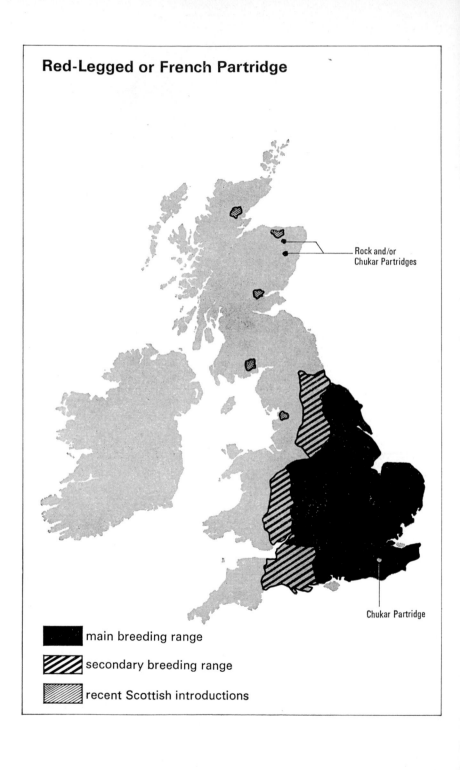

Red-Legged or French Partridge

Rock and/or
Chukar Partridges

Chukar Partridge

main breeding range

secondary breeding range

recent Scottish introductions

Monsieur Léon Dorez, of the Bibliothèque Nationale, has kindly communicated to me two letters written in 1673, which throw light on the question as to the date when the red-legged or French partridge was introduced into England. They were discovered at the Bibliothèque Nationale [Mélanges Colbert, vol. 166, fol. 276]. Daniel, in his *Rural Sports* published in 1801, says that the French partridge had been known in England since the time of Charles II. These letters make it certain that King Charles did his best, at any rate, to obtain some birds; and it is hardly possible to doubt that he was successful.

Holkham, March 14, 1921.   Leicester.

Paris, 2^n [November] 1673.

Sir,

The King my master, has sent me to France for some red partridges to introduce into England. Monsieur de Colbert, ambassador in England, has promised the King to procure for him as many as he wishes to have. I send you a letter which he sends you on the subject. I beg you, Sir, to have the goodness to let me know your instructions, which I am expecting as soon as possible, because of the time of year which obliges me to be quick. I am staying at Paris at the house of Monsieur Crettien, in the street of the Feronnerie, near St. Innocent, at the sign of the Town of Vernon, and I am, Sir, your very humble and very obedient servant.

Favennes De Mouchant.

To Monsieur Saumery, at Chambord.
[Saumery was Keeper of the Château de Chambord in the *département* of Loir-et-Cher.]

London, the 9th October, 1673.

Sir,

Mr. de Mouchant, Gamekeeper of the King of England, visiting France to procure some red-legged-partridges to fill the parks at Windsor and Richmond. His Majesty the King of Great Britain has desired that I should give him this letter to beg you very humbly, Sir, to be good enough to allow the said Mr. de Mouchant to take a quantity of the said partridges in the park or in the neighbourhood of Chambord. And the fact is that I could not avoid granting the wish of His Britannic Majesty, being assured that the King our master will not disapprove of your doing him the pleasure. And I take advantage of this occasion also, if you please, to assure you that I am always, Sir, your devoted, humble and very obedient servant.

Colbert.

[Charles Colbert, Marquis de Croissy (1625–96), was ambassador from Louis XIV to Charles II, from 1666 to 1673.]

From the lack of further reference to red-legged partridges at Windsor and Richmond, it seems clear that as Daniel wrote: 'they are supposed to have mostly perished'.

Fitter records two other early and apparently unsuccessful attempts to naturalize red-legged partridges in England. A brace of 'curious outlandish partridges' sent to the 9th Earl of Rutland at Belvoir Castle in Leicestershire, in November 1682, were probably of this species. At some time probably between 1712 and 1729, the 2nd Duke of Leeds reared some red-legged partridges from eggs on his estate at Wimbledon in Surrey, where they 'were, after increasing for a time, all destroyed by some disobliging neighbour'.

In 1883 J. E. Harting wrote: 'In or about the year 1770 several noble sportsmen appear to have combined in importing eggs from France, and hatching them out under hens on their estates in different parts of the country'. These 'noble sportsmen' included the Earl of Hertford (Ambassador to France from 1763–5) and Lord Rendlesham at Sudbourne and Rendlesham* Halls near Woodbridge in East Suffolk: the Duke of Northumberland at Alnwick Castle; and the 4th Earl of Rochford† at St Osyth in Essex. The birds in Suffolk quickly became established, probably because the light sandy soil and open heathland provided an ideal habitat for them, and were reported to be especially common on Dunmingworth Heath and between Aldeburgh and Woodbridge. In 1777 Daniel saw a covey of fourteen red-legged partridges (three brace of which he shot) near Colchester, which came presumably from St Osyth. Twenty-one years later he shot two and a half brace at Sudbourne – the first he had seen since 1777. George Montagu (1751–1815) in his *Ornithological Dictionary* (1812), quoted Daniel's *Rural Sports* that red-legged partridges were common around Oxford, and cited another source that they were frequently to be seen near Ipswich. Since about 1828 the red-legged partridge has outnumbered the grey partridge in most parts of East Anglia.

In 1776 Sir Harry Fetherstonhaugh of Uppark in Sussex imported a quantity of red-legged partridge eggs from France, which he reared in the park and in the walled gardens of nearby Harting Place. Although these birds failed to establish themselves in the neighbourhood on a permanent basis, Harting, writing in 1883, was of the

*An albino red-legged partridge – now in the British Museum (Natural History) – was shot at Rendlesham in 1905.

†Lord Rochford, ambassador in Turin from 1749 to 1753, also introduced the first seedling of the Lombardy Poplar (*Populus italica fastigiata*) to England in 1758.

opinion that they were the original source of the species in Sussex. Fitter, however, considers it more likely that the red-legged partridges of Sussex are descended from two broods which were hatched under domestic hens and liberated at Kirdford, 4 miles (6 km) north-west of Petworth, in 1841 – (see below).

In 1823 Lords Alvanley and De Ros imported large numbers of red-legged partridge eggs from France, which were successfully hatched on estates at Culford, Cavenham and Farnham near Bury St Edmunds in Suffolk, from where by 1839 the resulting offspring had crossed into Norfolk and Lincolnshire.

Harting has recorded in some detail the subsequent spread of red-legged partridges in England. In the *Zoologist* (1874; p. 4224) it was said that; 'This bird is becoming quite common in that part of Southern Lincolnshire bordering the Wash and opposite the county of Norfolk. I am told it has of late years gradually extended in that district.' From western Lincolnshire red-legged partridges crossed over into the neighbouring county of Nottinghamshire, where, according to J. Whitaker in *Notes on the Birds of Nottinghamshire* (1907), the first bird was recorded in 1851. There appears to have been little spread from the Duke of Northumberland's introduction at Alnwick; it is possible that the few breeding in Westmorland in the 1860s may have come from this source, although the birds would have had to cross the alien habitat of the northern Pennines. John Cordeaux in *A List of British Birds of the Humber District* (1872) wrote: 'The red-legged partridge is fortunately only an occasional wanderer in Northern Lincolnshire. I have seen birds that were shot in the neighbourhood of Ashby, near the River Trent.' In October 1835 a brace were shot at Ulceby west of a line between Hull and Grimsby. In their *Handbook of Yorkshire Vertebrata*, Clarke and Roebuck describe the species as 'resident in various parts of York-shire [*e.g.* around Doncaster] but in extremely limited numbers, and only very occasionally shot'. Dr A. G. More (*Ibis*, 1865; p. 428) writes that the red-legged partridge was 'breeding very rarely in West Yorkshire', where it had possibly arrived from Westmorland.

A number of unsuccessful attempts were made to naturalize red-legged partridges in Derbyshire (*e.g.* at Donnington Park by the Marquess of Hastings), and Fitter refers to an abortive introduction made in 1820 at (Great) Witley, 9 miles (14 km) north-west of Worcester. Garner, in his *Natural History of Staffordshire* states that the species was introduced to Teddesley Hall, 5 miles (8 km) south of Stafford by Mr Edward John Littleton, but scathingly adds that it was 'no desideratum for the sportsman'. By 1883 it was, according to Harting, breeding occasionally in Cambridgeshire (where the first bird was shot near Anglesea Abbey on 11 September 1821), Huntingdonshire (where it had been known in some numbers since 1850), around Lilford in Northamptonshire and in Rutland, but had not penetrated as far as Cheshire, Lancashire, Shropshire or north Wales. Red-legged partridges were recorded by Harting as being resident in Berkshire (where one was shot from a covey of grey partridges near Newbury in 1809); Buckinghamshire (where they were first recorded in 1835); Oxfordshire (where a pair was seen near Stokenchurch in 1835; soon afterwards the species was described as 'resident and becoming common'); and on Royston Heath in Hertfordshire, where they may have been the descendants of those introduced to St Osyth in Essex by Lord Rochford in about 1770. Fitter records that one was shot at Ippollitts near Hitchin as early as 1815, and by 1877 the species was common in that part of Hertfordshire.

Nearer London, Harting saw red-legged partridges at Stanmore, Elstree and Brockley Hill; in September 1880 he found a dead bird at Northaw near Barnet. The first bird in Middlesex was seen in 1865.

South of the Thames, red-legged partridges were recorded as breeding occasionally in parts of Kent (where a few were shot prior to 1823) by the late 1860s. Knox, in his *Ornithological Rambles in Sussex*, gives details of the introduction at Kirdford, referred to briefly above.

In July 1841 two coveys of red-legs were hatched under hens, and turned down on a manor in the parish of Kirdford in the Weald of Sussex. They were observed there for nearly a fortnight, when they suddenly disappeared. During the following September a small covey was sprung near Bolney, about twenty miles [32 km] further west, and a brace shot, probably a remnant of the Kirdford birds.

Harting shot red-legged partridges at Three Bridges, Frant, Uckfield, Hellingly, Hassocks, West Grinstead, Midhurst, Harting and Chichester. Gilbert White (1720–93) makes no reference to the species in Hampshire in *The Natural History of Selborne* (1789), but less than a hundred years later Harting saw red-legged partridges on the chalk of Butser Hill south of Petersfield, and at Liss, Holybourne,

Alresford and Thruxton. On the Isle of Wight two birds shot in Harting's time – one at Newchurch, the other at Freshwater – had presumably either flown across the Solent from Hampshire or over the Channel from France.

Of the red-legged partridge in Wiltshire the Revd A. C. Smith of Calne was able to write: 'It is our good fortune in Wiltshire to know but little of this bird. A few stragglers from time to time have made their way into the county, and have been shot at Winterslow, Draycot Park and Winterbourne Monkton.' Rowe mentions that 'a few specimens have been procured around Marlborough'. The first record for the county dates from 1861. According to Dr More (see above) the species also failed to become naturalized in the neighbouring county of Dorset. Fitter refers to a bird killed at Upwey between Weymouth and Dorchester before 1799 but was of the opinion that this may have been a natural immigrant from France.

The red-legged partridge does not appear in Dr. More's 1830 *List of Devonshire Birds*, nor in Bellamy's *Natural History of South Devon* (1837). The first record of it in the county, in 1844, occurs in *The Birds of Devon* (1892) by W. S. M. D'Urban and M. A. Mathew. Harting refers to one shot 'many years ago' on Waddle's Down near Whitstone, 6 miles (10 km) south-east of Bude in Cornwall; another at Newton St Cyres, 4 miles (6 km) north-west of Exeter, and a third killed on 3 October 1882 at Okehampton. Rodd makes no mention of red-legged partridges in his *Birds of Cornwall*, but in his *History of Cornwall* (1816) Richard Polwhele (1760–1838) quotes from the *Monthly Magazine* for 1 December 1808 (*The Literary Repository of Cornwall and Devon, Quadrupeds, Birds and Fishes*) from information received from a Mr James of Manaccan: 'Partridge Plenty: I have been told that Charles Rashleigh, Esq. of St. Austle [*sic*] procured from abroad some of the Red-legged, and turned them loose, and that they have multiplied'*. This release presumably took place on the Menabilly Estate near Fowey (overlooking St Austell Bay) seat of the Rashleigh family which has been prominent in the county since the sixteenth century.† The success was apparently short lived. From the mid-1850s Mr Augustus Smith turned down some red-legged (and grey) partridges on St Martin, and later on Tresco, in the Isles of Scilly, where they had all died out by 1864.‡

---

*I am grateful to Mr Roger Penhallurick for drawing my attention to this record.
†Rashleigh Papers: County Record Office, Truro. Account Books, Ref. DDR 4542.
‡'Birds of the Isles of Scilly', *Zool.* 1906.

Red-legged partridges are not referred to by Cecil Smith in his *Birds of Somerset*, published in 1869, but in September 1879 Mr C. Fry Edwards of the Grove, Wrington, shot a brace – and was told of others which had been seen – at Compton Bishop, 3 miles (5 km) south of Glastonbury. In a letter to Harting, Smith wrote:

On December 14 last [1882] I saw at one of the poulterer's shops in Taunton a red-legged partridge, which had been shot that morning at Kingston about three miles [5 km] off. . . . A small covey of four or five had been seen at Nynehead near Wellington. . . . In all probability they wandered [to Taunton] from some distance; for I have not heard of any being reared or turned out in this neighbourhood. . . . In the east of Somerset near the Mendips, I am informed 'that the red-legged partridge was introduced at Cheddar about sixty-six years since [*i.e.* in about 1816] by Mr. Cobley. . . . The birds spread and drove the grey birds till they became so strong that to preserve any of the old species it was resolved to exterminate the foreigners. This was done and the grey birds were restored, but some of the red-legs may have escaped.' . . . My father had hatched some [red-legs] from some eggs he brought with him from Paris in 1835; but these were kept in an aviary, and none escaped. So far as I remember also they did not breed.

Red-legged partridges spread to Gloucestershire probably from Wiltshire, though possibly also from Cheddar. On 7 October 1881 Mr Ernest Armitage wrote to Harting from Dadnor, near Ross-on-Wye in Herefordshire; 'A friend of mine shot a red-legged partridge in this neighbourhood last week. . . . No one can remember ever having heard of one being killed here before.'

Although the exact number of introductions in England is unknown, Fitter traced more than forty different attempts since 1830 in twenty-six counties. It is, however, well-nigh impossible to differentiate between 'introductions' in the sense discussed here, and 'rearing' for sporting purposes.

In Wales the red-legged partridge has always been, at most, of only very local distribution. According to E. Cambridge Phillips,

About six or seven years ago a young bird of this species was killed at Seethrog near Brecon. . . . About six months afterwards another was caught in a garden at Brecon. . . . Mr. Alfred Crawshay of Talybont turned out a couple of red-legged partridges about a year previously, and they must have hatched a small brood. In the autumn following there were four or five young ones.*

Fitter traced at least twelve introductions in seven Welsh counties, and records that the first red-legged partridge in Carmarthenshire was not seen until 1919.

The only record of a red-legged partridge in Scotland by the time of Harting's survey in 1883 was that of a single bird shot in a covey of grey partridges in January 1867 two miles from Aberdeen. Baikie and Heddle in *Historia Naturalis Orcadensis* (1848) state that in 1840 the species was introduced to Orkney by the Earl of Orkney. Fitter traced eight separate attempts at introduction in six counties in Scotland, where by 1920, when the species was apparently flourishing only in Fife, Professor Ritchie described 'the pretty red-legged partridge' as 'another introduction which has met with little success in Scotland. . . .' None were recorded in Fife during the 1968–72 survey for the British Trust for Ornithology's *Atlas*.

During the 1970s red-legged partridges have been more or less successfully introduced to a number of estates in Scotland. In March 1970, thirty were turned down at Rosehall in Sutherland, where in the following spring they were reported to be flourishing: in the same year others were introduced to Cullen in Banffshire and at Cortachy in Angus, and a single bird was seen on 10 April at Fingland Hill in Kirkcudbrightshire. In 1972 red-legged partridges were reported to be commonly seen in the Buckie district of Banffshire, where several hundred had been released: it is possible, however, that these birds were chukar partridges (see below) or chukar/red-legged hybrids (which were also reported from Dumfriesshire and Kirkcudbrightshire in 1974). Single birds were seen at Murthly in Perthshire on 23 June, and at Yetholm (Roxburghshire) and Southerness (Kirkcudbrightshire). In 1973 red-legged partridges were breeding at Redgorton, Dupplin and around Scone (about forty pairs) in Perthshire and in Dumfriesshire. They have also been turned down with limited success since 1973 on the Earl of Dundee's Birkhill estate near Cupar in Fife. Four pairs with nineteen young were released in that year at Killearn in west Stirlingshire, and the species was recorded in Aberdeenshire, Kincardineshire, Angus (Strathcathro and Cortacy), and Galloway (Cairnsmore). In 1974 three birds

*'Birds of Brecon', *Zool.* 1882.

were seen at Coupar Angus in Perthshire on 13 February, and a pair of adults with young at Aberfeldy in July; others were observed in June near Dunblane, where the species has been introduced. One hundred red-legged partridges were released in Caithness in the summer of 1974, some of which were still around in the following December. The increase in records in recent years is believed to be largely the result of such game-management releases.*

In Ireland, red-legged partridges were twice introduced to Co. Galway by a Mr Gildear a few years prior to 1844, and a single bird was shot near Clonmel on the border between Co. Tipperary and Co. Waterford on 4 February 1849. The species is no longer present in Ireland today.

Since its introduction to England over two hundred years ago, the red-legged partridge has had to contend with a considerable amount of prejudice from shooting-men, who have apostrophized it for lacking 'sporting' qualities because of its tendency to run rather than to fly, and when in flight to scatter instead of to pack. Until well into the present century red-legged partridges were treated as vermin in many parts of the country, where it was feared that they would oust the native species. The grey partridge has, indeed, suffered a serious setback in England in recent years. This, however, has been due not to the infiltration of the alien bird, but to a combination of circumstances. A series of cool, wet nesting seasons in the 1960s and 1970s, coupled with the increasing use of agricultural pesticides which has caused a shortage of *Symphyta* (sawfly) larvae, together resulted in a high rate of chick mortality; and the elimination of many miles of 'bullfinch'-type hedges surrounding farmland has drastically reduced the number of suitable nesting-sites. Shooting-men have reason to be thankful for the adaptability and hardiness of the red-legged partridge, whose strongholds today are the light, sandy soil and heathland of parts of East Anglia, and the chalky downland of south-central England. In the west, it is thinly distributed as far as Somerset, Dorset and eastern Devon, where small breeding populations were found in three areas in 1971–2. In

Scottish Bird Report, 3(7), p. 128, 1971; p. 347, 1972; 8(4), p. 234, 1973. *Perthshire Bird Report*, 1974.

1972 birds were seen at Venn Ottery, Aylesbeare Common, Souther-ton, Harpford Common and Zeal Monachorum. In 1973 some birds were released at Clinton – where more had been turned down in previous years – and others were reported from Venn Ottery, Hawkerland, Dotton, Stoneyford, near Dawlish, Powderham, Brampford Speke and Shebbear.* It will be interesting to see whether the birds will survive in this most westerly outpost of their range in England. Elsewhere the red-legged partridge ranges north-west to Shropshire, and north to northern Yorkshire – its range being principally governed by the incidence of precipitation. It occurs locally even further north and west as the result of game-releases (*e.g.* at Treales near Preston and on the Bank Hall Estate in Lan-cashire),† but appears to be unable to maintain its numbers without artificial reinforcements, principally due to the dampness of the climate. Red-legged partridges decreased in numbers in Sussex, Kent and Essex between 1950 and 1967, and in Suffolk from 1945 to 1959. In recent years there has been a marked increase in Suffolk, and some evidence of recovery in the south-eastern counties. During work from 1968–72 for the British Trust for Ornithology's *Atlas,* red-legged partridges were reported from the Isle of Man; recent records for the island include a pair seen at Poyllvaaish on 29 April 1973 and a single bird at Balladoole near Arbory on 12 May 1974; the popula-tion appears to be small and possibly not firmly established. In Wales, red-legged partridges are found in parts of Glamorganshire, Breck-nockshire, Radnorshire, Montgomeryshire and Denbighshire close to the English border.

*

Several attempts have been made over the years to naturalize both the EUROPEAN ROCK PARTRIDGE (*Alectoris graeca*) and the closely related CHUKAR PARTRIDGE (*A. chukar*); the latter ranges from Cyprus and the Balkans (Greece and east Bulgaria), eastwards through Iran and India (including the Himalayas) into China. Both are allied to, and closely resemble, the red-legged partridge, from which they are mainly distinguished by their slightly larger size, greyer plumage, and different call-notes.

*H. P. Sitters, *Atlas of Breeding Birds in Devon,* 1974.
†*Lancashire County Bird Report,* p. 10, 1971.

In the late 1920s or early 1930s Mr F. R. S. Balfour introduced some chukars to Dawyck Park, Stobo, Peeblesshire: they were also kept unpinioned between the wars by the Duke of Bedford at Woburn Abbey; at Whipsnade, and by Mr Alfred Ezra at Foxwarren Park, near Cobham in Surrey. At none of these places, however, were they ever properly established, although some were still breeding until the late 1940s. Following the decline of the grey partridge, and in order to see whether they might be more easily established than the red-legged partridge, a number of commercial game-farms have, in recent years reared both chukars and rock partridges; as a result, the former now live wild in small numbers on the South Downs in Sussex, and the latter (or both) in parts of Aberdeenshire and neighbouring counties where hybrids may be involved.*

*Scottish Bird Report, 8(4), p. 234, 1973.

Note: For Red-Legged or French Partridge see also Addenda.

38. Bobwhite Quail

(Colinus virginianus)

The bobwhite or Virginian quail is a native of the eastern United States north to the Great Lakes and the Dakotas, and west to eastern Texas, Colorado and south-east Wyoming. It is a dumpy reddish-brown bird, measuring 8–9½ in. (20–24 cm) in length – midway in size between the European quail (*Coturnix coturnix*) and the common partridge (*Perdix perdix*). The underparts are mainly chestnut or brick red in colour, the flanks being barred with chestnut crescents. The cock has a white throat and a stripe of the same colour runs from the forehead back above the eye and past the lores to the nape of the neck. In the hen these areas are an overall buff. The crown of the head is brown, the legs yellow, and the tail grey.

The bobwhite is a gregarious bird, living in coveys of up to thirty in deep brushwood, abandoned fields and open pinelands (known in the south-eastern States as 'boondocks' or 'booneys'), where it not infrequently perches in trees, although it is never seen in thickly forested areas. It builds its nest in late spring or early summer in thick undergrowth or in the cover of hedge-bottoms, where the hen lays between fourteen and sixteen dull or creamy unmarked matt-white eggs. Bobwhite quail eat mainly insects, grain and seeds: their call is a plaintive repeated whistle – 'bob-bob-white'. In North America the bird is greatly valued as a destroyer of insect pests and grubs, and has been much transplanted for sport:

. . . the bobwhite quail is everyman's gamebird, and it is no unusual sight to see country youths . . . in search of bobwhite coveys with a beautifully trained and co-ordinated pair of setters or pointers working ahead: 'No gamebird is more highly esteemed for its flesh than the quail, and its recreational value is a tonic to the bird-dog enthusiast.'

*J. N. P. Watson, 'Where Hunters Still Come First . . .' *Country Life*, pp. 1098–1102, 23 October 1975.

2D

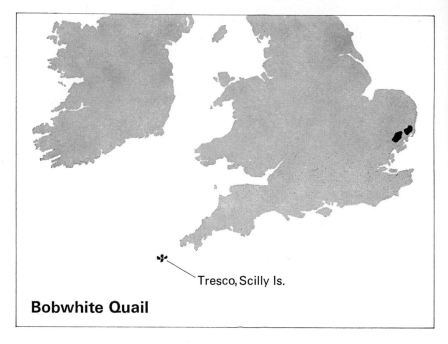

Tresco, Scilly Is.

Bobwhite Quail

In the nineteenth century bobwhite quail were turned down for sporting purposes on over a dozen separate occasions, principally in England, but also in Scotland (three times), Ireland (once) and Wales (twice).

The earliest of these introductions was one made to Ireland, prior to 1813, by General Gabbit. A few years later the Earl of Leicester introduced a considerable number to Holkham in Norfolk, where they apparently soon settled down and spread. It was presumably to one of Lord Leicester's birds that the Revd John Burrell, Rector of Letheringsett near Holt (11 miles (18 km) east of Holkham) was referring when on 11 November 1825 he wrote to Mr H. Denny:

I had yesterday a bird brought to my parlour which was shot here; it was given me as a new addition to the British Fauna; at first sight I thought it a quail, a bird I never saw. I examined it by Shaw's *General Zoology* and from some of the habits, which I have learnt from the sportsman since I received it, I am willing to accord it the nomenclature applied to it by the sportsman, the *Maryland Partridge* of Pennant's *Arctic Zoology*. It is not, however, such a novelty as my neighbour conjectured; I have a specimen previously set up; it was brought to me last year, when I contented myself with a bird's-eye view, and joined other students in Natural History in having hitherto confounded it with the quail. It is now *quite a colonized*

creature, and numerous are the covies, which report says that the poachers cannot destroy, its manners are so watchful and shy of man. It was too much shot for preservation, and therefore I not once thought of sending it to Norwich.

Bobwhite quail were still to be found between Holkham and Holt twenty years later.

In the early 1840s, Prince Albert turned down a considerable number of bobwhite quail from the United States in Windsor Park; some of these appear to have brought off young successfully, and several were shot in the home counties (*e.g.* a male near Chelsham Court, Godstone, Surrey in 1845, and a female at Rotherfield near Tunbridge Wells in Kent on New Year's Day 1850) until the early 1850s.

In 1867 a consignment of bobwhite quail from Canada arrived in Norfolk: Prince Albert's son, the Prince of Wales, later Edward VII, turned down three cocks and four hens at Sandringham in the west of the county, and others were liberated at Aylsham, 12 miles (19 km) north of Norwich, and on a number of other East Anglian estates, including that owned by the Maharajah Duleep Singh at Elveden in Suffolk. None appear to have survived for long.

In April 1860 sixteen bobwhite quail from Canada were turned down at Heron Court in Hampshire by Lord Malmesbury on behalf of the Acclimatisation Society: a further nine quail were received from Mr F. J. Stevenson in Canada, some of which were turned down by the vice-president of the Society, the Hon. Grantley F. Berkeley.

In 1870 and 1871 Lord Lilford and the Duke of Buccleuch attempted to naturalize bobwhite quail in Northamptonshire: they bred successfully for about ten years – some being shot in the neighbouring county of Rutland and in Nottinghamshire – but then apparently died out. In 1898 a number of bobwhite quail were shot in north Wales, two of which came from around Bettws-y-Coed near the Caernarvonshire/Denbighshire border. At about the same time Mr Edward John Littleton turned some down at Teddesley Hall in Staffordshire.

In recent years a number of attempts have been made to establish bobwhite quail in Britain, with a view to supplementing the depleted stocks of the common partridge.

In 1956 Major Pardoe turned out about sixty bobwhite quail at Dunwich on the coast of Suffolk between Southwold and Aldeburgh: they did well in the wild until the hard winter of 1962–3 drastically reduced their numbers. In about 1957 the late Lord Tollemache brought over by air from the United States about a hundred bobwhite quail eggs, all of which hatched successfully, the young birds being put out in coveys on the Helmingham estate near Stowmarket; they appeared to do well for two or three years, but then all disappeared, probably due to the lack of cover and heavy clay soil.

In 1961 a single bobwhite quail spent the spring and summer on the Royal Society for the Protection of Birds' Minsmere Reserve near Westleton, some 2½ miles (4 km) south-west of Dunwich. In the following year two pairs were seen at Minsmere where they may have bred, although no young were seen. In the spring of 1963 a single bird was observed in bushes near the coast, and three pairs were seen during the summer, one of which had chicks. In May 1964 bobwhite quail were heard uttering their mournful call in many areas around Westleton, and at least four pairs are believed to have bred. Captain Stuart Ogilvie reported that in June two pairs with about five young were released on the Sizewell portion of his estate, some 2 miles (3 km) south of Minsmere. No introductions had been made on the Scott's Hall estate (which includes Minsmere); birds seen there must, therefore, have arrived of their own volition presumably from Dunwich.

In 1965 at least four pairs were present at Minsmere, two of which raised young successfully. During the winter of 1965–6 a covey of twenty birds was in residence on the reserve, out of which probably eight pairs bred in the following spring, two coveys of a dozen birds being seen in September. These birds wintered well, and five coveys were present in the following summer. In January 1968 the keepers estimated that there were about forty birds on the estate, of which apparently only about five pairs nested, this number of fledged coveys of between four and six birds being seen. These birds spent the winter on the estate, and eight pairs nested in the following spring, but none of the young appeared to survive. A covey of three birds was present until the end of March 1970, and four coveys were seen in August. After an especially cold period in late December 1971, only one or two birds seemed to be around, but two pairs nested and a covey of no less then sixteen was seen on 9 June 1972. Between August 1971 and February 1972, when it was found dead after a cold spell, a solitary cock developed the habit of feeding with reared pheasants, waiting for the keepers on their rounds. In 1973 a second lone cock was observed from late May to mid-June, and a third was seen on 14 February 1974.

In 1961 and 1962 a number of bobwhite quail were released in Norfolk, and in the latter year birds were reported from Holkham on the north coast; Taverham, 6 miles (10 km) north-west of Norwich; Hainford, the same distance due north of Norwich; near Wymondham, 8 miles (13 km) south-west of Norwich, and at Somerton, 7 miles (11 km) south of Bury St Edmunds.*

According to the *Cornwall Bird Report* for 1968 (p. 58) bobwhite quail were 'Introduced to Tresco, 1966; bred in 1967 and 1968. Thought to have spread to Bryher.' In fact it was in 1964 that a clutch of six bobwhite quail eggs, imported from the Avon and Airlie Game Farms at Chippenham in Wiltshire by Lieutenant-Commander and Mrs Thomas Dorrien-Smith, was first successfully hatched on Tresco in the Scilly Isles, the young being released in late summer. In the following year at least two broods were successfully reared in the wild, and were supplemented by the release of a further six birds. The population has remained fairly static since then, and in the summer of 1975 several pairs nested successfully in the wild on Tresco.

A small number of bobwhite quail have been turned out in the

Norfolk Bird Report, 1962.

1970s in various parts of England – in Gloucestershire, Herefordshire (where one landowner shot a number of what he took to be a new species of partridge) and Wiltshire for example – but although one or two birds usually remain for some time, the main coveys soon disappear. According to the *Perthshire Bird Report* for 1974, bobwhite quail are 'thought to have been released on an estate near Dunblane where one was heard on 9 July and two were calling on 17 July'.

The principal reasons for the apparent failure of bobwhite quail to become established for any length of time on mainland Britain appear to be twofold: they seem not to be hardy enough to survive a cold and wet spring, especially on heavy clay land and, as they nest largely on road-side grass because of a shortage of more suitable thicker cover, they fall an easy prey to ground predators.

A report of the British Ornithologists' Union Records Committee for 1974 includes the following statement by Dr J. T. R. Sharrock:

Bob-white Quail *Colinus virginianus*. In view of continued breeding in the Minsmere area of Suffolk and on Tresco, Isles of Scilly, both England, the question of transferring this species to Category 'C' was assessed. It was unanimously decided to retain the species in Category 'D', but to reconsider the position in 1979.*

*Category 'C' includes introduced feral species which have been admitted to the British and Irish List: Category 'D' is an appendix for species worth recording but not yet given formal admission to that list.

Note: See also Addenda.

III Amphibians and Reptiles

39. Alpine Newt
(Triturus alpestris)

The alpine newt* is more widely distributed in Europe than its name might suggest: it is found from northern and eastern France westward to the Russian border, and from Denmark in the north, south to central Italy and northern Greece. A separate population lives in the Cantabrian Mountains of north-western Spain; the species may possibly also occur in parts of central Spain.

The male alpine newt is normally grey or blackish above with darker markings; the female tends to be browner and more uniformly coloured. The sides are marked with numerous small spots and sometimes with a whitish stipple, frequently set, in the males, on a pale lavender background. The underside is almost invariably a uniform shade of orange or red. In the breeding season, the male grows on his flattish back a low smooth-edged, yellowish crest, spotted or barred with black.

The alpine newt is nearly always found in the vicinity of water. In the south it is montane in its habits, sometimes being found at altitudes of between 7500 and 9000 ft (2286 and 2743 m). In the north of its range it has a much wider distribution, normally occurring in cold and nearly plantless waters, in forests, ponds, lakes and slow-moving streams in the mountains, but also in open and shallow waters in low-lying districts. Alpine newts emerge from hibernation in February, and after an elaborate courtship the female lays her eggs, singly and on the leaves of water-plants, between March and the end of May: metamorphosis is complete after three or four months. When living at high altitudes, the summer may be of too short a duration for the larvae to metamorphose in a single season. The food of the alpine newt consists mainly of insects, worms, centipedes, snails, *Daphnia*, and small tadpoles.

*The name 'newt' is derived from the Anglo-Saxon 'evet', 'effet', or 'eft' ('efeta').

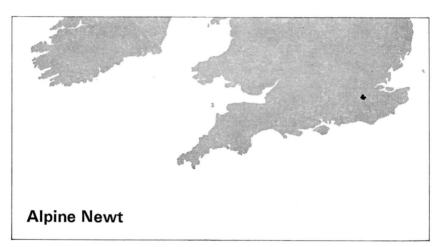

Alpine Newt

In Britain, a single colony of the alpine newt has survived for many years in a garden in Surrey; this population, which is well established and contains several hundred individuals, is in the neighbourhood of the Beam Brook Aquatic Nursery at Newdigate in Surrey, from which the colony originated.

40. Midwife Toad

(Alytes obstetricans)

The midwife toad is a native of western Europe as far south as the Alps and the Iberian Peninsula, eastwards to western Germany. Small and plump, it measures up to 2 in. (5 cm) in length, and is usually grey, olive or brown in colour, sometimes marked with darker or greenish patches. In a good light the pupil of its prominent eye appears, like that of a cat, as a vertical slit. Its voice, a high-pitched and musical bell-like 'poo, poo, poo', is normally uttered at night and frequently away from water. The favourite habitats of the midwife toad are gardens with drystone walls, quarries and rockslips, where it feeds mainly on small invertebrates. On the continent it is found at altitudes of up to 6000 ft. (1829 m). The midwife toad is a largely nocturnal animal, spending the day hidden from sight, sometimes in burrows which it digs for itself.

In 1919 in the course of an address to the South London Entomological and Natural History Society entitled 'British Batrachians', Mr G. A. Boulenger said:

The Midwife Toad has established itself, no one knows how, in a former nursery garden in Bedford; it has been there for many years, and a friend of mine found it still in plenty last summer. Its presence is revealed by its whistling note, which suggests the sound of a small bell, or a chime when uttered, as is usually the case, by a number of individuals, and is probably chiefly in the evening and at night. This so-called toad, a member of the very distinct family *Discoglossidae* furnishes an interesting example of parental solicitude, the male taking charge of the eggs, which are large and few [20–100] and strung together like a rosary, immediately after oviposition on land, not in the water as in most other Batrachians. After extraordinary contortions . . . the male fastens the string of eggs round its

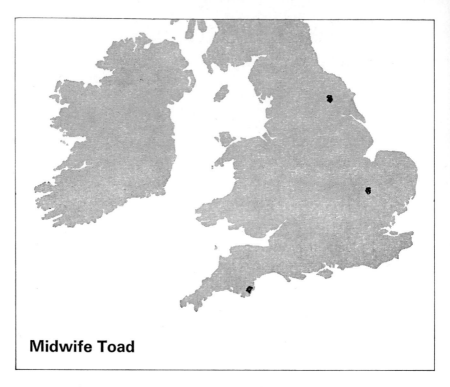

Midwife Toad

hind limbs and carries them for a period of about six weeks, when he betakes himself to the water for the purpose of releasing his progeny, which escapes from the egg capsules in the tadpole condition.

Some doubt exists as to whether it was in 1878 or 1898 that the toads forming this colony arrived in Bedford. As they first appeared and were heard calling in about 1903, the later date seems the more likely. The nursery garden belonged to the firm of Horton & Smart, by whom the toads are believed to have been introduced accidentally in a consignment of ferns and water-plants from southern France. In 1922 part of the garden was acquired by the Bedfordshire County Council, and the toads' breeding pond was filled in. About a dozen were removed by Mr W. S. Brocklehurst to his private, one acre (·4-hectare) walled garden about $\frac{1}{4}$ of a mile (400 m) away. In 1950 another colony was discovered in a neighbouring garden, and a small population was found to have survived on the original nursery garden site; all three colonies are believed to be still in existence, although the toads appear to have made no attempts to expand their range.

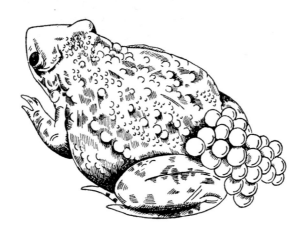

In 1947 Mr Robert Brocklehurst brought five toads and a dozen tadpoles from his father's garden in Bedford to his own home, Woodsetts Grange, near Worksop in Nottinghamshire. In 1933 another small colony was established in a private garden near York, which did well for many years but has recently shown some signs of declining. In about 1954 a colony of midwife toads was started by Viscount Chaplin at Blackawton near Totnes in south Devon with two egg-carrying males from the London Zoo; this population was apparently thriving until at least 1964, and a few individuals may still survive.

41. African Clawed Toad

(Xenopus laevis)

The African clawed toad, which is sometimes also known as the platanna or smooth spur-toed toad, is widely distributed throughout tropical Africa, where it ranges from Ethiopia south to the Cape of Good Hope.

This species has a smooth skin marked around the body with fairly distinctly defined lines of stitch-like swellings. In colour it is darkish brown above with a whitish underside; some individuals are uniformly coloured, while others are marked with brown spots on the belly. The head and mouth, the upper jaw of which contains a number of small teeth, are blunt, and the eyes have a circular pupil. The three inner toes of the webbed hind feet each have a sharp claw- or spur-like nail, from which the creature derives its name. The African clawed toad, an almost exclusively aquatic species, feeds on small water animals which it pursues and captures beneath the surface in its unwebbed front feet. In Africa the female clawed toad lays large eggs one by one in August and September. Three days after hatching the tadpoles develop a pair of barbels which hang down from each corner of the mouth.

In 1955 some metamorphosed tadpoles of the clawed toad were liberated in Kent, but they were almost certainly killed after a severe thunderstorm. It is, in any case, doubtful if they would have survived a severe winter in this easterly county.

In 1967 Mr Frank Boyce turned out a number of African clawed toads in some small ponds above the cliffs at Brook, about 3½ miles (5-6 km) from Freshwater on the south-west coast of the Isle of Wight.

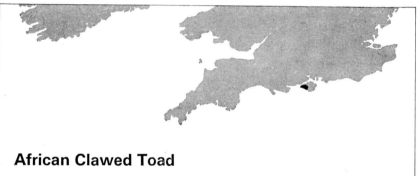

African Clawed Toad

Tadpoles were found in the ponds in 1970, and in 1974 a number were discovered in another pond nearby. In 1976 the population of this apparently thriving colony of African clawed toads was estimated at between forty and fifty individuals. The success of this colony is presumably due to the comparatively mild climate of the Isle of Wight.

42. European or Green Tree Frog

(Hyla arborea)

The European or green tree frog* is distributed over most of Europe apart from the north, the Balearic Islands and, probably, parts of southern France and the Iberian Peninsula. It also occurs in Asia Minor and southern Russia as far east as the Caspian sea.

The tree frog measures about 2 in. (5 cm) in length, and its smooth skin is usually bright green or yellow to dark brown in colour. (Tree frogs share with chameleons the ability to change their colour quite extensively). In continental European populations a dark line, often edged with cream, extends back from the eye past the ear-vent and along the sides as far as the groin, in front of which it almost invariably branches upwards and forwards. The limbs are also marked with prominent longitudinal stripes. In the breeding season, and possibly at other times, the males have a brownish or yellowish vocal sac beneath the chin, which becomes enlarged and spherical when they call; the song, uttered at night during the breeding season, is a harsh, swift 'krak, krak, krak', starting quickly and gradually decelerating, and sounding at a distance not unlike the quacking of a duck. The tree frog is the only northern and central European frog which has disc-like adhesive pads on its toes; these enable it to clamber about – usually at night – in trees, bushes, reeds and other herbaceous vegetation, where it feeds mainly on insects and spiders. In May the female lays about one thousand small eggs, normally in walnut-sized clumps, and metamorphosis is complete after about three months.

The earliest recorded introduction of the tree frog to Britain took place in the 1840s, when a number were turned out on the Undercliff

*Although thus named, the tree frog is in fact allied to the toads *Bufonidae* rather than to the frogs *Ranidae*.

at St Lawrence on the south coast of the Isle of Wight. In 1906 Lord Walsingham liberated some more at the same site, where they bred in a small pond for several years; however, as no frogs have been reported from this area since 1963 this colony may well be extinct. At about the same time a number of tree frogs escaped from a greenhouse at Freshwater Bay on the opposite side of the island, where they bred in a pond until it was filled in during the First War. On Lundy Island in the Bristol Channel a number of tree frogs escaped in 1933 when their container was over-turned by a bullock; in 1939 at least one remained alive and at liberty. Tree frogs have also been introduced with varying degrees of success to the Scilly Isles; at Paignton in south Devon, where fifty turned out in 1937 soon disappeared (a more recent attempt to found a colony there also failed); and in the garden of a Cambridge University college. Mr. T. B. Rothwell established a non-breeding colony at his aquatic nursery at Newdigate in Surrey. Some tree frogs which had been brought back to England from Cannes in the Alpes Maritimes were turned out many years ago in a walled garden in Suffolk, where they failed to breed. These were presumably the more southerly stripeless tree frog (*H. meridionalis*) which has no dark line on the flank and a slower call than *arborea*. The introduction of this species, which ranges from southern France through Spain and Portugal to North Africa, may explain why some other attempts to naturalize tree frogs in England have failed. Another reason may be that frequently few, if any, females are turned out. Tree frogs are usually collected at their breeding ponds, where the males remain throughout most of the breed-

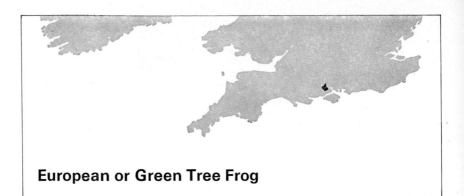

European or Green Tree Frog

ing season; the females, however, enter the ponds only to deposit their eggs, and then depart. Thus for most of the year the population of a pond is predominantly male.

In 1952 Mr Oliver H. Frazer liberated about fifty European tree frogs near his home, Mottistone Mill, at Brighstone on the Isle of Wight; two clumps of spawn were found in the following year but the frogs subsequently disappeared. Another attempted introduction in the Freshwater Bay district also apparently failed. The origin of a single frog caught on Hayes Common in Kent in May 1952 is unknown. In 1954 Lord Chaplin liberated some tree frogs in his garden near Totnes in south Devon, where it is possible that some remain today. At Boxley, north-east of Maidstone, in Kent, two dozen tree frogs were turned out in 1955; in May of the following year a single male was heard calling on two occasions.

In 1962 a colony of European tree frogs was discovered in a pond on the edge of the Beaulieu Abbey Estate in the New Forest in Hampshire, where local enquiries revealed that it had apparently been established by a Mr Jones about fifty or sixty years previously.* (Another colony has also been reported a few miles away, but there may have been some confusion here over locality names). Dr J. F. D. Frazer heard five or six males calling when he visited the pond one evening in May 1963. In the summer of 1973 Mr David Bird counted fourteen tadpoles in the pond, and on a sultry evening in June 1975 he heard two males calling. In recent years no more than three individuals have been seen at any one time, and Mr Bird estimates that

*Jones had reputedly been living for some years in North Africa: as this colony is undoubtedly *arborea* he presumably collected the original stock during his journey home through Europe. It is said that the introduction was made to a pond some distance away from the present site, to which the frogs in course of time migrated.

the entire colony may number only about a dozen or so: this, however, appears to be enough to maintain a viable population. The pond, which measures about 80 feet (24 m) in diameter and is surrounded on three sides by low trees and shrubs, is in an exposed position on the top of a hill; the water, which dries out in late summer, measures between 2 and 4 ft (61 and 122 cm) in depth, and in summer is appreciably warm to the touch; this may be an important factor in the success of this colony, which contains the only certainly extant population of wild European tree frogs in Britain.

43. Edible Frog
(Rana esculenta)

The edible frog is a native of Europe, from France eastwards to western Russia, north to southern Sweden and south as far as Italy, Corsica, Sicily and the northern Balkans.

The colour of the edible frog is subject to considerable variation, but is usually some shade of green (or occasionally brown), normally covered liberally with black markings and sometimes with a pale yellow line running down the spine. The back of the thighs is usually marbled with yellow and black. The male has a whitish, pea-sized sac at each corner of his mouth which is inflated when – between May and August and usually at night – he utters his rasping call, choruses of which can be heard at a distance of up to 1 mile (1–2 km). The edible frog is slightly larger than the common frog (*R. temporaria*), measuring about 3 in. (7–8 cm) in length, and has a narrower head and a more pointed nose.

Edible frogs emerge from hibernation in mid-April or May, and between 5000 and 6000 yellowish eggs are laid by the female in a number of separate clumps during May, June and July: the mating embrace, during which the male fertilizes the eggs as they are laid, may last for from several hours to a day or more. Metamorphosis is normally complete after from twelve to sixteen weeks. Somewhat surprisingly, the edible frog loves to feel the heat of the sun's rays on its back, and it may frequently be seen sunbathing during the summer months. The food of edible frogs consists mainly of insects, small aquatic invertebrates, fish-fry, tadpoles, and small frogs, which are frequently seized from a point of vantage on a water-lily leaf or some other piece of floating vegetation.

The hind-legs of the edible frog are much esteemed in France as a culinary delicacy, being said to resemble chicken in flavour. The

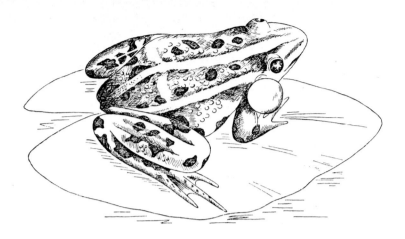

domestically reared *grenouilles de parc* are considered by most gourmets to have a superior flavour to wild-caught *grenouilles de pêche*. It is not, however, from this gastronomic habit that the French have acquired the nickname of 'Froggie': as early as the sixteenth century Michel de Nostredame (Nostradamus; 1503–66) was referring to the French in his *Centuries*, published in Lyons in 1555, as *crapauds* (toads). John Guillim (1565–1621) in his *Display of Heraldrie* (1610) described the arms of the city of Paris (at one time a quagmire named Lutetia ('mud-land') as consisting of 'three frogs [or toads] erect saltant'; he bestowed upon the Frenchman the derogatory nickname of 'Johnny Crapaud'. '*Qu'en disent les grenouilles?*' was a question current at Versailles at the time of the French Revolution, when members of the court and aristocracy were speculating on the intrigues and machinations of the *citoyens*.

It is conceivable that the edible frog inhabited southern Britain during the period of the climatic optimum around 4000 B.C., but no fossil evidence has so far been discovered to support this theory. The species may well have been imported by the Romans with the edible dormouse, and kept in domestication as a source of food, but again there is no evidence in archaeological remains to suggest that edible frogs ever became naturalized in the British countryside.

In the second edition (1849) of his *History of British Reptiles*, the

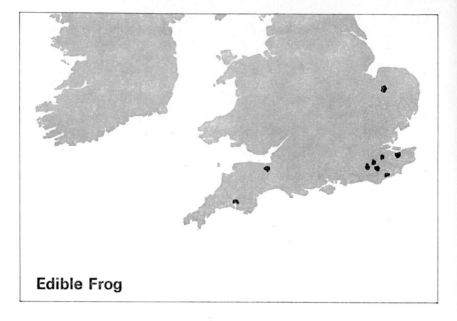

Edible Frog

zoologist and dental surgeon Thomas Bell (1792–1880) quotes F. Bond, who six years previously had written in the *Zoologist*:*

> I have the pleasure of recording in the *Zoologist* the capture of the Edible Frog (*Rana esculenta* Linnaeus) for the first time, I believe, in this country. Two specimens were taken by my friend C[harles] Thurnall, Esq. of Duxford, in Foulmire [Foulmere] Fen, Cambridgeshire, in September last.

Bond expressed surprise

> that they were never seen before, their croaking being so very different from that of the common frog; the sound is more of a loud snore, exactly like that of the barn owl *Strix flammea*. The whole fen was quite in a charm with their song . . . this Frog is a very timid animal, disappearing on the least alarm, and it is not very easy to catch. It seems to be entirely a water reptile, never coming on the land; at least I never could find one out of the water, like the common species.

Bell himself continues:

> The very remarkable and sonorous croak belonging to this species had procured for the Frogs of this neighbourhood the names 'Cambridgeshire

*2, pp. 393, 677, 1843.

Nightingales' and of 'Whaddon Organs'; and I have often heard my father, who was a native of those parts, say that the croak of the frogs there was so different from that of others, that he thought they must be of a different kind.

'Dutch Nightingales' and 'Boston Waites' were additional local names for the frogs recorded by Pennant in 1776; this, taken in conjunction with the reference to his father by Bell (who makes it clear that he is referring to a period some eighty years previously), presupposes, as Fitter points out, that edible frogs were established in Foulmire at least by the 1770s. Thomas Bell senr was proved correct in his belief when the two frogs captured by Charles Thurnall were subsequently identified at the British Museum. In 1844 Bond and William Yarrell each presented two edible frogs (presumably from Foulmire) to the museum. Two years later Mr Henry Doubleday liberated some edible frogs from Foulmire in a pond in Epping Forest in Essex; they soon moved to another pond nearby, where they apparently remained for only a short time.

The earliest recorded introduction of the edible frog into the British Isles was made by Mr George Berney of Morton Hall in Norfolk:

I went to Paris in 1837 and brought home two hundred edible frogs and a great quantity of spawn. These were deposited in the ditches in the meadows at Morton [about 10 miles (16 km) north-west of Norwich], in some ponds at Hockering [about 3 miles (5 km) south-west of Morton], and some were placed in the fens at Foulden near Stoke Ferry. They did not like the meadows and left them for the ponds. I found them in a pond at the top of Honingham Heights. I have measured the distance on a map and find it to be in a straight line one and three-quarter miles and forty yards [about 3 km]. In the whole distance there is not one drop of water. In 1841 I imported another lot from Brussels. In 1842 I brought over from St. Omer 1,300 in large hampers made with plenty of tiers and covered with water-lily leaves that the frogs might be comfortable. These were dispersed about in the above mentioned places and many hundreds were put into the fens at Foulden and in the neighbourhood.

A few years later Mr Berney compared the state of his colony with that of the diminishing one at Foulmire Fen: 'I believe that my thriving colony is on its way to such semi-extinction.' Nevertheless, in 1884, two edible frogs from Foulden were presented to the British

Museum, and in the previous year Hans Gadow heard the mating call of male edible frogs at Hickling Broad, some 18 miles (29 km) east of Morton; others were reported from Wroxham Broad, 6 miles (10 km) west of Hickling until as late as the outbreak of the First World War. Edible frogs appear, therefore, to have survived at Foulden for over forty years, and the descendants of those liberated at Morton for perhaps more than seventy years.

In 1853 Alfred Newton, F.R.S. (1829–1907) and his brother discovered a colony of edible frogs – a number of which he presented to Norwich Museum – in a pond between Thetford and Scoulton. Newton reported that the frogs were 'exceedingly noisy, puffing out their faucial sacs, like so many dwellers in the cave of Aeolus'.* Twenty-three years later Newton and Lord Walsingham (1843–1919) found another colony at Stow Bedon in the same district, where they were told that the frogs had been present for a dozen years or more. Lord Walsingham, who presented a specimen to the Norwich Museum, reported, after making local enquiries, that 'the species is pretty generally diffused in a south-westerly direction from the place where we found it'. In 1884 Lord Walsingham presented seven edible frogs from Stow Bedon to the British Museum. In 1898 Mr G. A. Boulenger claimed that edible frogs had apparently been at Stow Bedon, where they were still common, for seventy years or more. Six years later Gadow stated that they were still sometimes to be seen at Foulden and were common in the Thetford/Scoulton district where, from enquiries made by Lord Walsingham, they were said to have existed since at least 1820. W. G. Clarke, writing in 1925, claimed that edible frogs were still to be found at Stow Bedon Mere – almost one hundred years after their first introduction to Hockering.

In about 1840 Colonel Durant introduced a number of edible frogs to ponds at Tong Castle near Shifnal in Shropshire; these had all apparently disappeared by the turn of the century.

Around 1884 Dr St George Mivart imported a quantity of edible (and marsh) frogs from Brussels, which he set free at Chilworth in Surrey. The Duke of Bedford's son, Lord Arthur Russell, brought some edible (or possibly marsh) frogs from Berlin at about the same time, which he turned out at Shere between Guildford and Dorking;

---

*According to Homer, Aeolus was the son of Hippotes and god and father of the winds (*Odyssey*, X: 1). Virgil wrote that he kept the winds imprisoned in a cave on one of the Aeolian islands off the north coast of Sicily (*Aeneid*: I: 52).

some of these later appeared on a marsh at the neighbouring village of Gomshall. Ten years later fifty more frogs from Berlin were set free at Shere where a small number survived for a further decade. A large colony of edible frogs was subsequently discovered in a pond at Ockham, 6 miles (10 km) north of Shere. Between 1905 and 1910 edible frogs were introduced by the Hon. Charles Ellis to ponds at Frensham Hall near Shottermill, where they bred successfully until the winter of 1939–40.

In about 1895 a Mr Cotton released a number of edible frogs into the river Itchen near Bishopstoke in Hampshire, where they had apparently disappeared by 1914. A colony discovered at Bolney in east Sussex in 1943 (which may have come from a collection formed at the Beam Brook Aquatic Nursery at Newdigate in Surrey 12 miles (19 km) to the north-west between 1925 and 1930) had vanished by 1948. In April 1950 Mr W. G. Ruffle discovered between sixty and seventy edible frogs in a pond at Buxted, also in east Sussex, which had reputedly been in the neighbourhood for at least seven years, having previously lived in a larger pond some 300 yards (270 m) away: this colony may also have come from Newdigate, 20 miles (32 km) to the north-west.

Mr G. E. Mason liberated a quantity of edible frogs in some ponds at Brandon on the west Suffolk/Norfolk border in about 1884; these were, however, never seen again. A number of frogs imported from Normandy in 1882 and 1892 were turned out at Blaxhall in east Suffolk where they survived for only a few years. Mr E. J. Rope introduced some edible frogs at Snape, about $2\frac{1}{2}$ miles (4 km) south of Saxmundham in east Suffolk in about 1930, where they bred for a number of years.

During the first half of the present century large numbers of edible frogs, mostly from Berlin, were introduced to ponds at Woburn where, although they bred, their numbers constantly had to be reinforced by new importations. Some edible frogs from Italy were turned out in Oxfordshire prior to 1897, where they apparently did not survive for long. In 1907 a few were placed in a pond near Queensferry in Midlothian, where they remained for about forty years.

The majority of the more recent introductions of edible frogs to England have taken place in and around the London area. From at least 1929 to 1961 they were to be found in gravel-pits at Ham in Surrey, whence they spread to Twickenham (1938), Teddington in

Middlesex (where they remained until 1956), Ham Common, Sudbrook Park, and Richmond Park (1931), where in May 1945 Fitter heard one croaking in Leg of Mutton Pond: none has been seen or heard there recently. A colony of twenty or more on Esher Common in 1958 and 1959 were not recorded in 1960 or 1961 after their pond had been drained for cleaning during the winter of 1959–60. Individuals were, however, seen or heard in 1962, 1963 and 1964, so this colony may still survive.

From 1942 until 1960 edible frogs were established in some disused brick ponds at Dunton Green and Otford, north of Sevenoaks in Kent; the source of these colonies is uncertain, but they too may have originated in the collection at Newdigate, some 30 miles (48 km) to the south-west. In 1950 three or four frogs seen in Dartford Creek may have been *esculenta*.

In north London, edible frogs were present in the various ponds on Hampstead Heath between about 1939 and at least 1965. These were derived from escapees from a private garden in the former year, and from an introduction of five frogs from Ham to ponds in Golders Hill Park in the summer of 1948. A dozen edible frogs from Ham (and sixteen marsh frogs from Romney Marsh) were introduced to a pond at Thorne in southern Yorkshire in the following year. Mr John Hillaby discovered colonies totalling more than a hundred frogs in the two Highgate Ponds and Viaduct Pond in May 1948, where by 1965 there were well over two hundred individuals. Alfred Leutscher told Fitter that a colony of edible frogs which he found in Whipps Cross Pond in Epping Forest, Essex, in 1948, had probably been there for at least four years; these frogs had apparently died out by about 1954. Between 1938 and 1958 there were edible frogs at Walthamstow, and others have been established at Snaresbrook and Leytonstone. In the late 1950s, Mrs M. G. be Udy of Bratton, between Warminster and Devizes in Wiltshire, told Fitter that the descendants of some edible (or marsh) frogs which she brought back from Cannes in southern France in 1938 or 1939 were still breeding annually in a pool in her garden; this population died out in the early 1960s. In 1959 and 1961 Mr Frank Boyce introduced three dozen edible frogs to his home at Freshwater on the Isle of Wight: here they bred for about four years, but by the winter of 1975–6 only one remained; Mr Boyce suspects that some may have migrated to nearby Freshwater Marsh. In 1966 nine edible frogs were discovered in a garden near Birmingham; their origin is unknown.

In his excellent book *The British Amphibians and Reptiles* (1951), Malcolm Smith wrote: 'the Edible Frog cannot be found in any part of Norfolk [or Cambridgeshire]'. In 1975 Alfred Leutscher was of the opinion that

to-day the edible frog is practically extinct in England. Apart from an occasional escape from a private collection here and there, it now seems to have vanished . . . whatever the cause, natural or man-made, it seems that this interesting addition to our fauna may now have to be struck off.

Fortunately for herpetologists, this gloomy prognostication has not yet come about. Edible frogs are at the time of writing (1976) established at a minimum of eight sites in three English counties. In Norfolk there are at least three small surviving colonies (one of which was rediscovered in 1960) in the general area of the introductions of the late 1830s and 1840s, although whether the present colonies, numbering dozens rather than hundreds of individuals, are the direct descendants of those original introductions is uncertain. Already pressure has been placed on these colonies by both local and (surprisingly, as the species is so common on the continent) foreign collectors, who have removed large numbers of plants and frogs, and also by herons who eat quantities of tadpoles. Some years ago half a dozen edible frogs were introduced into a garden in Surrey, which they promptly deserted in favour of a pond on a nearby common. What is probably the largest naturalized colony of edible frogs in Britain today – estimated to number well over a thousand individuals – is established in the seventy or so ponds on the 22 acres (9 hectares) of the Beam Brook Aquatic Nursery at Newdigate, previously owned by Mr T. B. Rothwell but at present under the control of Mr Gilbert Taylor. In Sussex there are four small colonies in at least two of which (one at Partridge Green between Worthing and Horsham and another which for the past ten years has occupied a series of fishing ponds at Washington 7 miles (11 km) north of Worthing), breeding occurs. In 1967 some edible frogs were introduced into a garden pond at Sittingbourne in Kent, where they have bred, although not annually. There are unconfirmed reports of colonies in Devon and Somerset, and others may exist elsewhere. There have, in addition, been many undocumented introductions of edible frogs over the years in various parts of southern England, some of which may still survive.

The reasons for the failure of the edible frog to become firmly established in southern and eastern England in spite of repeated introductions, compared with the success of its close relation the marsh frog, remain something of a mystery. Probably a combination of various factors are involved. The unpredictable British temperate maritime climate may be more unsuitable than the extremes commonly experienced on the continent. It seems unlikely that the British winters are too cold; a continuous hard winter often helps the survival of a hibernating amphibian, which might otherwise awaken during a thaw and become trapped and suffocate under the ice when freezing conditions return: a series of cool summers, on the other hand, might well affect an amphibian such as the edible frog which spawns as late as July, and there is believed to be a high mortality rate among edible frog tadpoles. It may be significant that the decline of the edible frog in East Anglia coincided with changes in agricultural management, and with the increasing use by man of inland waterways which the reduction of the use of Norfolk reeds for thatching purposes helped to silt-up. Unsuitable habitats thus formed and inadequate supplies of food may well have been largely responsible for the comparative failure of the edible frog to become an established member of the British herpetofauna.

44. Marsh Frog
(Rana ridibunda)

The marsh frog is a native of two separate areas of continental Europe; one of these comprises Spain, Portugal and southern France; the other stretches from Denmark and Germany eastwards into Russia and as far south as the southern Balkans.

In Britain the marsh frog is basically olive brown in colour, the head and back sometimes being suffused with light green. The back is covered with warty protuberances and usually – though not always – with black spots which may vary considerably in shape, size and number. The legs are marked with dark-green bars, the back of the thighs being marbled with olive or whitish-grey. The underside is either uniformly white or is speckled and spotted with black or dark green. The iris of the eye is golden yellow with black vermiculations. This species is the largest native European frog, sometimes reaching 5 in. (13 cm) or more in length.

Marsh frogs normally emerge from hibernation in early April and mate in May and June; between 5000 and 10000 eggs are laid by the female and metamorphosis is completed by August. At each corner of his mouth the male possesses a greyish, pea-sized vocal sac which he inflates when in the water to utter his loud call, which has earned for the animal the alternative name of 'laughing' (*ridibunda*) frog. The song of the male was described by Aristophanes (450–385 B.C.) as 'Brek-ek-ek-ek and 'Kroax-kroax'; it also produces a repeated and plaintive 'kee-oink, kee-oink'. The food of the marsh frog consists mainly of small crustaceans and invertebrates, adult newts and frogs, small fish, and even occasionally mice and fledgeling birds. Like the edible frog, the marsh frog much enjoys basking in the sun, usually on the banks of streams and dykes, or on a floating lily pad, from which it takes to the water with a noticeable 'plop'.

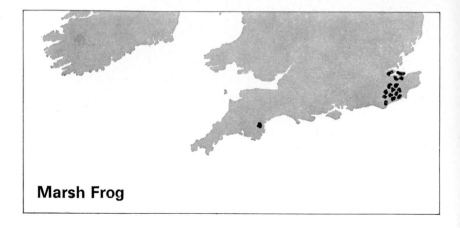

Marsh Frog

Fitter was able to trace only one pre-twentieth century attempt to introduce marsh frogs to Britain, when in about 1884 Dr St George Mivart turned some out at Chilworth in Surrey.

The most important introduction was that made by Mr E. P. Smith (Edward Percy, the playwright) who described it as follows:

In the winter of 1934–35 I introduced twelve specimens of the European edible frog *Rana esculenta** (the Hungarian variety) into a pond beside a running stream in my garden at Stone-in-Oxney, East Kent. The site is an old sea-creek which, in Tudor times, formed a small landing stage; but now, since the sea has receded five miles [8 km] or more, abuts upon a tract of land where the flats comprising the Walland, Romney and Denge Marshes usurp the place of the salt water. It is a maze of dykes, canals, meres and streams intersecting rich pastures.

In June 1935 two of the frogs migrated overland to a mere about ½ mile (800 m) distant, where they were joined by the remainder in the following October. This mere had dried out two years previously, and was thus free of large aquatic predators. Here the frogs became so excessively noisy that the local villagers were roused into complaining to their local Member of Parliament and to the Minister of Health, as a result of which a Sunday tabloid appeared with the startling headline 'Twelve Frogs Keep Sleepy Village Awake'.

*The marsh frog, here described, used to be regarded by some herpetologists as a subspecies of the edible frog (*R. esculenta*) rather than a full species. These particular specimens had been sent from Debrecen in eastern Hungary for purposes of scientific research at University College, London, a fate from which they were rescued by Mr Smith.

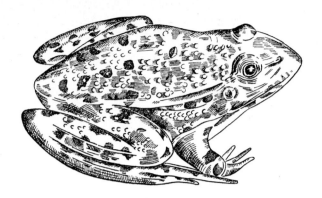

By the autumn of 1936 the mere had become colonized by the frogs, which later in the year appeared 3 miles (5 km) away at Appledore which is connected to the mere by a circuitous series of dykes and sewers. 'In May 1937', wrote Smith, 'began what I might call the Great Year of Rana esculenta.' Large quantities of spawn appeared, followed in due course by numerous tadpoles and small frogs. A few managed to reach Great Lake at Bedgebury, 14 miles (22 km) from the mere and unconnected to it by water, presumably having been carried there by human agency. At this time marsh frogs also first appeared at Lydd, the same distance from the mere but in the opposite direction, and with a direct if serpentine water connection. Following further complaints about the frogs' vocal activity, a number were captured and despatched to Scotland, Essex, Hertfordshire and Sussex. In 1939 seventy frogs from Hungary were turned out by a son of Professor (later Sir) Archibald Vivian Hill in Reach Lode which runs into the Cam at Upware near Wicken Fen in Cambridgeshire, but were never seen again.

By 1938 marsh frogs had colonized an area of Romney Marsh covering some 28 square miles. Smith, who had been studying the creatures during the previous three and a half years, came to the following conclusions about them: they migrate principally in June and October; they are very powerful swimmers and jumpers (one specimen, with 11 in. (28 cm) long hind legs, leaped a distance of 4 ft 9 in. (145 cm); they neither breed nor sing until they have reached their second year, and are fully grown after four or five years; they are shy but curious; their voice has a ventriloquial quality, and although a heron flying overhead will cause them to become silent, another noise will appear to stimulate them to vocal competition.

By the outbreak of war marsh frogs had occupied part of the Royal Military Canal, and were beginning to expand their range into many other districts of Kent and east Sussex. In 1943 they crossed the south bank of the river Rother at Thornsdale, Iden, and by 1946 had spread as far upstream as Maytham and downstream to near the coast at Rye.

In 1958 marsh frogs first appeared at Baynham on White Kemp Sewer, and by 1956 had travelled 4½ miles (7 km) along the New Sewer from Appledore as far as Brenzett. Brackish water was clearly no barrier to their advance; the first specimens were observed in the harbour at Rye in 1951, and by the following year they had penetrated to the Pett Level and Winchelsea Marsh.

Until the 1950s the main spread of the marsh frog was by way of the waterways which intersect the area of its introduction. Thereafter it extended its range even more rapidly until it had colonized virtually the whole of Romney, Walland and Denge Marshes and much of the Pett Level. It soon extended from west of Hythe to south-west of the Pett Level near Hastings, and as far up the Rother and Brede valleys as Newenden and Brede Bridges. The lower valley of the Tillingham and the land between the Royal Military Canal and the coast (with the exception of portions of Denge beach) were also colonized.

By 1975 marsh frogs were widely but patchily distributed over an area of more than 100 square miles of Romney Marsh and the Rother Levels, extending as far west as the eastern end of Pevensey Levels near Bexhill. They are also found in the southern half of the Isle of Sheppey and on the Iwade marshes in north Kent: new areas are still being colonized, although more slowly than before, and there is some evidence of a decline in numbers in recent years.

In 1948 Alfred Leutscher turned out some marsh frogs in Wanstead Park in Essex, and in the following year sixteen were set free at Thorne between Doncaster and Goole, in Yorkshire; in 1958 others were liberated in Devon and near Minehead in Somerset, and in 1961 a pair were put out in a sewage farm at Beddington (near Croydon) in south London. In about 1954 Mr W. L. Coleridge introduced about a dozen marsh frogs from Romney Marsh to his sunken water garden, containing three small lily-pools and one larger pond, at Bishopsteignton near Newton Abbot in Devon; here they have bred almost every year – especially successfully in hot summers such as those of 1975 and 1976 – although many young are killed and eaten by their parents. In the autumn of 1975 the population included five adult females, two adult males, and a large nomber of juveniles.

On 6 September 1974 a single marsh frog was discovered on a roadside verge at St Breward on the edge of Bodmin Moor in Cornwall, where it was claimed that the species had been seen on a number of occasions in previous years, thus suggesting the possibility of a hitherto unsuspected population.

As a colonizer, the marsh frog is undoubtedly the most successful British alien amphibian. A number of factors have helped it to become established, probably the most important of which are the suitability of the habitat into which it was introduced, and the abundance of the available food supply.

Romney Marsh – an area given over almost entirely to sheep-farming – consists mainly of small fields divided by ditches and sewers edged by strips of ungrazed grass, with no need for hedges. This provides an ideal environment and source of food for the marsh frog, which is not only considerably more prolific than the common frog (which lays only between 1000 and 2000 eggs), but may also consume the latter's tadpoles. The aggressiveness of the male marsh frog and the species' migrating instinct help considerably in the colonization of fresh territories.

The future of the marsh frog, however, in its present stronghold in south-east England, is almost entirely dependent on the continuation of sheep-farming. Should this ever be abandoned in favour of arable agriculture, which would result in the infilling of the waterways and their replacement by hedges, the marsh frog population would probably soon become extinct.

2F

45. Wall Lizard

(Podarcis muralis)

The wall lizard* is found in continental Europe as far north as France, southern Belgium and the Netherlands, the valley of the Rhine, Bavaria, Czechoslovakia and Rumania, south to central Spain, southern Italy, and the Balkans. It occurs on a number of islands in the Ligurian Sea and on those off the Atlantic coast of France, Spain and Portugal, including the Channel Island of Jersey where it lives on the cliffs of the north-east coast between Gorey Common and Bonley Bay, seldom more than 200 yards (180 m) inland. The wall lizard also inhabits north-western Asia Minor.

The body of this lizard is flatter and its legs are longer than are those of the common or viviparous lizard (*Lacerta vivipara*) and the sand lizard (*L. agilis*), from which it may also be distinguished by the smoother edge to the collar on the underside of the neck. The colour and pattern of its skin are subject to much individual variation; typical specimens may be greyish or olive brown, sometimes tinged with green, and frequently marked on the sides of the tail with conspicuous black and white bars. The females often have darkish sides and occasionally a dark vertebral stripe running between pale streaks or spots. The males are sometimes similar in appearance, but may also have light spots on the sides and bolder markings on the back. The underside of both sexes is occasionally whitish or pale buff in colour, but the males often have an orange-red belly and a white chin and throat marked with rust-red blotches and dark spots. This bright underside colouring is often absent in the females who may, however,

---

*The classification of the wall lizard has, in the past, caused considerable confusion. It is now generally accepted that there are fourteen subspecies of *Podarcis*, most of which are confined to the Mediterranean region. The two most frequently introduced to Britain are *P. muralis* and the Italian wall lizard, *P. sicula*. Two forms of *P. muralis* are imported, *P. m. muralis* (usually brown) and *P. m. brueggemanni* from north-western Italy (in which the males are normally green-backed with very heavy markings beneath). The usual form of *P. sicula* is *P. s. campestris* from northern Italy.

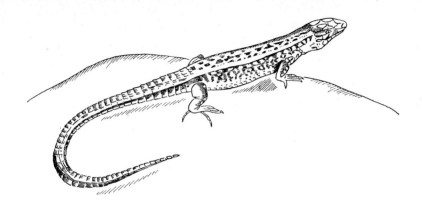

have the throat and chin markings. Both sexes measure up to 3 in. (7–8 cm) in length, excluding the tail; in May and June the females lay clutches of from two to four eggs in a small pit, which hatch in from six to eight weeks.

Wall lizards frequent garden- and field-walls, rock-faces, piles of stones and the trunks of trees, often quite close to human habitation, where they climb about swiftly and actively in search of the small invertebrates (mainly insects, spiders and worms) upon which they feed. The wall lizard is usually found in drier, sunnier and less grassy areas than are frequented by the common lizard, although in the south of its range – where it is often encountered in mountainous districts at heights of more than 6000 ft (1829 m) – it may inhabit damper and shadier places.

The earliest recorded attempt to naturalize wall lizards in Britain took place in May 1932, when a dozen (*P. m. muralis*) were turned out in a walled garden at Farnham Castle in Surrey; these were reinforced a year later by the addition of a further pair. Young were seen in October 1933 and a gravid female in the summer of 1934. In 1951 Dr Malcolm Smith announced the rediscovery of this colony on an old wall in a neighbouring private estate, where it continues to survive. A small colony, which was at one time established at the Beam Brook Aquatic Nursery at Newdigate, no longer exists.

In 1937 two hundred wall lizards (species unknown) were set free at Paignton in south Devon, where they apparently remained in any numbers for only about a decade or so, although until the 1960s sightings of occasional individuals were reported.

Wall Lizard

In 1954 a stock of fifteen wall lizards (possibly *P. m. nigriventris* which is found near Rome and Naples and is black with a few green spots, but more probably *brueggemanni*) was liberated by Lord Chaplin in his walled garden near Totnes in south Devon, where they have bred annually ever since: this thriving population now numbers well over a hundred individuals. The young in this colony normally emerge during the third week of September (it is interesting to note that in the exceptionally hot summers of 1975 and 1976 the young hatched out in early August). A number of *P. s. campestris* put out with them soon disappeared, as did a dozen of the same subspecies liberated in the area in 1961 or 1962.

Two flourishing colonies of wall lizards (species unknown) were discovered in about 1962 on the Isle of Wight: one of these, which was originally introduced to a walled garden near Ventnor, continues to survive, and has spread outside the garden to an area behind the La Falaise car park.

In 1964, a number of wall lizards (species unknown) were turned out on East Burnham Common near High Wycombe, in Buckinghamshire; it is not known how these fared.

It has been suggested that the success or failure of individual introductions of wall lizards is largely dependent on the suitability of the habitat for the species liberated; it seems probable, however, that temperature requirements are at least an equally critical factor.

*

Other Amphibians and Reptiles

Compared to our knowledge of other alien vertebrates living at large in the British Isles, information on exotic amphibians and reptiles is meagre indeed. This is principally because they are such small and inconspicuous animals that their presence is easily overlooked; it may, therefore, be that there exist a number of hitherto unrecorded colonies of the foregoing species. It is also possible that other reptiles and amphibians may be living in so far undetected small and isolated numbers in various parts of Britain.

The EUROPEAN POND TORTOISE or TERRAPIN (*Emys orbicularis*) ranges over most of continentinal Europe except in parts of the north. It also occurs on the islands of Sicily, Corsica and Sardinia in the Mediterranean, and in western Asia and north-west Africa.

The carapace of the pond tortoise, which is basically oval in outline, is blackish or brownish in colour, usually marked with light, often yellowish, streaks and spots. The plastron or ventral part of the shell has a transverse hinge which allows a small degree of movement to the fore section. The pond tortoise favours still or slowly flowing waters plentifully supplied with aquatic plants and overhanging undergrowth, where it can often be seen basking in the sun on a rock or log. In the water it sometimes swims with only its head and neck raised above the surface. It is also found in marshes, swamps, ditches, dykes and even, on occasion, in brackish water.

Dr Malcolm Smith has pointed out that the pond tortoise inhabited

England and continental Europe as far north as Denmark after the Ice Age (fossil remains have been discovered in East Anglia which date from around 3000 B.C.), but was apparently unable to survive for long possibly because a succession of cool, damp summers prevented the eggs from hatching, thus resulting in the loss of an entire generation. England and Denmark are several hundred miles from the nearest present-day breeding areas of the pond tortoise in north-eastern Germany and central France, and even there the eggs may fail to hatch in an inclement season. Given suitably warm summers, however – such as those of 1975 and 1976 – there seems to be no reason why this species should not be able to survive in the wild in southern England.

The earliest documented introduction to Britain of the European pond tortoise appears to have taken place in 1890–1 when some were put down by Lord Arthur Russell at Shere near Guildford in Surrey. In 1894–5 a number were turned out at Blaxhall and Little Glemham south of Saxmundham in Suffolk; a considerable quantity of young tortoises were seen in 1929, and in the spring of 1932 a sub-adult was reported: two years later another was discovered in the neighbouring village of Snape. Lord Walsingham turned some out in Old Park pond at St Lawrence on the Isle of Wight in about 1906; in the following summer one was seen in the millpond at Carisbrooke, 8 miles (13 km) to the north in the centre of the island, and in 1952 another was found on Chale beach, on the other side of St Catherine's Point from St Lawrence. Considerable numbers of European pond tortoises were liberated at Frensham Hall near Haslemere in Surrey between about 1905 and 1910, and others were set free on Lambay Island north-east of Malahide, Co. Dublin, in 1906, and at Woburn Abbey in Bedfordshire where they were still to be found in 1950. Nine pond tortoises were captured in a number of small ponds in north Surrey in 1948, which may possibly have been the descendants of those introduced at Frensham forty or so years before.

The European pond tortoise has for long been one of the most popular imported reptile pets, and over the years there must have been innumerable unrecorded attempts at naturalization in addition to those mentioned above, as well as many accidental escapes from captivity:* it may very well be, therefore, that unknown colonies of

---

*As there have also been of another favourite pet, the common Greek tortoise (*Testudo graeca*) which has, however, seldom if ever successfully bred in the wild in Britain.

this species exist today in parts of southern England, although as *Emys*, like many other chelonians, lives to a great age, a population may exist for decades without breeding.

The TESSELATED* or DICE SNAKE (*Natrix tesselatus*) is found at heights of up to 8000 ft (2438 m) in Italy and south-east Europe north to southern Switzerland, Austria, Czechoslavakia, Hungary, south-western U.S.S.R. and south-western Asia. Isolated populations also exist on the rivers Rhine and Elbe. This snake which is brownish in colour and measures about 3 ft (91 cm) in length, is a largely aquatic species, feeding principally on small fish. The female lays between four and a dozen eggs in June and July.

The tesselated snake is another extremely popular imported pet which frequently escapes from captivity or is deliberatley released, and there is at least one record of suspected breeding in the wild at Holme-upon-Spalding Moor 6 miles (10 km) south-east of York in 1971. Dr Ian F. Spellerberg writes: 'several researchers have said that this species could become established (if this has not already happened) as it requires a similar habitat to our largest snake, the grass snake'. The tesselated snake is, however, much more aquatic in its habits than *N. natrix* and needs a milder climate for its survival and successful breeding than does the native species.

*Regularly chequered.

A small colony of the SOUTHERN or ITALIAN CRESTED NEWT (*Triturus cristatus carnifex*) – a subspecies of the indigenous crested or warty newt – has existed in the wild for a number of years in the grounds of the Beam Brook Aquatic Nursery at Newdigate in Surrey.

In about 1954 Lord Chaplin turned out some YELLOW-BELLIED TOADS (*Bombina variegata*) in his walled garden containing an orchard with a pond in the middle near Totnes in south Devon: ten years later these toads – which are distributed throughout continental Europe except in the extreme north – were breeding and thriving, and

it is possible that some may still survive. In about 1964 Mr W. L. Coleridge brought five yellow-bellied toads back from Switzerland to his
garden at Bishopsteignton near Newton Abbot in Devon, where
they have bred annually ever since and by 1976 had increased to about
thirty individuals.

The FIRE-BELLIED TOAD (*Bombina bombina*) of central and
eastern Europe and Asia has in the past been naturalized at the Beam
Brook Aquatic Nursery at Newdigate. Numerous attempts have been
made since the turn of the century to establish it at Woburn Abbey
in Bedfordshire, but none have been successful. It is possible, however,
that other so far unknown colonies of this toad may exist elsewhere.

*

The GREEN LIZARD (*Lacerta viridis*) ranges from northern
Spain to the Channel Isles. It is longer (12–15 ins; 30–40 cms;) than
the wall lizard and green in colour. 'Green' and 'Guernsey' lizards
mentioned by Gilbert White as early as 1769 were presumably either
the indigenous sand lizard (*L. agilis*) or the introduced *muralis*, with
both of which *viridis* can be confused. The earliest recorded introduction of *viridis* was to the Ynysneuadd Woods, Ebbw Vale,
Monmouth in 1872. In 1899, 230 were set free at St Lawrence, Isle

of Wight, where some survived until 1936. From 1905–10 a number were liberated at Frensham Hall, Haslemere, Surrey. At Portmerion, Merioneth twenty turned down in 1931 lived until 1935, while 100 liberated at Paignton, Devon in 1937 survived until 1946. Eight males and seven females turned loose on the Burren, Co. Clare in 1958 were still there in 1962. Several records from Gloucestershire in the late 1960s revealed no breeding colony. Even in southern England *viridis* is on the northern edge of its range, although it might well breed here successfully in exceptionally hot summers.

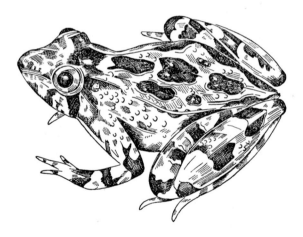

In 1947 L. A. Lantz wrote that he released young PAINTED FROGS (*Discoglossus pictus*) – which are natives of southern Europe and north-west Africa – in the Manchester area, and collected adult specimens in later years. It is possible that some of the descendants of his original stock may still survive, as may those of a small number turned down in the early 1960s in north London.

IV Fish

46. Rainbow Trout

(Salmo gairdneri)

The natural range of the rainbow trout is the coastal drainage area of western North America – especially in the tributaries and associated mountain streams of the Sacramento River – from Mexico in the south as far north as Alaska.

In general shape and appearance the rainbow trout resembles the native brown trout (*S. trutta*): the colour is a light to dark olive-green on the upper sides, speckled on the flanks with innumerable small dark spots and bluish-green blotches. The underside is a silvery gold. Along the lateral line runs a broad purple band from which the fish derives its name. The pectoral and pelvic fins are dark with light borders, and the caudal-fin is spotted with black. Rainbow trout eat a wide variety of food including small insects, snails, caddis larvae, freshwater shrimps, worms, beetles and, on occasion, small fish. In Britain they normally spawn in the spring or autumn, when the female lays between 1000 and 5000 eggs. In North America rainbow trout sometimes reach a length of over 2 ft (61 cm) and a weight of 15 lb (7 kg). They are a hardy fish and are able to tolerate a wider range of water temperature and a greater degree of pollution than the brown trout: they also make use of a greater variety of food and grow faster than the native species.

The scientific name *Salmo gairdneri* covers two forms of rainbow trout – the steelhead which, like the sea-trout, migrates to the sea, and the sedentary rainbow, which lives always in fresh water. Attempts in the past to classify rainbow trout according to their place of origin have resulted in such scientific names as *S. irideus*, which has been applied to fish which spawn between February and May, and *S. i. shasta* from the McCloud river in California, which spawn in November and December. As, however, it is now believed that autumn spawning is the direct result of selective breeding, the names *irideus* and *shasta* no longer have any taxonomic value, and

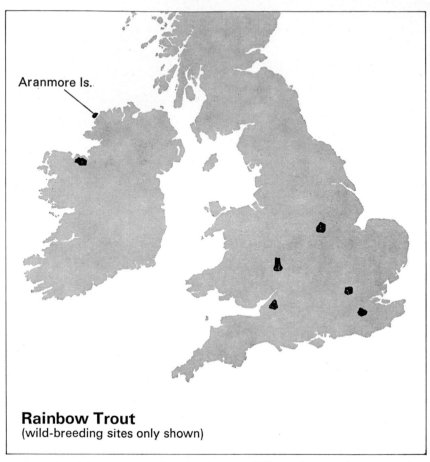

Aranmore Is.

Rainbow Trout
(wild-breeding sites only shown)

are used solely among fish culturists to differentiate spring-spawning from autumn-spawning fish.

Dr E. B. Worthington has traced in some detail the story of the naturalization of the rainbow trout outside its natural range. In 1872 Mr Livingstone Stone founded, on behalf of the United States Government, the Baird Trout Hatchery on the McCloud river which flows from Mt Shasta into Lake Shasta east of the Sacramento river in northern California. Eight years later the first consignment of rainbow trout eggs was despatched from Baird to New York and to Wytheville, Virginia, for the establishment of further Government hatcheries. In 1882 more eggs were sent from Baird – and in subsequent years from Wytheville – to a number of eastern states.

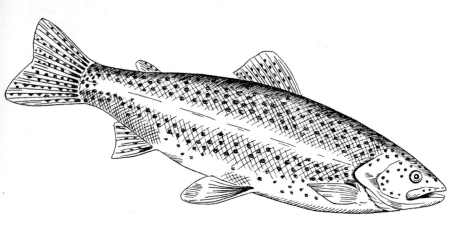

Rainbow trout appear to have been first successfully acclimatized in the eastern United States by Mr H. J. Maynard in the Spring river in south-west Missouri in 1888. Mr G. A. Seagle at Wytheville, Superintendent of the United States Fish Commission, later recorded successful acclimatization of rainbows in rivers in Georgia, Indiana, Michigan Montana, North Carolina, Virginia and Wyoming. In 1898 Mr G. Schnitzer reported that in 1885 fish from 10000 Baird hatchery eggs – and from double that number in the following year – had been liberated in the Laramie river in south-east Wyoming, where by 1895 some were caught weighing 9 lb (4 kg). For every successful attempt to acclimatize rainbow trout in the United States, however, many unsuccessful introductions were made in, for example, the New England states where the native brook trout (*Salvelinus fontinalis*) is found and the brown trout has been successfully introduced.

The first consignment of rainbow trout ova to Europe was made in 1882, when 1200 were despatched, presumably from the Baird Hatchery, to Herr von Behr, President of the *Deutsche Fischerei Verein*, in Germany, where they were all subsequently reported by Herr Jaffé to have died. These were followed four years later by a further shipment of 25000 eggs – from either Baird or Wytheville – to Moisburg in Germany. Fish hatched from these eggs – reinforced with further consignments from the United States – formed the basis of much of the subsequent stock of European rainbow trout.

The first shipment of rainbow trout eggs to England, amounting to about 3000, arrived at the National Fish Culture Association at Delaford Park, near Iver in Buckinghamshire, on 14 February 1884. These were supplemented in the following year by further consignments of 10000 and 5000 ova. Quantities of the resulting fish were despatched to a number of fish-breeders in England, and some were also sent to Sir James Maitland's hatchery at Howietoun in Stirlingshire, where they were reported to be breeding by 1887. These early shipments from the McCloud river were probably all of the *shasta* strain: subsequent consignments from hatcheries in the United States were probably either *irideus* or hybrids.

In almost every winter between 1888 and 1905 rainbow trout eggs from the United States arrived in Britain and were sent to hatcheries at Bridgnorth in Shropshire and Malvern Wells in Worcestershire; to Inishannon and the river Bandon in Co. Cork and Ballymena in Co. Antrim; to Loch Uisg and a number of small lochans near Lochbuie on the Island of Mull, and to Inverness. Professor James Ritchie considers that when in 1898 over a thousand fish (reputedly rainbow trout) were introduced to the river Buchat, a feeder of the Aberdeenshire Don, the fish released were not in fact rainbows but American brook charr. In 1890 rainbow eggs arrived in England from Herr Jaffé's hatchery at Osnabrück in Germany: subsequent shipments from Europe may have been migratory 'steelheads' which have been much introduced on the continent.

No rainbow trout eggs were imported from the United States between 1905 and 1931, but in the winter of 1931–2 Mr D. F. Leney of the Surrey Trout Farm and United Fisheries imported – not directly from Baird but through the United States Bureau of Fisheries and from the Government hatchery at White Sulphur Springs, West Virginia – 50000 ova, followed by the same number in 1938–9. These were all spring spawners, probably *shasta* type.

When, at the request of the National Association of Fishery Boards, Dr Worthington, then Director of the Freshwater Biological Association at Wray Castle on Windermere, conducted his survey of rainbow trout in Britain (the findings of which were published in 1940 and 1941), he gave particulars of waters in which rainbow trout were known to have bred.

Year of Introduction	Source	Locality
About 1900 (stock died during the drought of 1933–5: lakes were re-stocked in 1936).	—	Two lakes near Great Missenden at the head of r. Gade (a tributary of the Colne), Hertfordshire.
1904 (as yearlings).	Mr Neville Grenville, Butleigh Court.	Blagdon Reservoir, Somerset.
About 1905.	Mr C. Maude of Glen House, Aran, probably *via* his brother, Captain Maude, bailiff to the Marquess of Conyngham and manager of Hanlon's Hatchery, Dungloe, Co. Donegal.	Lough Shure (*Loc a cabanaig* – 'Loch of the Shepherd') a peaty moorland lough measuring about $\frac{1}{4}$ mile (402 m) wide by $\frac{1}{2}$ mile (804 m) long at a height of about 600 ft (183 m) on the western side of the Island of Aranmore.
1909–10 (possibly as early as 1890, or not until 1912).	Ashford Lake Ashford Hall, Bakewell (or possibly from Blagdon Reservoir).	Derbyshire Wye, from Ashford, 1 mile (1–2 km) up river of Bakewell, down past Haddon Hall to the confluence with the river Derwent – a stretch of about 6 miles (10 km).
1917–18	—	r. Misbourne, a tributary of the Colne, Buckinghamshire.
Before 1920	—	r. Chess and Bury Lake, Chesham, Hertfordshire.
1920. (Perhaps before the First World War).	—	The upper reaches of the r. Cam above Shelford; the Granta through Barbraham, Linton and Bartlow: r. Rhee through Harston and Barrington.
1930	—	r. Garron (a tributary of the Welsh Wye), Herefordshire.
1931	The Piscatorial Society.	A monastic fish-pond near Newbury, Berkshire, fed by the r. Lambourn.

2G

Year of Introduction	Source	Locality
May, 1934	Mr H. E. Bennett, Ditton Court, Maidstone.	r. Malling Bourne (a tributary of the Medway), Maidstone, Kent, over a length of about 4 miles (6 km).
1935	The Trent Fish Culture Co., Mercaston	The upper reaches of the r. Guash above Empingham, Rutland.
c. 1936	—	Lake at Newstead Abbey, Nottinghamshire.
?	Viscount Hampden	r. Mimran, in the Chiltern Hills.

Dr Worthington also recorded successful breeding of rainbow trout up to the outbreak of the Second World War in five small lakes near Haslemere and in the river Wey in Surrey; in a stream near Chichester in Sussex; in a flooded quarry pit in Kent, and in Norfolk. Mr A. R. Peart told Fitter that he knew of rainbow trout breeding in the river Beane in Hertfordshire, in the Wye in Buckinghamshire, and in the Anton and possibly other streams in southern Hampshire. It is of interest to note that, with the exception of Loch Shure on Aranmore where the water is peaty, nearly all the waters in the British Isles where rainbow trout breed successfully are spring-fed and alkaline.

Dr Worthington also gave some examples of waters in which rainbow trout failed to become self-perpetuating: in the river Coquet where a number were introduced by the Northumbrian Anglers' Federation; in the Welsh Dee where some were turned out in the early 1900s; and in the Lambourn in Berkshire where large numbers were liberated between 1921 and 1927. From the Lambourn rainbow trout (perhaps migratory 'steelheads') dispersed to the Kennet, where they were outnumbered by native brown trout, which were said to be more sporting fish and, with a dark pink flesh, to provide better eating. (It is interesting to note that today in such waters on the Kennet as Avington (Lord Howard de Walden), Craven (Earl of Craven) and Wilderness (Sir Richard Sutton's Settled Estates) – where rainbows are extensively reared – they now outnumber brown trout and provide better sport and superior eating to the latter species, which frequently assumes a distinctly muddy flavour. The same is true on the nearby river Loddon on the Duke of Wellington's estate at Stratfield Saye). In 1938 some rainbow trout escaped into the river Test at Romsey in Hampshire, where they failed to become established. Some fish introduced to the river Tamar near the Devon/ Cornwall border at about the same time apparently also soon disdisappeared. Since the war further attempts have been made to naturalize rainbow trout in a number of Cornish streams.

In all, Dr Worthington recorded rainbow trout as being present in 1940 in between fifty and fifty-five waters in the British Isles, of which only one was in Ireland. In the 1960s a number of rainbow trout, which are believed to have come from a small lochan in the valley of Blane Water – the principal tributary of the river Endrick which is regularly stocked with brown and rainbow trout fry – were caught in Loch Lomond between Stirlingshire and Dunbartonshire.

In 1967 and 1969 the late Dr T. A. Stuart submitted reports to the Scottish Department of Agriculture and Fisheries about the experimental rearing of rainbow trout in the rehabilitated hatchery at the Lake of Menteith in Perthshire. At the end of April 1967, 11 217 yearling fish which had been bred in the hatchery, were turned out in the lake, and had soon spread out over its 650 acres (263 hectares). In the autumn of 1967 a run of (all male) rainbow trout was reported in inflowing burns. On 26 and 27 March 1968, during a heavy spate, a second run took place: fifty ripe male fish were caught in the Malling Burn, and fifteen ripe males and one ripe female in the Portend Burn, where on 22 April a further eleven ripe males and a single ripe female were taken: on the same day seven redds with ova were discovered in the lower reaches of the Malling Burn about 40 yards (36 m) from the lake, and a week later redds were found in the Portend Burn. The first eggs hatched on 30 May, and a number of fry were subsequently seen. In October 1968 rainbow trout were observed making their way (with brown trout) up the Malling and Portend Burns, in the former of which the complete pattern of spawning behaviour by the rainbows – territory guarding, courting, reddcutting and oviposition – was observed. Although it is possible, as Dr Stuart suggested, that some of these rainbows were naturalized fish, this is impossible to prove because the proprietors of the lake –

in response to the increasing demands of anglers – regularly supplement whatever natural spawning there may have been by further introductions of artificially reared fish.

In 1969 Dr Winifred E. Frost was asked by the Salmon and Trout Association to conduct a follow-up survey to the one carried out by Dr Worthington thirty years previously. Much of what follows is drawn from her report which was published in the summer of 1974.

In 1971 (the year in which the survey was made) there were a total of 491 waters in the British Isles which were known to contain rainbow trout, of which 462 are in England, Scotland and Wales, 21 in Northern Ireland and 8 in the Republic. In addition there were about 67 waters which had recently held a stock of rainbows but where the position in 1971 was uncertain. The survey also revealed that in the decade between 1940 and 1950 there were few introductions of rainbow trout into British waters, and none into those of southern Ireland. Some introductions were made in Britain between 1950 and 1960 (in the Irish Republic from 1955 to 1964) after which the number of waters stocked with rainbows steadily increased; since 1965 more and more waters have been, and are currently continuing to be, stocked with rainbow trout – a reflection of the ever-increasing demand for trout fishing in Britain.

Dr Frost's enquiry revealed that of the various waters mentioned by Dr Worthington as containing stocks of breeding rainbows in 1941, only five – Blagdon, the Derbyshire Wye, Lough Shure, the Misbourne in Buckinghamshire, and the Wey in Surrey – certainly contained a breeding stock thirty years later. Only Lough Shure and the Derbyshire Wye held self-perpetuating populations unassisted by artificial stocking in both 1941 and 1971. The 1971 survey further revealed that rainbows were breeding in some forty waters in the British Isles, only five of which – the Derbyshire Wye and a tributary, the Lathkill; a tributary to the Leigh Brook in Worcestershire; Lough Shure, and Lough na Leibe, Ballymote, in Co. Sligo – held self-perpetuating stocks.

A 10 mile (16 km) stretch of the river Wye in Derbyshire today holds a stock of successfully reproducing rainbow trout: this stretch comprises 5½ miles (9 km) of the Haddon Hall Water from the village of Rowsley – at the confluence with the river Derwent – upstream to Ashford Hall Lake, 1 mile (1–2 km) above Bakewell, and the 4½ mile (11 km) reach on the Chatsworth Water (the Duke of Devonshire) from Ashford Hall through the Marble Works at Ashford to Cressbrook Mill and Monsal Dale, where rainbows were apparently 'introduced in the early 1900s and have been there ever since [and] the population is maintained by natural propagation'. A 1½ mile (2–3 km) stretch of the river Lathkill upstream to Alport Mill has also contained since 1959 a stock of self-perpetuating rainbow trout. In a letter to Dr Frost in 1971, Mr Fraser of the Haddon Hall Estate Office wrote that the river Derwent, immediately upstream and downstream of its confluence with the Wye at Rowsley, showed 'increasing signs that rainbow are colonizing and indeed spawning in it in good numbers'. The Monsal Dale Fisheries reported that they believed that 'rainbows are also breeding in part of the Chatsworth Fisheries (r. Derwent) below One Arch Bridge, Beeley, and Smelting Hall Brook, Rowsley'.

The self-perpetuating population of rainbow trout in a tributary of the Leigh Brook at Acton Mill Farm (owned by Mr M. Narbeth) at Suckley, on the Herefordshire/Worcestershire border, was brought to the notice of Dr Frost by Mr J. D. Kelsall, the Severn River Authority Fisheries Officer. Here there is a small, shallow ornamental pool (about 20 ft [6 m] by 100 ft [30 m]) with a bottom of mud and sand, fed by a tributary of the Leigh Brook, which measures only about 4 ft (125 cm) wide and has a gravel bottom: Mr Narbeth owns some 200 yards (180 m) of water containing a number of deep holding pools formed by the erection of several small weirs. In about 1962 both pool and brook were stocked with approximately fifty 6–8 in. (15–20 cm) rainbow trout (? *irideus*) from a hatchery in Derbyshire, and successful spawning has taken place in the early spring (usually in February) of each year since then. There have been no subsequent introductions. Several large rainbows have been caught downstream of the Acton Mill Water, and a certain amount of upstream movement has been observed in the winter when there is a greater depth of water. The movement downstream, which takes place out of the breeding season, is believed to be caused by overcrowding. Some rainbows are stocked downstream of the Acton Mill Water, and it is possible that these may travel upstream to breed with Mr Narbeth's self-perpetuating population.

There has been no other introduction of rainbow trout to Lough Shure since that of 1905 which, because the fish spawn in February, is believed to have consisted of *irideus*. In 1971 Dr Frost was told that the angling for rainbows there was 'still good'.

The self-perpetuating stock of rainbow trout in Lough na Leibe near Ballymote in Co. Sligo was drawn to the attention of Dr Frost by Dr Michael Kennedy of the Inland Fisheries Trust in Dublin, by which the water was originally stocked in 1955. This lough, which has a limestone bottom, covers an area of between 10 and 12 acres (4–5 hectares), and is mainly spring fed. The outflow is a narrow stony stream which after about ½ mile (804 m) disappears in a limestone fissure in a small pool. 'In this stream and in the stony shallows of the lough near its outflow', wrote Dr Kennedy, 'we found redds and subsequently post alevins.' There appear to be enough spawning sites available to support a naturally regenerative stock of rainbows, but in the interests of angling, and because there is a greater supply of food than the self-maintaining fish could consume, every other year or so a stock of takeable-sized fish is introduced. That a nucleus of self-perpetuating rainbows exists is shown by the results of the examination of scales of some catches by Dr Kennedy, 'which showed that some of the fish belonged to the "unstocked" year classes and therefore these had to be native fish'. Dr Frost concludes 'that the rainbows when first introduced into Lough na Leibe established a self-perpetuating population which still exists contemporaneously with the stocked rainbows, which are added infrequently and in small numbers'.

Somewhat similar conditions to those obtaining in Lough na Leibe occur in White Lake, Co. Westmeath, which was also stocked with rainbows by the Inland Fisheries Trust in 1955. These reproduced themselves naturally, but because of the limited number of suitable spawning sites the quantity of fish was inadequate to meet the demands of anglers, and both rainbow and brown trout were extensively introduced. 'It would seem', writes Dr Frost, 'that if there is any spawning going on it may well be that the brown trout dominate the much limited spawning grounds . . . this would suggest that on White Lake, the small natural breeding population, at first established, has not been able to survive the predation of the large stock fish (rainbow and brown) on any small native rainbows and the heavy angling pressures on any native specimens which may have survived to become larger size.'

In May 1975 Mr A. F. Walker of the Freshwater Fisheries Laboratory at Faskally discovered a self-maintaining colony of rainbow trout in a 3 acre (*c.* 1 hectare) hill lochan at an altitude of 1400 feet (425 m) near Loch Loyne in Inverness-shire. The lochan, which is slightly acid, is said to contain only rainbow trout (Mr Walker obtained a single cock fish $10\frac{1}{2}$ in. (26 cm) in length), and according to the present owners was last stocked over seventeen years ago. Although both the inflowing and the top of the outflowing burn (the latter falls steeply into Loch Loyne) appear suitable for spawning, no trout were observed in either.

In the early 1950s a viral disease of young salmonids – subsequently named infectious pancreatic necrosis (IPN) – made its appearance in the eastern United States, from where it later spread west throughout the country. It first arrived in continental Europe in the early 1960s, where it was said to have been introduced to France in a consignment of contaminated rainbow trout ova from the United States. From France the disease soon spread to Denmark and then to other European countries before crossing the Channel to Britain, where it was first reported in Scotland and Ireland in 1971 and in England (Hampshire) in 1972. In the next few years the virus spread to Canada and Japan, and it is now believed to have become distributed worldwide. In both North America and in continental Europe IPN is now regarded as endemic.

Dr Barry Hill of the Ministry of Agriculture's Fish Disease Laboratories believes that the virus has probably been present in fish stocks throughout the world for many years, and that it has only recently been discovered because of the enormous expansion of intensive trout farming. Dr Hill also points out that the chronological order of the discovery of the virus in different countries almost exactly coincides with the establishment in those countries of laboratories with diagnostic capabilities.

In 1975 about eight outbreaks of IPN occurred in the British Isles, the first of which are believed to have been at trout farms at West Acre and Narborough on the river Nar in Norfolk. From these two farms fish suspected of carrying the disease were in the same year introduced to sixty-two waters (including four other fish farms) in the area controlled by the Anglian Water Authority.

The IPN virus, which is easily transferable by the movement of either infected fish (as carriers or infected fry) or contaminated eyed ova, kills the young feeding fry of both rainbow and brown trout, whose resistance increases with age until, at between three and five months, they are believed to become immune. The disease appears in several strains with different degrees of virulence; some cause few if any deaths, while others have a mortality potential of 90 per cent. So far the strains in Britain seem to be less powerful than those in continental Europe, but are nevertheless capable of causing death-rates of over 50 per cent. IPN is now believed to be endemic in parts of Ireland and well established in some Scottish waters; it may well be too late to prevent it from becoming endemic throughout the British Isles. Its future depends on several factors, among which are whether more virulent strains make their appearance; what effect, if any, it may have on the fry of coarse fish in the wild; and – not least – the conscientiousness with which fish farmers notify the Ministry of Agriculture of outbreaks in their hatcheries, as they are by law required to do under the Diseases of Fish Act (1937).

The reasons for the comparative failure of the rainbow trout to establish more viable breeding populations in the British Isles – in spite of the fact that the fish reproduce to some extent in about forty different waters – remain obscure. The species is, indeed, a classic example of an animal which easily becomes acclimatized* but is difficult to naturalize. The temperature levels and chemical character-istics of the majority of waters holding rainbow trout appear to be suitable for successful reproduction and the establishment of self-perpetuating populations. One reason for the failure may be inter-specific competition: the fry of the brown trout hatch earlier in the year than do those of the rainbow, and are thus able to dominate and harass the latter in their search for food and territory at a critical stage in their development; many of those rainbows which survive to become yearlings doubtless fall prey to larger introduced rainbows and brown trout: thus, heavy artificial stocking, which in many cases may be accompanied by severe over-fishing, could well be another factor inhibiting the establishment of more self-maintaining populations of rainbow trout in the British Isles.*

*In recent years rainbow trout weighing up to 2–3 lb (*c.* 1 kg) or more have been caught in the tidal Thames as far up-river as Kew and Strand-on-the-Green. These fish may possibly have been migrants from the Kennet, but are more likely to have been locally introduced.

47. American Brook Trout or Brook Charr
(Salvelinus fontinalis)

The American brook trout or brook charr* is, as its name implies, a native of North America, where it is found – usually in clear streams and lakes – from Quebec to South Carolina.

In general appearance the brook charr resembles the brown trout or common charr (*S. alpinus*). The body dorsally is greyish-blue or green flecked with yellow, the sides are silvery-green dotted with purple, red or black small x-shaped marks, and the underside is usually a soft pink tinged with green or gold. The leading edges of the pectoral, pelvic and anal fins are white outlined by a black margin. The head is pointed and ends in large and powerful jaws. In Britain brook charr seldom exceed 15 in. (38 cm) in length or $1\frac{1}{4}$ lb (550 g) in weight, although in North America they grow to a considerably larger size.

Brook charr favour especially the upper reaches of cool, clear, swiftly-flowing spring-fed streams and oxygen-rich lakes. Their food is much the same as that of brown and rainbow trout. They spawn between October and March, when the hen fish lays some 2000 yellow eggs in the coarse gravel bottom of moving and standing waters. Various (invariably sterile) hybrids have been artificially produced in hatcheries, *e.g.* brook charr × salmon par: brook charr × Lochleven trout, which have been termed 'leopard-trout'; brook charr × common charr ('Struan-trout'); and brook charr × brown trout ('zebra-trout' or 'tiger-trout'). These last have also been taken in the wild in a chain of pools in Surrey, and in Wise Een Tarn† –

*In North America this fish 'is called the "brook trout", "speckled trout" or "speckled charr". In Britain it has most commonly been called "American brook trout", but since as a *Salvelinus* it is properly a charr rather than a trout (*Salmo*), the names "speckled charr" or "brook charr" seem preferable.' P. S. Maitland, *Key to British Freshwater Fishes*. 1972.

†When originally constructed, Wise Een covered about 14 acres (5–6 hectares): by 1975 it had been reduced to about 11–12 acres (4–5 hectares).

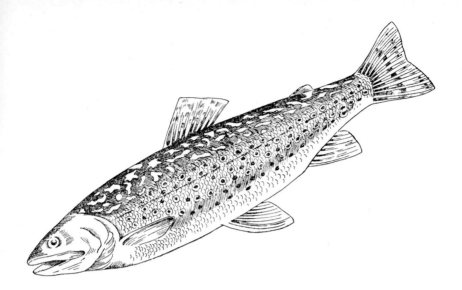

which brook charr colonized from the 2 acre (about 1 hectare) Scale Tarn after it had been stocked in about 1925 – between Windermere and Esthwaite Water in the Lake District of Lancashire: these fish, which were between three and four years old and were certainly natural hybrids, showed at least as good a growth rate as the brown trout in the tarn.

The first shipment of American brook charr ova to reach England* was despatched from North America in the late winter of 1868 and reached its destination early in the following spring. Mr Livingstone Stone, then owner of the Cold Spring Trout Ponds at Charlestown, New Hampshire, recorded that 'one lot was sent to England to Mr Frank Buckland, and was favourably noticed in the London *Times*'; this is confirmed by Buckland himself, who in 1871 wrote:

American brook trout brought over by Mr Parnaby of Troutdale Fishery, Keswick . . . the first specimens ever seen in this country were sent to me beautifully packed with moss in tin boxes, by some friends in

*American brook charr were first introduced to Norway in 1870; France and Germany (1879); Austria (after 1879); the Netherlands (before 1883); Switzerland (1883); Czechoslovakia (before 1890); Italy (1891); Sweden (1891–2); Finland (1898); Poland and Rumania (late nineteenth century); Denmark (about 1900); Hungary and U.S.S.R. (before 1914); Bulgaria (1930); and Spain (1934): they have also been successfully introduced to South America, Asia, Africa, Australia and New Zealand.

American Brook Trout or Brook Charr

America. The parent fish were obtained from Lake Huron, in Canada.* Since that time the import of eggs of *fontinalis* has become a regular business.

(John Parnaby, who owned the Troutdale Fishery in partnership with J. J. Armistead, had been employed in an advisory capacity on piscicultural matters by the Government of Canada).

This initial introduction was followed on 20 November 1871 by a second shipment of which Stone wrote:

10,000 trout eggs were packed in sphagnum moss in a common wooden box about a foot [30 cm] square at Charlestown, N.H. They went from Charlestown to Boston, 120 miles [193 km] by rail on the same day. They remained at Boston overnight and the next morning were put on board the ocean steamer which sailed that day. They made a long passage of eighteen days to Liverpool and a considerable journey by rail from Liverpool to Keswick. At the end of the journey two-thirds were found to be in good condition, although some had hatched on the way and died, and the byssus† generated by these and by some of the eggs that were killed during the first part of the trip made great havoc in places.

These are generally accepted as the earliest introductions of brook charr to Britain. In a somewhat petulant letter to *The Times* of 28 October 1885, however, Mr Parker Gillmore claimed that in 1866–7 he had been the first to suggest the introduction of this and other species of North American fish, and that towards the end of the latter year he had travelled to North America and had collected some brook charr, as well as some salmon and oysters, which were shipped to England early in 1868, where they were apparently distributed before his return and without his knowledge. He complained in his letter that

the Prince of Wales had a pond stocked with *fontinalis*. When I visited it the person in charge was ignorant of my being the introducer, as was

---

*The migratory instinct, which was strongly implanted in the brook charr imported to Britain during the nineteenth century, may have been derived from stock which had been migratory in the wild in North America, where some brook charr are known to migrate between the Great Lakes and their inflowing rivers: the Nepigon River, for example, which is reputed to hold the largest brook charr in the world, flows into Lake Superior. Armistead (1895) records that some brook charr in Britain were caught 'in salt water in some of our bays and estuaries' – the implication being that they had migrated there from the upper reaches of the inflowing rivers.

†A filamentous fungoid growth.

also His Royal Highness, until I informed him by letter. . . . at the Zoological Gardens, I also saw *fontinalis* in a glass tank with another person's name given on the label as presenter and introducer.

Poor Mr Gillmore went on to bemoan the fact that 'I have not only been left a pecuniary loser, but also have never been credited with my work'.

Writing from the Crystal Palace Aquarium to the *Zoologist* on 19 September 1876, Mr John T. Carrington stated that at Mr James Forbes's fishery at Chertsey Bridge on the upper Thames he had seen

a large number of beautiful specimens of several ages, of the American brook trout (*Salvelinus fontinalis*). Of this last named truly handsome fish Mr. Forbes has many fine examples . . . we observed some about a pound [450 g] in weight and others were a year and a half old . . . I cannot help thinking that owing to Mr. Forbes's efforts, this fish will soon obtain permanent hold in the Thames.

In his *British and Irish Salmonidae* (1887) Day wrote of the brook charr as follows:

It has, during the last twenty years, been acclimatized in this country, and thrives in some of the places where it has been turned out, in south, east or west. Although in Norfolk it has been stated to have done well, and grown twice as quickly as the brook [brown] trout this is denied, and Mr. Southwell informed me (December 6th. 1886) 'The remark that *S. fontinalis* "thrives well" in Norfolk (among other places) is quite incorrect; of all those which have been introduced, I doubt whether any survive.' At Howietoun [to which Sir James Maitland imported 10 000 ova in 1878] it has done fairly well, but does not often seem to live over its fifth year. . . . in Mr. Andrews' ponds [at Guildford], in Surrey, and [in the River Wey] it is said to have done well, while in Bagshot Park it is likewise stated to have thriven. . . . In one of Mr. Basset's ponds, at Tehidy, near Camborne, Cornwall, Mr. Cornish (*Land and Water*, May 1st. 1886) tells us that a 9¼ lb. [4 kg] one was captured in April 1886; also that with this species 'Mr. Basset stocked his ponds some nine years since . . .'. The Maclaine of Lochbuie has acclimatized this fish in a moor loch about a thousand feet [304 m] above the sea, near Loch Uisk, in Mull: in 1884 on was captured 2½ lbs. [1 kg] in weight, and they are said to have attained 5 lbs. [2 kg].

In their *Fauna of Argyll and the Inner Hebrides* (1892) Harvie Browne and Buckley record that brook charr 'have been introduced by Mr M'Fadyen into the lochs of the Cuilfail district [above Kilmelford near Oban] over many years. They have also been successfully placed in the Lochbuie lochs on Mull. The Maclaine writes "they have done better here than (so far as I can learn) any other part of the United Kingdom".' In the 1880s the Duke of Sutherland introduced a number of brook charr into Loch Brora and the Kintradwell Burn in Sutherland, and in 1889 a hundred were turned out in the river Eden at Appleby in Westmorland. As mentioned in the previous chapter, Professor Ritchie believed that over a thousand fish, reputed to be rainbow trout, which were introduced in 1898 to the Buchart, a tributary of the Aberdeenshire Don, were in fact brook charr.

In 1901 Scott and Brown wrote that brook charr had been

introduced in many small lakes throughout the [Clyde drainage] district, also in Loch Lomond where it still maintains its identity, but has not thriven. Has been distributed throughout Renfrewshire and Ayrshire, and is thriving in the Rivers Ayr and Irvine, and in the Waters of Borland, Kilmarnock, Cessnock, Carmel and Alnwick . . . it has been widely distributed throughout Scotland, even in islands such as Mull.

Apart from the foregoing references, information on brook charr in British piscicultural literature is meagre indeed. After the death of John Parnaby, his partner, J. J. Armistead moved to New Abbey near the banks of the river Nith on the Solway Firth, where he founded the Solway Fishery. After Armistead's death this fishery was purchased by Mr F. G. Richmond, on whose death in 1935 it was acquired by Mr D. F. Leney of the Surrey Trout Farm at Haslemere: here a thriving population of brook charr was established, which is still in existence today.

In 1936 Mr Leney introduced between two and three hundred brook charr, as well as some brown and rainbow trout, to a chain of spring-fed pools – at one time watercress-beds – on the Lower Green Sand at Deepdene near Haslemere, the home of Sir Hildebrand Harmsworth, Bt; here the brown and rainbow trout have virtually died out, but before they did so a small number of brown trout × brook charr hybrids ('zebras') were produced; the successful establishment of this population of brook charr (which has never been reinforced by further stocking) is believed by Mr Leney to be due to the absence of lime in the water.

In the late 1950s or early 1960s Mr Leney sold a number of brook charr and hybrids to the Bristol Waterworks Company for stocking their reservoirs at Chew, where although the conditions were apparently suitable, the fish failed to become established. The lake at Newstead Abbey Park near Nottingham at one time held a thriving and self-maintaining population of brook charr (as well as rainbow and brown trout) which were introduced by the owner of the estate, Mr Jobling: all the fish spawned naturally, although the young were artificially protected by a dam underwater and by wire-netting overhead. The trout and charr were all exterminated following pollution during the last war and later subsidence from coal mining.

On the Claife Heights above the western shores of Windermere in Lancashire, lie three artificial man-made tarns, managed by the Freshwater Biological Association at Ambleside, each of which contains small numbers of brook charr. In 1962 Dr Winifred Frost of the F.B.A. told Mr Maurice de Bunsen that

Scale Tarn (near Wise Een) which was built in 1910 was the pond where *fontinalis* were first planted – the fish were hatched and reared in Wraymires Hatchery [near the tarns], where the eggs were obtained from I do not know. The fish grew well. During 1914–18 the Scale Tarn stock nearly died out, but one male and one female were stripped by the hatchery manager and reared – and from these all *fontinalis* in the tarns on Claife Heights (i.e. Wise Een, Wraymires and Scale) are descended.

In another Tarn nearby [Moss Eccles], (made about 1906) *fontinalis* were also present in 1910–13. How they came there we do not know, but probably from the Wraymires Hatchery.

Today [1962] Scale Tarn has few if any *fontinalis*, Wraymires few, and Wise Een not many; their numbers have diminished in the last ten years – now we are conserving them.

All these tarns (which were constructed by a Mr Fowkes) contain brown trout – and Wise Een also holds rainbows – with which the brook charr have to compete. Dr Frost believes that between 1970 and 1975 the brook charr populations in Wise Een and Wraymires declined considerably from the numbers present between 1965 and 1970. In both waters, however, the surviving fish spawn naturally on the rocky and gravelly shores, and although there have been one or two reinforcing introductions during the last fifteen years or so, these populations are undoubtedly self-perpetuating.

One of the most successful British populations of brook charr is that in Llyn Tarw (or Llyn-y-Tarur) on the Plas Dinam estate of Lord Davies at Llandinam, near Newtown, Montgomeryshire.* When and how this population became established is unknown, but it is entirely self-perpetuating – spawning taking place in the lake itself since there are no feeder streams. A creel census (representing minimum catches, as only a 10 per cent return was made by anglers) revealed that 452 fish were caught in 1970; 430 in 1971, and 371 in 1972, suggesting that the total population may be considerable.

In the early 1960s, when anglers were complaining of the small size of the fish in Llyn Tarw, the Newtown Angling Association sought advice from the Fisheries Department of the Severn River Authority. As a temporary measure the minimum size limit was removed, which resulted in the capture in 1972 of a then British rod-caught record brook charr of 1 lb 14 oz (845 g). In 1973–4 a twelve-month monitoring programme was carried out in Llyn Tarw; after unsuccessful attempts with seine nets, in the autumns of 1973 and 1975 a number of gill nets were set to catch gravid fish in order to obtain eggs for incubation, of which about 3000 were secured: these were successfully fertilized and were placed in the Authority's hatchery.

Following this survey, Mr A. Churchward, District Fisheries Officer of the Severn-Trent Water Authority, supervised the stocking of two waters in the Forest of Dean in Gloucestershire with fry from Llyn Tarw. A small hill water, Llyn Bugail, and a minor tributary of the upper Severn were also stocked with takeable fish from the same source; in the former Mr Ayton reports that at least twenty-five fish have been caught; in the river there was a mass migration down-stream (incidentally over a large waterfall), and a number of fish were captured several miles away from their place of introduction. In December 1975 approximately 12000 eggs were obtained from gravid fish in Llyn Tarw to provide both brood stock of a 'wild' strain and surplus fish for stocking purposes. Mr Ayton is currently (1976) engaged in stocking two upland waters in central Wales (one

*For information on this colony I am indebted to Mr Warwick J. Ayton, Principal Fisheries Officer of the Welsh National Water Development Authority.

2H

with brook charr only, the other also with brown trout) from the Chirk Fishery in Denbighshire, which in 1964 imported 10000 ova from the Paradise Trout Company of Cresco, Pennsylvania. 'It will be of interest to determine', writes Mr Ayton,

how the brook trout face competition . . . more important is to establish whether this species is capable of providing alternative and more productive fisheries in upland, base deficient lakes. Their allegedly faster growth rate and less specific spawning requirements compared with brown trout is encouraging in this respect.

The oldest colony of brook charr in Scotland* appears to be one which was established towards the end of the nineteenth century in a 2 acre (about 1 hectare) hill loch at the head of Kirkton Glen north-north-west of Balquhidder in Perthshire. This water, Lochan an Eireannaich ('The Loch of the Irishman' – so-named from a nearby rock known as 'The Irishman's Leap'), was last stocked shortly after the First World War by the then owner, Mr James Carnegie of Stronvar, from whose son George the estate was purchased by the present owner, Mr P. D. Ferguson, in 1958. This loch contains no other fish and the brook charr, which appear to be firmly established, thus suffer from no inter-specific competition. In 1974 the loch yielded the present British rod-caught record brook charr which weighed 2 lb 9 oz (1·1 kg).

The population of brook charr in Monzievaird Loch on the Ochtertyre estate near Crieff in Perthshire was introduced early this century by Sir Patrick Keith Murray, Bt, who in about 1890 transported them by pony-and-trap from the Tayfield estate of Dr John Berry at Newport-on-Tay, Fife. At Ochtertyre they soon became firmly established in St Serf's Water (or the Serpentine as it is known locally), a small, muddy, tree-girt lochan above Monzievaird to which it is connected by a short silted burn. St Serf's Water also contained carp, tench and eels, while Monzievaird held mainly pike and perch. The inflow to the former water is only a tiny muddy trickle, the outflow to the latter being controlled by a sluice-gate, and it is uncertain how the brook charr were able to maintain themselves so well.

*For information on brook charr in Scotland I am largely indebted to Mr A. F. Walker of the Department of Agriculture and Fisheries for Scotland and to Mr R. N. Campbell of the Nature Conservancy Council.

In order to improve the game fishing in Monzievaird, both it and St Serf's Water were treated with derris in 1965 which, of course, destroyed all the fish. Some of the brook charr were removed beforehand, and these fish were subsequently re-introduced to both waters. In 1973 St Serf's Water was drained, and Mr Walker considers that it is doubtful if it now contains any fish. Monzievaird, however, still holds a good quantity of both brook charr and brown trout and now – under new ownership – is also being extensively stocked with rainbow trout. The brook charr appear at the moment to be firmly established (Mr Walker caught thirty in 1975), but their future must be in some jeopardy. The Freshwater Fisheries Laboratory at Faskally has recently stripped some Monzievaird brook charr and plans to rear the fry for comparison with their own stock, with a view eventually to re-stocking St Serf's Water.

The original population of brook charr at Tayfield, which became established in a pond and stream on Dr Berry's estate prior to 1890, was exterminated by disastrous pollution in 1934, following which the pond became silted up and the sluice fell into disrepair. In the late 1950s or early 1960s Dr Berry attempted to re-establish this colony; the pond was cleared and enlarged and the sluice-gate repaired; eighteen fish were obtained from Monzievaird and more were introduced in 1974 and 1975; so far however – possibly because the spawning-beds may be too heavily silted – the fish have failed to become established.

In 1960 Mr Niall Campbell of the Freshwater Fisheries Laboratory removed a number of brook charr from Monzievaird; from these fish were hatched two broods of fry, which were used to stock a number of neighbouring waters: two hundred (as well as many brown and rainbow trout) were placed in the spawning burn of Fincastle Loch (Mr de Bunsen), a limestone rich, man-made water 3 miles (5 km) north-west of Faskally, previously stocked with brown trout: in 1970 some brook charr were introduced by Mr Campbell into a small limestone lochan above Fincastle, where a few were caught by local anglers: no brook charr remain in either water today. In 1961 Mr Campbell introduced thirty brook charr fry to Loch Dunmore, an artificial woodland pond of 4 or 5 acres (about 2 hectares), on land owned by the Forestry Commission near Pitlochry: some of these fish reached a weight of 2 lb (907 g), and as a number of young fish have subsequently been caught it is assumed that successful spawning has taken place, although the only inflowing burn seems too small for this purpose and Mr Walker has searched it without discovering either ova or fry. Loch Dunmore is currently badly over-grown with water-plants, but some brook charr remain, as in 1975 Mr Walker caught one there weighing 1½ lb (679 g). Fish from Monzie-vaird have also been used to stock Lindores Loch near Newburgh in Fife, and several small lochs controlled by the Milton Park Hotel, Dalry, Kirkcudbrightshire.

In 1968 the late Dr T. A. Stuart of the Laboratory at Faskally stocked a stream – which at one point widens into a small pond – running through his garden near Pitlochry, with forty two-year-old brook charr. This stream, which was at one time a mill lade, is adjacent to the aptly named Moulin Burn which runs through the town, and has screens and sluices at head and foot. The brook charr spawned successfully every October and soon became firmly estab-lished in both pond and stream, where in the former one specimen grew to a weight of 6 lb (about 3 kg). Several ½–1 lb (224–453 g) fish have been caught downstream in the river Tummel, but there is no evi-dence to show that any have become established outside the mill lade system. After Dr Stuart's death in 1972 the bulk of the fish were removed to ponds at the Laboratory in order to breed from them for experimental stocking purposes. A number of mature fish were introduced in 1972 to Loch Bhac, 3 miles (5 km) north-west of Fincastle, where they did extremely well. In the spring of 1974 and 1975 Mr Walker introduced a further 1000 brook charr fry to this loch, where they have grown well and may soon become established. In 1975 some fry were reintroduced to the late Dr Stuart's water, although Mr Walker believes that a stock of brook charr from the original introduction still remained there: the screens are not now cleaned as frequently as they used to be, and it remains to be seen whether the brook charr will be able to compete with the increasing numbers of brown trout which are entering the water.

Four small hill lochs between 3 and 15 acres (1 to 6 hectares) in area, on land owned by the National Trust 2150 ft (655 m) high in the Torridon Mountains of Wester Ross also hold firmly established populations of small 8–9 in. (20–23 cm) brook charr; in one of these lochs the fish co-exist with the indigenous common charr. Mr Niall Campbell, who has visited these lochans, was told locally that the brook charr were originally introduced in the 1890s.

In recent years the Freshwater Fisheries Laboratory has experi-mentally stocked six lochs in Sutherland, one in Angus and eleven in Perthshire with brook charr; it has also given a number of fry to fish-farms in Morayshire, Fife, Wester Ross, Stirlingshire and East Lothian, and is shortly to present brood stock fish to the Highlands and Islands Development Board.

In his 1962 paper, Maurice de Bunsen listed the following waters – in addition to those mentioned above – into which brook charr have been introduced but from which they have subsequently disappeared:

Date	Water	Remarks
1870 and 1890	River Nene, Northamptonshire.	British Museum.
1871	?	Specimens presented to Queen Victoria by Mr Frank Buckland.
1883	Northamptonshire.	British Museum.
before 1885	?	Specimens presented to the Prince of Wales, possibly by Mr Parker Gillmore.
1887–9	River Esk, Midlothian.	Musselburgh Angling Improvement Association.
1889	Cardiganshire.	British Museum.
1890–2	Fallodon, Northumberland.	Introduced by Viscount Grey, from the Howietoun Hatchery.
1891	Glastonbury, Somerset.	Brook charr × brown trout hybrids. (British Museum).
before 1895	Loch Lomond, Stirlingshire.	—
before 1895	New Abbey, Dumfries, (loch 3 miles [5 km] from the Solway Firth).	By Mr J. J. Armistead.
1924	Loch Coulin, Ross and Cromarty (and another small neighbouring lochan).	British Museum.
1925	Yew Tree Tarn, Westmorland.	Introduced by Mr J. Maitland from his hatchery at Monk Coniston, to a new artificial water on the west

Date	Water	Remarks
		side of the Coniston–Ambleside road.
1931	An artificial water on the Wiltshire/Dorset border.	British Museum.
1936	Kinlochewe, Ross and Cromarty.	British Museum.

American brook charr have apparently never been introduced into Ireland.

The American brook charr – which appears to favour acid or only mildly alkaline waters within a temperature range of 11° to 16°C – provides another example of a species which readily acclimatizes in Britain but rarely becomes naturalized. This is unfortunate because, possessing as it does a number of advantages over the brown trout, it would be a welcome addition to any trout fishery; it has a quicker growth rate (up to two times during the first two years), provides better sport and superior eating, and accepts inferior spawning conditions than the native species. Yet, as Mr Walker writes,

stocking trials are reinforcing the view that brook trout very seldom become established in the presence of appreciable numbers of brown trout. Precisely why this is so is uncertain, but it could be due to competition between the species in the spawning streams. Brook trout can spawn very effectively in flowing water in artificially created situations, where brown trout are scarce or absent. They also spawn in still water and, judging from the North American fisheries literature, we might expect them to be more successful in this respect than brown trout. Nevertheless despite their widespread introductions throughout Britain in the past and early this century, naturalization has been very rare.

It will be interesting to see whether the latest attempts to establish brook charr in Scotland succeed in materially altering this situation.

48. Common Carp

(Cyprinus carpio)

Recent research carried out independently by Mr Alwyne Wheeler of the British Museum in London and by Mr Eugene K. Balon of the Royal Ontario Museum in Canada has revealed that, contrary to previous belief, the common carp is not a native of China, but rather a central Asiatic species which spread naturally east to Manchuria and west to the rivers of the Black Sea.

In colour the common carp varies from brown or brownish-green on the back through sides touched with gold to a yellowish belly. The dorsal fin is usually dark – the other fins sometimes having a reddish tinge. In southern England the common carp spawns in shallow water in May and June, when the female lays between 60 000 and 70 000 eggs on various species of aquatic vegetation: a minimum water temperature of 64°–68°F (18°–20°C) is necessary for the eggs to hatch successfully, which they normally do after from three days to a week. Common carp favour rivers and lakes with slowly moving or still waters with a muddy bottom and surrounded by dense vegetation; here they feed mainly on water-fleas, midge larvae, water insects, snails, crustaceans, worms, algae, and the seeds of aquatic plants. In domestication the common carp may reach a maximum length of about 3 ft (91 cm) and a weight of around 60 lb (27 kg) at the age of about forty years. Several domesticated varieties (with humped backs and scale variations) such as the mirror or king, golden, and leather carp, have been developed on the continent; they were first introduced to ornamental waters in southern England well over a hundred years ago.*

*The crucian carp (*Carassius carassius*) is an allied but separate species which may possibly have been introduced to Britain; however, the discovery in 1975 of the jaw-bone of a crucian carp in Roman remains at Southwark suggests, according to Mr A. C. Wheeler, that it is much more probably indigenous. It is found today mainly in parts of southern England (especially in the Thames Valley) and in East Anglia: a variety is the Prussian carp (*C. c. gibelio*) from which the goldfish (*C. auratus*) probably evolved.

[*Footnote continued on page 441*

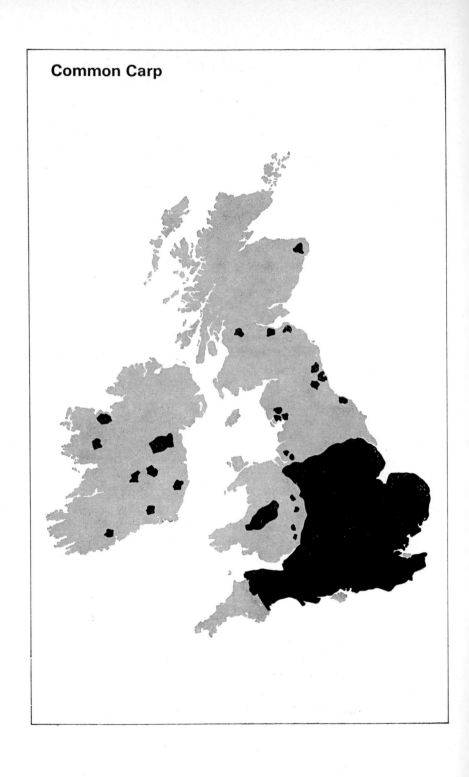

Common Carp

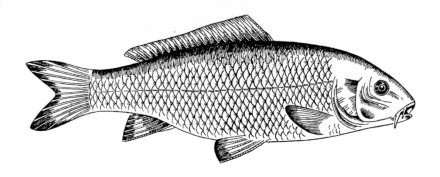

Balon suggests that the original home of the wild common carp was in central Asia (eastward from the Caspian Sea), from where it spread naturally during the last post-glacial period east into China and westward to the Black and Aral Seas, first appearing in the Danube some 8000–10000 years B.P. That it did not arrive in these areas earlier is suggested by the fact that during the Pleistocene, before the great continental glaciers retreated, the species had not reached Scandinavia, the British Isles, or North America, nor had it arrived in the Far East by the end of the Pliocene.

Except for the river Danube and its tributaries, it seems unlikely that the carp was found naturally in Europe as early as the beginning of the Christian era. Decimus Magnus Ausonius (*c.* 310–95 A.D.) made no mention of carp in the Rhine and Moselle in his descriptive poem *Mosella*. The most westerly known natural occurrence of rheophilic (stream-loving) wild carp in the Danube is at the mouth of the Morava River where, near Devin in Bulgaria, the famous Amber Road crossed the Danube. From this area – where immense numbers of wild carp doubtless spawned annually in the flooded water-meadows – the fish were transported to European waters outside the Danube river-system by the Romans who traded along the Amber Road. Evidence for the natural occurrence of wild carp in the Danube is provided by the different names given to the fish by the inhabitants of the region; from the use of the same Celtic name outside the western limit of its natural range; from medieval manu-

Mr D. F. Leney writes that Japanese golden carp (*Hi Goi*), which have been bred on the continent for many years, were first introduced to England in the early 1900s; but although they have bred here they have never become established. During the 1960s and 1970s large numbers of Japanese 'coloured' carp (*Koi* or *Coi*) have been imported direct from Japan by air, and have successfully bred in many ornamental pools.

scripts; and from present-day field-work carried out among the rheophilic wild carp of the Danube.

Flavius Magnus Aurelius Cassiodorus (*c.* 490–585 A.D.) provides evidence in his *Variae* for the Danubian origin of western domesticated carp, when he refers to the *Carpa* of the Danube as an expensive delicacy served to King Theodorus (490–526 A.D.) at Ravenna. The common carp is extremely tenacious to life, and can survive for a surprisingly long time out of its natural element and without food, and lives happily in brackish water. Having been transferred by the Romans from the Danube to Italy, carp were kept in domestication there in reservoirs (*piscinae*) designed by Sergius Orata in the first century B.C. From these they naturally escaped from time to time into the nearby rivers, where they became naturalized. After the fall of the Roman Empire and the establishment of Christianity, they began to be reared for fast-day food in monastic stews. Quantities of wild carp endemic to the Danube may subsequently have been transferred directly to European waters. Albertus Magnus ('Albert of Cologne': ? 1206–80) makes what is probably the earliest reference to the breeding in captivity of domesticated carp in his commentary on the *Books of Sentences* written by Peter Lombard, Bishop of Paris, (*c.* 1100–60).

Balon suggests the following stages in the domestication of Danubian wild carp: 1. Transported west of the Danube during the first to fourth centuries A.D.; 2. Sporadic western introductions and artificial rearing in the fifth to sixth centuries; 3. The beginning of rearing on a mass scale with continued introductions from the seventh to thirteenth centuries (to Germany and France by 1258); 4. Continued breeding and mass culture in the fourteenth to sixteenth centuries (in Sweden by 1560); 5. Intensification of (selective) breeding and further introductions (*e.g.* to Denmark by 1660) and outside Europe (*e.g.* to North America in 1831; to Australia (1860), and in 1896 to South Africa), from the seventeenth century onwards. The domestication of wild carp in China began independently some five-hundred years earlier than in Europe, but there is no evidence to suggest that European pond-culture was in any way derived from this earlier Chinese domestication

The Carp is the Queen of Rivers: a stately, a good, and a very subtle fish, that was not at first bred, nor hath been long in *England*, but is now naturalized. It is said, they were brought hither by one Mr. *Mascal*, a gentleman that then lived at *Plumsted* in *Sussex*, a County that abounds more with this fish than any in this nation.

You may remember that I told you, *Gesner* says, there are no Pikes in *Spain*; and doubtless, there was a time, about a hundred, or a few more years ago, when there were no Carps in *England*, as may seem to be affirmed by Sir *Richard Baker* [1568–1645], in whose Chronicle [of the King's of England] you may find these verses.

Hops and Turkies, Carps and Beer,
Came into England all in a year.

So wrote the great Izaak Walton (1593–1683) in his delightful treatise of fishing, *The Compleat Angler*, first published in 1653. Baker's distich, however, post-dates the arrival of carp in England by at least a quarter of a century. Hops – from which beer, as opposed to ale which is made from malt liquor, is produced – were only introduced into southern England from Flanders in about 1520–4, and turkeys, which were discovered (in domestication) in Mexico in 1518, were unknown in Europe until about 1530, having probably been imported *via* North America by Sebastian Cabot (1476–1557) or one of his contemporaries.

In his *History of British Fishes* (1859) William Yarrell writes:

Leonard Mascall [*c*. 1514–1589; kitchen-clerk to Matthew Parker (1504–75), appointed Archbishop of Canterbury in 1559] takes credit to himself [in *A Booke of Fishing with Hooke & Line* . . . (1590)] for having introduced the Carp . . . ; but notices of the existence of the Carp in England occur prior to Mascall's time, or 1600. In the celebrated *Boke of St. Albans*, by Dame Juliana Barnes, or Berners [or Bernes *b.c.* 1388], the Prioress of Sopwell Nunnery . . . Carp is mentioned as 'a deyntous fisshe, but there ben but fewe in Englonde, and therefore I wryte the lesse of hym.' . . . in the Privy Purse expenses of King Henry the Eighth in 1532, various entries are made of rewards to persons for bringing 'Carpes to the King'.

Dame Juliana's *Boke of St. Albans* was first published in 1486; the edition which refers to carp – printed ten years later at Westminster by Wynkyn de Worde – contains a 'Treatyse of fysshynge wyth an Angle' (said to date from about 1450) which was not included in the earlier edition.

In *The Compleat Angler* Izaak Walton quotes a number of earlier writers on the common carp:

. . . in some ponds Carps will not breed, especially in cold ponds; but where they will breed, they breed innumerably: *Aristotle* [384–22 B.C.] and *Pliny* [23–79 A.D.] say, six times in a year. . . .

'Tis said by [Paulus] *Jovius* [1483–1552], who hath writ of fishes [in *History of His Own Time*], that in the lake *Lurian* in *Italy*, Carps have thrived to be more than fifty pound [23 kg.] weight;

The age of the Carps is by Sir *Francis Bacon* [1561–1626], in his *History of Life and Death*, observed to be but ten years, yet others think they live longer. [Konrad von] *Gesner* [1516–65] says [in *Historia Animalium*], a Carp has been known to live in the *Palatinate* above a hundred years.

Janus Dubravius . . . says, that Carps begin to spawn at the age of three years, and continue to do so till thirty: he says also that in the time of their breeding, which is in the Summer, when the sun hath warmed both the earth and water, and so apted them also for generation; that then three or four male Carps will follow a female, and that then she putting on a seeming coyness, they force her through weeds and flags, where she lets fall her eggs or spawn, which sticks fast to the weeds, and then they let fall their melt upon it, and so it becomes in a short time to be a living fish; . . . much more might be said out of him, and out of *Aristotle*, which *Dubravius* often quotes in his *Discourse of Fishes*.

The history of the introduction of the carp to Ireland, which has been described by Dr Went, can be traced in some detail.

The Journal Book Entry of the Royal Society for 29 April 1663 records that 'Mr. Boyle was desired to communicate his papers, concerning the manner how my Lord of Corke, his ffather, had Carpes transported into Ireland, where they were not before'. In the course of his address to the Royal Society, Robert Boyle (1627–91), the seventh son and fourteenth child of Richard Boyle (1566–1643) the 1st 'Great' Earl of Cork, read the following extracts from his father's diary:

3 Sept. 1634: I wrott to the Lord President of Mounster, and as he desired, gave order to Sir John Leek to deliver his Lop. 20 yonge carpes . . .*
26 Sept. 1640. I wrott by Badnedg to Sir John Leek to furnish Sir Phillip Percival with 40 young carpes, Mr. Henry warren second Remembrancer of Exchequer with 20 Carpes and my daughter the countess of Kildare, with 20 carpes to stoar the new pond withall.†

*Lismore Papers, 1st series, 4, p. 44.
†Ibid., 5, p. 161.

7 Oct. 1640. Letters to Sir John Leek to deliver to Sir Philip percival 40 young carpes . . . towards the storing of his ffishe pond.*

Charles Smith in his *History of Cork* (1750), commenting on a part of Edmund Spenser's *Faërie Queene* (1590), claims that carp (and tench) were to be found in the river Awbeg in Co. Cork during the reign (1603–25) of James I, and that both species were commonly kept in captivity in ornamental waters, although tench were less plentiful. In his *Natural History of Dublin* (1772) Rutty repeats the suggestion that carp were originally introduced to Ireland during the reign of James I.

William Tighe, in his *Statistical Observations Relative to the County of Kilkenny* (1802), wrote: 'The tench (*Cyprinus tinca*) is found in the Barrow together with the *Cyprinus carpio* or carp; it is said they first came there upon the breaking down of some ponds at Low-Grange by a flood'. In his *Rural Sports* (1807) W. B. Daniel mentions that the 'Lakes of Killarney' contained large numbers of fish, including carp – a statement which was repeated in 1849 by James Windele in his *Historical and Descriptive Notices of the City of Cork and its Vicinity*.

The following extract from the register of a guest-house at Clifden, Corofin, on the edge of Lough Inchquin, appears to be the earliest reference to carp in Co. Clare:

23rd. June 1815. . . . About 25 years ago his [William Burton's] father put into the lake 24 Carp. It is surprising that after such a lapse of time in so fine a piece of water, so prolific a fish as *Cyprinus carpo* has not multiplied to a sufficient degree to afford a proof of their existence therein, not one of them has been taken.

John Templeton of Cranmore, in *Irish Vertebrate Animals* (published in the *Magazine of Natural History* by his son Robert in 1837) describes the carp as a 'naturalized species'. Finally, William Thompin his *Natural History of Ireland* (1849–56) wrote:

This fish [the carp] which was introduced into the British Islands has long been in Ireland. Localities noted: Montalto and Killyleagh, Co. Down; and Markethill, Co. Armagh (Mr. J. Sinclaire); Co. Dublin (Dr. Ball); Counties of Galway and Sligo (Mr. R. Barklie).†

*Lismore Papers, 1st series, 4, p. 44.
†Introduced by the Great Earl of Cork in the South of England – *vide* Robert Boyle in a paper to the Royal Society: R. Ball.

The common carp is today fairly widely distributed in slow-flowing or still waters throughout central and southern England, in parts of Ireland, in southern Wales, and to a lesser extent in northern Wales and Scotland. Even in southern England, summer water temperatures in some waters are often barely high enough to ensure successful spawning and an adequate supply of young fish. In many places carp populations only survive through reinforcement by artificial stocking.

49. Goldfish

(*Carassius auratus*)

The natural home of the goldfish ranges from eastern Europe* into Asia as far east as China.

In appearance the goldfish is very similar to the closely related crucian carp (*C. carassius*)† from which it is principally differentiated by its shallower body and larger scales. The spines of the anal and dorsal fins are deeply serrated – the rays on the anal fin normally being five in number. Goldfish are subject to considerable colour variation; most specimens in the wild are olive-green or brown on the back, golden or silvery-white on the sides, and whitish-silver on the underside, with no dark spot in front of the tail-fin. Goldfish seldom exceed 12 in. (30 cm) in length or 2 lb (907 g) in weight.

In the wild goldfish live in thickly weeded rivers and lakes; they spawn in June and July when the hen fish lays between 160000 and 380000 eggs, which normally hatch after about one week. In the East the numbers of the sexes are usually fairly equal; in many parts of Europe, however, the population consists entirely of females, and reproduction is by gynogenesis.

Several occidental authors have claimed that the goldfish,‡ which in its domesticated form is known in China as *Chin Chi-yü*, was being kept in captivity as early as the time of the T'ang Dynasty

*According to legend, Great Schutt Island in the upper reaches of the Middle Danube was surrounded in ancient times by 'a great number of golden carp (*Cyprinus auratus*) that enabled even the poorest people to make a living; yes, there were times when the fishermen gave them away as gifts'. A. Khin: *Hausen on Great Schutt Island and its Fishing*, 1930.

†Some authorities consider that the goldfish is a subspecies (*C. c. auratus*) of *C. carassius*; others believe that the variety of the crucian carp native to eastern Europe, the Prussian carp (*C. c. gibelio*), is a subspecies of the Asiatic goldfish (*C. a. auratus*).

‡For the history and development of the goldfish I am indebted to the excellent monograph on the species by G. F. Hervey and J. Hems (see Bibliography).

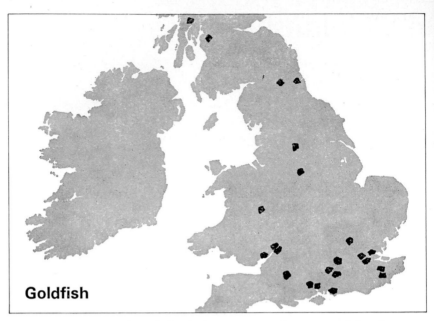

Goldfish

(618–907 A.D.), basing their theory on two circumstantial references in contemporary literature. It is not, however, until the early years of the Sung Dynasty (960–1279 A.D.), when they are mentioned by three separate authors, that firm evidence for their domestication occurs. Li Shih-chên wrote in about 960 A.D., in his *Pên Ts'ao Kang Mu (Materia Medica)*; 'Formerly in ancient times they [goldfish] were little known . . . no one kept them domesticated before the [period of] the Sung.' In his *Wu Lei Hsiang Chih (Account of the Mutual Influence of Things)* the monk (Kao) Tsan-ning (918–99 A.D.) wrote: 'If goldfish eat the refuse of olives or soapy water then they die; if they have poplar bark they do not breed lice.' In *Ch'ien T'ang Hsien Chih (Description of the District of Ch'ien T'ang)* we read: 'Only since the Sung [Dynasty] have there been [goldfish] breeders.'

In about the year 1030 Su Tzŭ-mei (1008–48) composed *Poem on the Pagoda* of the Six Harmonies*, in which the following couplet occurs:

> At the Pine Bridge I waited
> for the Golden Chi,
> All day I lingered, alone.

*This pagoda, named Liu Ho T'a, was constructed at Zakow near Hangchow in 971 A.D.

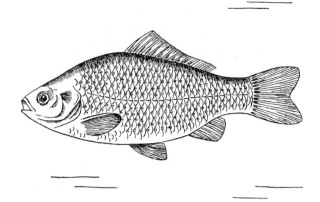

A contemporary poet, Su Tung-p'o (1036–1101) wrote:

At first I did not understand these words, until I became Deputy of Ch'ien-T'ang, and then I knew that there were these fish coloured like gold in the pool at the back of the monastery. Yesterday I strolled again above the pool and threw in cakes and dumplings for a long time. They came out indeed a little, but did not eat, and went in again so that I could not see them again . . . the present time is separated from [that of] Su Tzŭ-mei by forty years, but they are swimming about as of old. This may be called long life indeed.

In the year 1089 or 1090 the same poet wrote:

> I love the Golden Chi
> fish of Nan-p'ing,
> Coming again I lean on the
> rail
> And throw them the crumbs,
> of my frugal feast.

Towards the middle of the same century Su Tzŭ-mei and his contemporary Wên T'ung (1019–79) wrote, in *Lines in Reply to Tzŭ-lü,*

> The Golden Chi at dawn
> is the beauty of fishes.

By about the middle of the Sung Dynasty the goldfish was well established in China as a popular pet. The author Yo K'o (1173–1240) wrote in his *T'ing Shih*:

At the present time there are fish breeders in Chung-tu [Peking] who can change the colour of fish to gold. The Chi [goldfish] is the most prized for keeping, and the Li [carp] comes next.

Liu T'ung and Yü I–chêng wrote, in their *Ti Ching Ching Wu Lüeh*, that the famous Goldfish Pool (originally named the Fish-Weed Pool) in Peking had been constructed as early as the period of the Second (Chin) Dynasty (1115–1234), and that goldfish were being bred in economic quantities at this time. According to Mêng Liang Lu, by the year 1276 silver, black, red, yellow and mottled varieties as well as golden ones were being produced by breeders and offered for sale in the market at Hangchow.

The earliest printed picture of a goldfish may be an illustration of the species in *San T'sai T'u Hui* (1607). The British Museum contains a painting of a woman and children playing with goldfish which, although not earlier than of the Ming Dynasty (1368–1644) and more probably of the eighteenth century, is clearly after the southern Sung court painter Su Han-ch'ên (fl. before 1127). Goldfish are frequently depicted in other branches of Chinese art by, for example, bronze-workers, wood- and stone-carvers, potters and tapestry-weavers – and especially in the fine porcelain produced during the Chin Dynasty.

In 1904 the well-known Japanese ichthyologist Mitsukuri, wrote: 'There is a record that about four hundred years ago – that is to say about the year 1500 – some common goldfish [Kingye] were brought from China to Sakai, a town near Osaka.' The German traveller and physician Engelbrecht Kaempfer (1651–1716), who was in the country in 1691–2, wrote in his *History of Japan* (1727): 'In China and Japan and almost all over the Indies, this fish is kept in ponds and fed with flies before their wings come out. The goldfish', he continues, 'is a small fish, seldom exceeding a finger in length, red, with a beautiful shining yellow or gold-coloured tail, which in the young ones is rather black. . . . Another kind hath a silver-coloured tail.'

In Japanese painting, goldfish are first depicted by Nobuharu of the Ashikaga Period (1335–1573); the earliest known Japanese print of the species is a woodcut by Utamaro (1754–1806) of a child upsetting a goldfish from a bowl of water.

In Europe, goldfish may first have arrived in Portugal as early as 1611, '. . . after the people of that country had discovered the route

to the East Indies by the Cape of Good Hope'.* The earliest mention
of the species in European literature appears to be that made by
Martin Martini (1614–61; stationed at Hangchow 1646–50) who in his
Description Géographique de l'Empire de la Chine wrote of 'the little
gilded fish ... which the Chinese have for this reason named Chin-yü,
because the skin glitters, being somehow interwoven with threads of
gold – the whole back sprinkled as it were with gold dust ... they
are a strange sight. The Chinese make much of them.' Samuel Pepys
(1633–1703) wrote in his *Diary* on Sunday 28 May 1665: 'Thence
home and to see my Lady Pen, where my wife and I were shown a
fine rarity: of fishes kept in a glass of water, that will live so for ever;
and finely marked they are being foreign.' These are commonly
assumed to have been goldfish, although Hervey and Hems cast
doubt on this attribution. The German ichthyologist M. E. Bloch
went a step further by claiming that he could trace the arrival of the
goldfish in England back to the reign (1603–25) of James I.

In his *Nouveaux Mémoires* (1696) Louis le Comte incorrectly
describes the goldfish which he claims to have seen on a visit to
China: 'They are commonly of a finger's length, and of a proportion-
ate thickness; the male is of a most delicate red ... the female is
white ... those who breed them, ought to have great care; for they
are extraordinarily tender and sensible of the least injuries of the air.
They [the Chinese] put them in great basins, such as are found in
gardens, very deep and large.'

Thomas Gray (1716–71) appears to have been the first English
author to mention the species when in 1742 he wrote his *Ode on the
Death of a Favourite Cat, Drowned in a Tub of Gold Fishes*, the animal
referred to being Horace Walpole's 'Selima'. The pond (Po-yang) in
Walpole's garden at Strawberry Hill was celebrated for its goldfish.
'They breed with me excessively and grow to the size of small perch',
wrote Walpole to George Montagu on 6 June 1752; on 16 August of
the following year he wrote again to Montagu: 'You may get your
pond ready as soon as you please, the goldfish swarm. Mr Bentley
carried a dozen to town t'other day in a decanter.'

There is circumstantial evidence to show that a number of goldfish were on board an East India Company ship which set sail from
Macao in September 1691 and berthed in London early in the following year. George Edwards in his *Natural History of Birds* (1745)
wrote: 'The first account of these fishes being brought to England

Loudon's Magazine of Natural History, 3, p. 478.

may be seen in Petiver's works, published about *anno* 1691.' This statement is repeated by Thomas Pennant in his *British Zoology* (1768–70), and by William Bingley (1774–1823) – 'Gold Fish are natives of China . . . they were first introduced into England about the year 1691' – in his *Animal Biography* (1813). The former author writes:

In China the most beautiful kinds are taken in a small lake in the province of Che–Kyang [south of Hangchow]. Every person of fashion keeps them for amusement, either in porcelain vessels, or in the small basins that decorate the courts of Chinese houses. The beauty of their colours, and their lively motions, give great entertainment, especially to the ladies, whose pleasures, from the policy of that country, are extremely limited.

James Petiver's book *Gazophylacium Naturae et Artis*, in which the author (who died in 1718) mentions two live goldfish imported direct from China, was however published not in 1691 but in 1711, the date apparently being misread or misprinted by Edwards,* who also mentions that Charles Lennox (1701–50), Duke of Richmond, possessed a large Chinese earthenware vessel full of goldfish which had been imported alive to England. Two goldfish are figured in Petiver, and there is evidence to suggest that the drawings for the illustration were made in 1705 and in any case not later than in the following year, when the plate was offered for sale. The species is not mentioned by Izaak Walton (1593–1683) in his *Compleat Angler*, first published in 1653, by John Ray in *Synopsis Methodica Piscium* (1710), or by Francis Willughby.

During the 1730s increasingly large numbers of goldfish were imported into England. Both George Edwards and the Dutchman Job Baster of Haarlem in his *Opuscula Subseciva* (1765) record that in 1728 some specimens – which had been given to him by Sir Matthew Decker, Bt, a Director of the East India Company – were introduced from China *via* St Helena by Captain Philip Worth of the Company's ship *Houghton*. Again however the date appears to have been misread, as the *Houghton* sailed from England towards the end of 1728 and returned two years later. In the same work Baster mentions that goldfish 'were placed in fish-ponds in England, and were increased, so that, sent into other parts of Europe, they became well known'. By the 1750s goldfish were being widely kept in domestication as pets, and Horace Walpole (1717–90), who in a letter dated

*This theory was first propounded by G. F. Hervey in *The Aquarist* in December 1946.

19 July 1746 to Henry Fox (1705–74) refers to the goldfish in the 'purling* basons' at Vauxhall, gave them away as presents to his friends.

The first goldfish in France were to be seen in Paris in about 1750, having been landed at the port of Lorient by a ship of the French East India Company. These may have been brought as a gift for the aptly named Jeanne Antoinette Poisson (Le Normant D'Etioles), Marquise de Pompadour (1721–64), although the de Goncourts – biographers of the mistresses of Louis XV – do not confirm this. The plate of the species in L. E. Billardon de Sauvigny's *Histoire Naturelle des Dorades de la Chine* (1780) was copied from a painting on a Chinese scroll which had been despatched to Paris from Peking by F. M. Martinet eight years previously.

The earliest reference to the introduction of goldfish (*Pesce dorato*) to Italy appears to be that contained in a letter of 6 May 1775 from Walpole to Richard Bentley (1708–82); 'I have lately given Count Perron some goldfish, which he has carried in his post-chaise to Turin: he has already carried some before. The Russian minister has asked me for some too, but I doubt their succeeding there.'

In 1753–4 a number of goldfish (*Goudvisch*) were introduced to ornamental waters belonging to Count Clifford and Lord of Rhoon in Holland, but they had not managed to breed there by 1765. Job Baster, who introduced sixteen goldfish to two ponds in Zeeland during the winter of 1759–60, met with better success: on 13 June of the latter year he saw 'some little fish 4–6 lines long, and of a blackish or swarthy colour. About six weeks later most of them developed silvery or white spots between the dorsal fin and the tail.'

Goldfish (*Guldfisk* or *Guldfirk*) were introduced to Scandinavia at some date prior to 1740, when Linnaeus received a specimen, which he later described in his *Systema Naturae* (1758), from the Swedish Ambassador to the Danish Court: Baster records that this specimen was subsequently preserved in alcohol and presented to the Swedish Academy 'because of its rarity'. On 6 November 1759 Linnaeus wrote to John Ellis that '. . . Mr Guy, who is famous for the cure of cancers . . . sent me a pair of Chinese gold fishes alive. . . .'†

According to Bloch, goldfish (*Goldfisch*) were introduced to Germany at least by 1780, when the German Ambassador to Holland,

*Rippling, or undulating.
†J. E. Smith, *A Selection of the Correspondence of Linnaeus* . . . vol. 1, p. 127, London, 1821. I am grateful to Mr A. C. Wheeler for drawing my attention to this record.

the Count of Heyden, brought some back from that country to Berlin. The Countess von Goes of Carinthia, Herr Grewe of Hamburg, and the Burgomaster of Bremen, Herr Oelrichs, were among others to import goldfish to ornament their waters.

In Russia, Gregory Alexandrovitch (1739–91) Prince Potemkin, the favourite of Catherine the Great (1729–96), gave a banquet in her honour on 1 April 1791 when goldfish (*Zolotoi ribki*) in silver bowls decorated the tables in the Winter Garden.

The goldfish appears to have been naturalized in parts of the United States before 1859, when Arthur M. Edwards wrote in his *Life Beneath the Waters, or the Aquarium in America*:

The gold carp (*Cyprinus auratus*) is one of the handsomest fish for an *Aquarium*, and at the same time it is easy of domestication. This beautiful creature is a native of China . . . it is now to be caught in the Schuykill River [in Pennsylvania] . . . into which stream it has escaped from some pond, and increased greatly in numbers, size and beauty – fish which breed in a semi-wild state always being of more brilliant color than those reared in confinement.*

Thirty years later at least one flourishing goldfish nursery had been established in the state of Maryland.

In the wild the coloration of the goldfish, as has already been said, does not differ from that of the crucian carp. The Chinese for a thousand years or so, and the Japanese for some four hundred years, have, by an intensive system of selective breeding, produced the domesticated variety which is now a popular pond and aquarium pet. They have also succeeded in propagating a large number of bizarre varieties, among the best known of which are the 'Veil-Tail'; the 'Telescope-Eyed' (which has a short nose, projecting eyes, no dorsal fin and a long three- or four-lobed caudal fin); the scaleless 'Shubunkin'; the 'Comet'; the 'Fantail'; the 'Moor' (an all-black variety); the 'Celestial'; the 'Lionhead'; and the 'Oranda', which is foreshortened and whose head is covered with wart-like protuberances.

In a wild state these weird varieties do not long survive, and the stock soon reverts to the original crucian carp type. Feral goldfish of the Prussian carp form can now be found living in suitable waters

---

*As the contrary of this last statement is in fact the case, it has been suggested by Dr James W. Atz that these fish were probably of very recent naturalization.

in many parts of southern and central Europe. Goldfish of the Asiatic type occur in isolated lakes and, to a lesser extent, in rivers in parts of southern England; they are extremely hardy fish, but breed most prolifically where the water is artificially heated (by, for example, electricity generating stations), as a minimum temperature of 20°C (68°F) is necessary for the eggs to hatch successfully. Mr Alwyne Wheeler of the British Museum considers that feral goldfish have for many years been well distributed in suitable enclosed waters in and around the London area; they are for example, established at Hampton Court; in the Round Pond in Kensington Gardens and in the Serpentine in Hyde Park; since at least 1923 they have been naturalized in several ponds in the dockland district of east London. During his work on the tidal Thames from 1968–74 Mr Wheeler caught goldfish on several occasions between Wandsworth Bridge and Barking; some were of the 'wild form' and coloration, while others possessed long fin-rays and were golden-red in colour: these may have been liberated pet aquarium fish, but as they were captured in small numbers during a five-year period and over a length of river, Mr Wheeler suspects that there is a small but established population of goldfish in the London Thames.

Goldfish are still being introduced sporadically all over the British Isles, and it is difficult to establish which records refer to self-perpetuating naturalized colonies and which to isolated individuals. Dr P. S. Maitland lists the following sites – mostly ornamental lakes on large estates – from which feral goldfish have been reported: Ben More Botanic Gardens, Argyllshire; Kelvingrove Park, Glasgow; Forth and Clyde Canal, Clydebank; Longleat, Warminster, Wiltshire; Almondbury, Huddersfield; Harlesthorpe Dam, Clowne, Chesterfield; Leazes Park, Newcastle; Bigg Waters, Wideopen, Northumberland; Raphael Park and Marshall's Lake, Romford, Essex; the river Mardyke, Aveley, Essex; Farm Pond, Tonbridge, Kent; Cooksmill Green, Roxwell, Chelmsford, Essex; the 'Blue Lagoon', Arlesley, Bedfordshire; Uckfield, Sussex; Totteridge Pond, north-west London; Mill Pond, Halifax; Keighley, Yorkshire; Edderton Hall, Welshpool, Montgomeryshire; Irrigation Pond, Newent, Gloucestershire; New Grange Pool, Dyneck, Gloucestershire; Cemetery Lake, Southampton Common; and 'Little California', Finchampstead, Berkshire. Feral goldfish have also been reported from the river Kennet between Hungerford and Newbury in Berkshire.

50. Bitterling

(Rhodeus amarus)

The bitterling* ranges over much of continental Europe, from the basins of the rivers Seine and Loire in the west as far east as those of the Black and Caspian Seas.

This small fish, which seldom exceeds 3 in. (7–8 cm) in length and has a disproportionately deep body, is greyish-green or dark blue on the back and silvery-white on the flanks and belly: the fins are a pale orange. Bitterling prefer densely weeded ponds and small lakes or slow-flowing streams with muddy or sandy bottoms: they feed largely on worms, *Enchytrae, Daphnia*, and the seeds of water plants.

The bitterling's method of reproduction is, perhaps, the most interesting of all naturalized British vertebrates. During the spawning season, from April to July, the colouring of the male takes on an iridescent hue, and a light blue bar appears in front of the caudal fin; the eyes and fins assume a reddish tinge, and around the former and on the upper lips appear small wart-like white papillae. The female develops, at the genital opening, a pendulous ovipositor or duct which may be almost as long as herself. The male chooses the shell of some species of unionid (Pseudanodonta) freshwater mussel – usually the pearl mussel, the swan mussel, the pond mussel, or the painter's mussel – into which the female, through her long egg-duct, lays single eggs among the gill-plates or mantle within the mussel's shells. The male then sheds his milt into the surrounding water, which is sucked between the shells through the mussel's inhalant syphon, and the eggs become fertilized. Because the means whereby they are deposited ensures a high survival rate, only between forty and one hundred eggs are normally laid. These develop and eventually hatch within the confines of the mussel's shells, where the fry feed on minute particles of food which are swept in by the movement of the water. After between three and four weeks the fry leave the safety of the mussel-shell as $\frac{1}{2}$ inch (1–2 cm) long fish.

*So called because of the reputedly bitter taste of its flesh.

The commensalism of the bitterling is, however, counterbalanced by the parasitism of its host, with those breeding season its own tends to coincide. The mussel lays its eggs in brood-pouches situated inside its outer gill-plates, where after about five months they hatch and tiny and highly mobile larvae or glochidia emerge. These float about in the water until they come into contact with a fish (frequently a bitterling) to which they become attached by a sticky filament: they then anchor themselves more securely to their host by small hooked teeth on the edge of their shells. The fish's tissues become irritated by the glochidium and form a cyst, in which the latter remains to feed on its host for up to four months; then – having grown a new shell – it falls from the cyst to commence a sedentary life on the pond- or river-bed, possibly in some entirely new area to which it has been borne by its unwilling host.

The exact date of the introduction of the bitterling to England is uncertain, but it has for many years been imported from the continent by fish-culturists for sale to coldwater aquarists. The earliest evidence for the suspected presence of these fish in the wild in England appears to date from the early 1900s, when Hardy records that the uncle of an official of the St Helens Anglers' Association, while a

Bitterling

boy, recalled catching them (under the name of 'Prussian Carp') in a pond in a field at Moss Lane, nearly opposite the ground of the St Helens Rugby League Football Club. As Wheeler and Maitland, however, point out, the lack of definite identification means that this record must be treated with some reserve, and it may well be that bitterlings were not established in the wild in England until the 1920s. Before the last war an angler fishing the Carr Mill Dam beside the East Lancashire (A580) road near St Helens noticed the distinguishing pale-blue stripe in front of the caudal fin on a number of fish – known locally as 'Pomeranian Bream' – which were being used by fellow-anglers as live-bait for perch: the source of these fish was the disused arm of a neighbouring canal at Blackbrook, and others are known to have been present in Leg of Mutton Dam, also near St Helens.

After the war the canal was found still to contain large numbers of bitterlings, and following an exhibition of some specimens by the Merseyside Naturalists' Association it was discovered that they were known also to be present in between ten and twelve other waters in south Lancashire, but that their numbers had declined in Leg of Mutton Dam. It is assumed that their original appearance in the wild was the result of aquarists disposing of unwanted stock, and that their spread is due to the release by anglers of surplus live-bait. In 1948 Mr H. Norman Edwards brought a dozen bitterling from Carr Mill Dam to use as live-bait for perch in Black Harry's Pit* at

*Specimens from Black Harry's Pit have been identified by Mr Alwyne Wheeler at the British Museum as apparently belonging to the European subspecies *R. sericeus amarus.*

Moreton near Wallasey in the Wirral, Cheshire, where a number were released. Six years later he transported some more to the Station Pond at Meols in Lancashire, where they thrived; after Black Harry's Pit was filled in, the stock of bitterling there was transferred to Meols. Writing in 1954, Hardy recorded bitterling as present in Collins Green Flashes; Duckery Flashes at Derbyshire Hill; part of the Southport Sluice; the rock hole at Bold; some ponds in the Haydock area; near the Black Horse Inn at Rainhill; and in the Knowsley area (where they had been introduced by Mr Edwards in 1950) – all within 4 miles (6 km) of St Helens. Local anglers used bitterling as live-bait when fishing for perch in the Flash at Pennington near Leigh, 8 miles (13 km) east of St Helens, and in colliery flashes at Wigan the same distance to the north-east, where previously bitterling were unknown. In 1953 large numbers of bitterling were transported by anglers on chub-fishing trips to Esthwaite Water in the Lake District of Lancashire, and to Rydal Water and Grasmere in Westmorland.

In Scotland, Dr John Berry released some bitterling in his pond in Fife in 1925, but by 1934 all had disappeared. A number were reported by Lansbury in a pond at Hadley Green in Hertfordshire in 1956. In the following year Orkin recorded that they were 'said now to be established in southern Yorkshire'; four years later Torbett repeated that bitterling were 'found wild in one or two small ponds in Yorkshire and Lancashire': these references to Yorkshire, however, may have resulted from confusion with waters in Lancashire.

The current status of the bitterling in England is uncertain and it seems likely that a number of introductions have failed and that overall the population may have declined in recent years. In 1965 Mr N. F. Ellison informed Mr Alwyne Wheeler that none had recently been seen in the Station Pond at Meols. A number of bitterling have been collected by hand-net in the Praes Branch of the Shropshire Union Canal on several occasions since September 1969, and Mr Wheeler received confirmation that they were still to be found there in 1974.

Note: See also Addenda.

51. Orfe
(*Leuciscus idus*)

The orfe or ide is widely distributed in slow-flowing and still waters throughout eastern and central Europe, and in the brackish waters of eastern Scandinavia bordering the Baltic Sea.

The back and upper parts of the sides of the orfe are greyish black and the lower sides and belly are silvery white; the eyes are yellow, and all the fins but the dorsal have a reddish tinge. An average length of about 1 foot (30 cm) is reached at the age of between six and nine years. An ornamental erythristic form known as the 'golden orfe' varies from pale yellow to orange in colour, often marked with dark patches.

The orfe is a migratory fish which between April and July (when the males become more brightly coloured) ascends small streams to spawn on stony and sandy beds, for which it requires a minimum water temperature of about 45°F (about 7–8°C). The cock fish usually reach the spawning redds first, where the females lay between about 40000 and 110000 sticky yellow eggs which become attached to aquatic vegetation, where they hatch after between a fortnight and three weeks. After spawning is completed, the fish migrate downstream – on the continent sometimes in immense shoals. In freshwater, orfe feed principally on aquatic insects (especially mayflies) and insect larvae – in brackish water on mussels, snails, and crustaceans. In continental Europe orfe tend to spend the winter in deeper waters.

The earliest known introduction of orfe to England appears to be one reported to Frank Buckland by Lord Arthur Russell in March 1874:

Ever since I first saw these splendid fish *Cyprinus orfus* in the ponds of the Imperial Palace, Laxenburg, near Vienna, I determined to introduce them if possible into England, and I was encouraged by your successful intro-

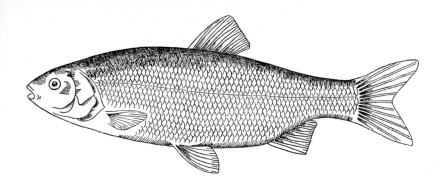

duction of the golden tench.* My first attempt a year ago was unsuccessful, two gold orfes only survived of the batch my brother [Lord Odo Russell, Ambassador at Berlin] had obtained in Berlin. They are still living in a pond at Woburn Abbey. My second and successful attempt has been accomplished with the assistance of Mr. Kirsch, Director of the Association for Pisciculture at Wiesbaden; he sent me one hundred and fourteen golden orfes of last year's breed, about two inches [5 cm] long each, and two large specimens. They travelled from Wiesbaden in two tin cans in charge of one of the clerks of the association, and favoured by the cold weather, they were all deposited, without a single loss, in one of the Duke of Bedford's ponds at Woburn Abbey.

Numerous other attempts have been made over the years to naturalize orfes in Britain, but few have succeeded in establishing a breeding stock. The poet and sportsman, Patrick Chalmers, believed that golden orfe seen from time to time in the Thames were probably the survivors or progeny of some introduced to the river Kennet by Frank Buckland. Writing in 1959 Fitter recorded breeding populations in ponds at Shottermill (near Haslemere) in Surrey; at Wivenhoe (near Colchester) in Essex; in Dumfriesshire; and in a ½-acre (0·2 hectare) pool at Owermoigne (between Dorchester and Swanage) in Dorset, where a stock of about 30–40 1½ lb (680 g) fish has existed since about 1910.

Mr Alwyne Wheeler, caught orfes in a forest pool at Loughton in Essex until 1961; the British Museum collection contains specimens from ponds at Westcott, near Dorking in Surrey (August, 1938);

*A variety of the common tench (*Tinca tinca*), the golden tench or schlei was first introduced to England either by Buckland or by Higford Burr of Aldermaston Park near Reading: a few small self-supporting colonies exist today in parts of southern England.

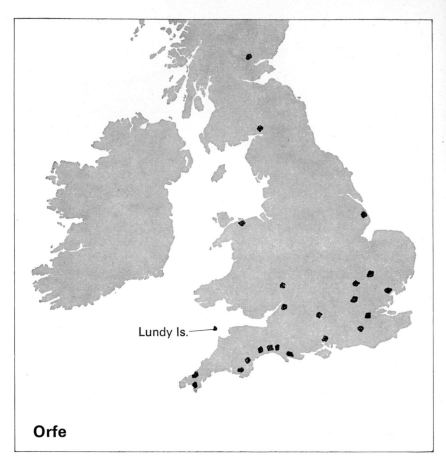

Lundy Is.

Orfe

from the river Kennet near Cookham in Hampshire (August 1938); and from the river Ember at Hampton Court, Middlesex (November 1968). In this last year a number of orfes were introduced to a shallow lake at Hayle Kimbro near the Lizard, and in 1971 to Tory Pond at Stithians (between Redruth and Falmouth) – both in Cornwall – where breeding appears to have occurred. The *Anglers' Mail* for 26 September 1973 reported the capture of a 4 lb (nearly 2 kg) orfe in the Winford Brook near Bristol, and the *Angling Times* of 1–2 January 1975 referred to two orfes, each weighing about 1 lb (453 g), caught in the Thames near Chertsey Bridge. In January 1976 Mr Brian Mills captured, in the river Test in Hampshire, a new British record rod-caught golden orfe which weighed 4 lb 3 oz (2 kg) and was subsequently identified by Mr Wheeler at the British Museum (*Anglers' Mail*, 4 February 1976).

Dr P. S. Maitland mentions the following additional places which are recorded as holding stocks of feral orfes: Oakington, Cambridgeshire; Hartford, Huntingdonshire; the Thames at Henley; the Ouse at Ely; Spetchley Lake, Worcestershire; Woburn Abbey, Bedfordshire; Lundy Island in the Bristol Channel; Moigne Combe Pond, Moreton, Dorset; Tortworth Lake, Falfield, Gloucestershire; Killiecrankie, Perthshire; Cemetery Lake, Southampton Common; Bullwants Pond, Mablethorpe, Lincolnshire; and Bodnant Gardens, Denbighshire.

According to the Duke of Bedford, the original stock of orfes at Woburn died out as a result of competition from other coarse fish in the lakes, and this may well be one of the principal causes for the species' failure to become naturalized in other apparently suitable waters in Britain.

52. Wels or European Catfish

(Silurus glanis)

The wels or European catfish is indigenous to continental Europe east of the river Rhine; it appears to be particularly common in eastern Europe, especially in the basin of the river Danube.

The slimy, scaleless, elongated body and broad, flat head containing a wide mouth give the wels a distinctly sinister appearance. The head, back and sides are usually some shade of greenish-black spotted with olive-green, and the underside is yellowy-white, with an indistinct blackish marbling; the head and back may sometimes be a deep velvety black and the sides occasionally take on a bronzy sheen. Two long barbels depending from the upper jaw, and four short ones from the lower jaw, help to give the catfish its name. There is no adipose fin, but an enormously elongated anal fin: a tiny dorsal fin is situated half-way between the bases of the pectoral and pelvic fins. The largest authenticated wels, taken from the river Dnieper in the Ukraine in the southern U.S.S.R., measured over 16 ft (5 m) in length and weighed 675 lb (306 kg); elsewhere in Europe and in England, however, the wels seldom exceeds 5 ft (152 cm) in length and 25 lb (11 kg) in weight.

The wels is a solitary species, living mainly in lakes and in the still waters of marshes and lagoons (sometimes in the lower reaches and backwaters of slow-flowing rivers) with a soft muddy bottom. It also occurs in the brackish water of certain parts of the Black and Baltic Seas. It spends the daylight hours skulking under stones or in the mud of the lake- or river-bed, or hidden in hollows or holes in the banks, only emerging after dark to forage for food, usually in shallow water. When adult the wels is a voracious predator, consuming – among other species – turbot, bream, crayfish, eels, frogs, roach, tench, ducklings, goslings and occasionally water-voles.* The wels

*According to Valenciennes. 'In the year 1700, on the 3rd of July, a countryman took one near Thorn [or Torun, Poland] which had the entire body of an infant in its stomach'; Grossinger relates that a Hungarian fisherman discovered the corpse of a woman in another 'having a marriage ring on her finger and a purse full of money at her girdle'.

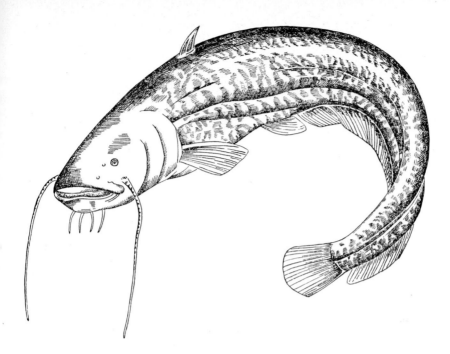

spends the winter in a torpid condition, emerging from semi-hibernation to spawn between May and July when the water has reached a minimum temperature of 68°F (20°C). The male scrapes a shallow hollow in the mud amidst thick vegetation close to the shore in which the female deposits about 15000 eggs per 1 lb (453 gm) of body-weight; these adhere to the vegetation which lines the nest, where they are guarded by the male until the young emerge after about three weeks.

In the Azov and Caspian Seas, in Lake Aral, and in those countries on either side of the river Danube, the wels is an important economic fish; in eastern Europe it is stocked in a considerable number of commercial fish-farms (especially in Hungary) where its scaleless skin is employed in the production of glue and leather, and its eggs are sometimes used to adulterate caviar.

John Fleming (1785–1857) includes *Silurus glanis* in his *British Animals* (1828) on the slender evidence of having discovered '*Silurus Sive Glanis*' in the *Scotia Illustrata* (1684) of Sir Robert Sibbald (1641 –1722) – who refers to *Atlas Novus* pt. 5; p. 148 – although Willughby, who also mentions *Silurus glanis*, does not list it as being found in

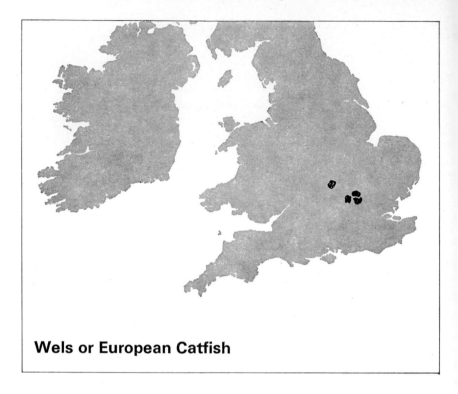

Wels or European Catfish

the British Isles. It seems possible that Sibbald was describing the superficially similar burbot (*Lota lota*), although a number of old authorities used the term *Silurus* when referring to the sturgeon (*Acipenser sturio*).

In his *Natural History of Ireland* (1849–56) Thompson, whose story is repeated by Yarrell and Day, describes the capture in Ireland of an unusual species of fish. According to Day,

an unique example of a fish which some have considered may be the *Silurus glanis* is stated to have been captured about 1827 or 1828 from a tributary of the Shannon, near its source, about three miles [5 km] above Lough Allen. A fisherman [William Blair of Florence Court] asserted that a fish at least 2½ feet [76 cm] long, and 8 or 9 lbs. [3–4 kg] weight, was seen struggling in a pool in the river as a flood subsided; that it had worm-like feelers to its mouth, while its appearance was so hideous that those who first saw it were afraid of touching it. The mouth of the figure of *Silurus glanis* in Yarrell's *British Fishes* [shown to Blair by Lord Enniskillen] was not considered large enough for that of the Irish specimen, but it must be observed that inquiries were only instituted in 1840. The captured fish was not eaten but adorned a bush for two or three years until the skeleton fell to pieces, and with it all evidence to connect *Silurus glanis* with Ireland.

In a footnote Day adds that

one or more, *fortunately unsuccessful*, attempts have been made during the last few years to introduce this hideous monster into British rivers. *Silurus glanis* has a voracious appetite, is a foul feeder, inferior as food, and almost rank when of large size, its presence would be of exceedingly questionable advantage.

The earliest of the attempts to naturalize the wels in Britain noted by Day appears to have been one referred to somewhat enigmatically by Mr Llewellyn Lloyd in his collection of reminiscences entitled *Scandinavian Adventures*, published in two volumes in 1854, in which he wrote: 'Through the indefatigable exertions of Mr George D. Berney of Morton, Norfolk, the silurus was last year [i.e. 1853] introduced into England. . . .' Since nothing further is heard of this introduction, it must be assumed that it was a failure.

Under the heading 'The Arrival of the Silurus Glanis in England', a second attempt was described with typically Victorian verbosity in *The Field* of 17 September 1864* by James Lowe, joint honorary secretary with Frank Buckland of the Acclimatisation Society, as follows:

That much desired fish, the Silurus, has at last been brought alive to this country, after various failures. The success is entirely due to the intelligent enterprise and perseverance of Sir Stephen B. Lakeman, who himself accompanied the fish all the way from Bucharest, a distance of 1800 miles; and on Thursday night I had the pleasure of assisting Mr. Francis Francis† in placing fourteen lively little baby-siluri in a pond not far from the fish-hatching apparatus belonging to the Acclimatisation Society on Mr. Francis's grounds at Twickenham.

When I state that Sir S. Lakeman had to change railway carriages more than thirty times during the journey, not to mention other vehicles, such as horse-carriages and steamers; that he started on the 23rd of August, and arrived in London with the fish on the evening of the 15th of September; and that during all that long journey he had to wage perpetual battle with the indifference and stupidity of officials, from station-masters down

*Although the contemporary account by Lowe clearly states that the date was 1864, Buckland gives 1862 or 1863 and Fitter gives 1865.
†Piscatorial Director of the Acclimatisation Society.

to porters (most of whom seemed to regard the fact of his travelling with a strange fish as rather a misdemeanour than otherwise), the reader will have some notion of the difficulties which have been overcome.

The fourteen little siluri (or siluruses) which have arrived are what remain of thirty-six of same species, which started from Kopacheni, where Sir S. Lakeman's estate is situated. This place is on the banks of the Argisch, a tributary of the Danube, and is about ten miles [16 km] from Bucharest. The Argisch abounds in silurus, and in all the other curious and almost unknown fish which swarm in the Danube, some of which (thanks to Sir S. Lakeman) we hope, at no very distant day, to reintroduce to their old friends, the siluri, in Mr. Francis's pond.

By way of preparation for the journey, the Siluri were placed in a water-cask, covered with a net, and placed in a large pond or lake of about 30 acres [12 hectares], belonging to Sir S. Lakeman; which pond abounds with fish, and yields silurus weighing up to 30 lb. and 40 lb. [13–18 kg], which may be caught with the line. . . . Of the thirty-six fish which started from Kopacheni, some were comparatively large (weighing up to 4 lb. [nearly 2 kg], and one of about 6 lb. [nearly 3 kg]), and some were mere fry. . . .

Sir S. Lakeman started (as I have stated) from Kopacheni on Aug. 23. He brought the fish, by Bucharest, to Giurgevo, a distance of fifty miles [80 km]; thence by steamer to Basias (in Transylvania), and so on by railway to Pesth, Vienna, Nuremburg, Cologne, Brussels, and Boulogne. The larger fish died first, all but the six-pounder, which endured to Vienna; and he only died there, it is supposed, because the servant in charge put his barrel into a stable, and it is likely that the ammoniacal atmosphere of the place disagreed with him. . . .

On arriving at Folkestone, there were fourteen survivors of the thirty-six which started from Kopacheni, and I am happy to say that every one of these reached Mr. Francis in the most lively and promising state. . . .

Immediately on his arrival in London, Sir Stephen Lakeman, with most praiseworthy public spirit, thought more of the fish than of himself; for without even driving to an hotel, he made his way to *The Field* office; and I need not describe with what delight he and his charge were welcomed. In a very short time we were on our way to Twickenham. . . .

So we got them safely down to Mr. Francis's, and on the brink of the pond turned them into a trough – fourteen little siluri, all alive and kicking, and as spry and frisky as possible. Their size varied from an ounce and a half to two ounces [42–56 g], for they are not more than three months old; but Sir S. Lakeman (who is well acquainted with the fish) declares that in a few weeks, when they have had the benefit of fresh water and plenty of food, their increase will be rapid and astonishing. When put into the water, they dived down to the bottom at once, with an easy vigorous movement, and waving their long barbels about, quite as if they knew their way about the pond which they then saw for the first time. From their flourishing condition, there is every reason to hope that they will increase and multiply. Indeed, I have now very little doubt that (with ordinary luck) this country has now acquired the *Silurus Glanis*. . . .

This is (so far as I am aware) the first time that this valuable fish has been brought to our shores; and the gratitude not only of the Acclimatisation Society, but of the country, is due to Sir Stephen B. Lakeman for the admirable manner in which he has effected the task which he unselfishly (and let me say patriotically) imposed upon himself.

Presumably at a later date, Buckland 'took down ten of them to my friend, Higford Burr, Esq.; Aldermaston Park, Reading, and turned them out into a large pond in front of the house. Some three years afterwards this pond was let dry – the silurus had entirely disappeared.'

In *The Field* of 8 September, 1894 Dr A. Gunther wrote:

A week or two ago, the gamekeepers [of Mr Nocton, of Langham Hall, Colchester] . . . caught a fish unknown to them, 4 ft. 3 in. [129 cm] long, and weighing over 30 lbs. [13 kg] in the River Stour at Flatford Mill, Suffolk. . . . Mr. Nocton suspected at once the real nature of the fish, and had it mounted by Mr. Gardener, of Oxford Street. How did the fish get into the River Stour? I am informed that Sir Joshua Rowley put, about twenty-nine years ago [*i.e.* in about 1865], young Siluri into a lake communicating with that river, and distant some six or seven miles [10–11 km] from the place of capture. There is, therefore, no doubt that this fish was a survivor of Sir Joshua's experiment. . . .

Fitter records that in 1872 the Marquess of Bath introduced a few small wels to his lake at Frome in Somerset, where they ate so many trout that, in 1875, the lake was drained to remove the three survivors, the heaviest of which weighed 28 lb (13 kg).

In 1881 Frank Buckland wrote:

There are several casts of Silurus in my museum in South Kensington.* The first I received from Mr. [T. R.] Sachs, in January, 1875. Herr Max von dem Borne was kind enough to catch this fish for us at Witsterwitz,

*The Fisheries Museum, founded by Buckland in 1865 in the Science Museum: expanded into the International Fisheries Exhibition in 1883.

near Berlin. He weighed 18 lbs. [8 kg] and measured 3 ft. 10 in. [116 cm] in length. . . .

In February, 1876, Lord Odo Russell, Ambassador at Berlin, kindly sent me, through Lord Arthur Russell, a magnificent specimen of the silurus. It measured 5 ft. 6 in. [167 cm] long, and weighed about 100 lbs. [45 kg]. A cast of this monster fish is in my museum. . . .

On the 27th. October, 1880, Lord Odo Russell brought to Woburn Abbey, from Berlin, seventy of the *Silurus glanis* in their second year. These Siluri were bred at Berneuchen, near Custrin, by Herr von dem Borne.

This is the largest and the most successful introduction of wels to Britain. (Fitter suggests that there may have been some subsequent importations to Woburn from the Danube.) The fish at Woburn flourished; when the lakes were netted in 1947 a specimen weighing 60 lb (27 kg)* was removed, and in 1951 several wels measuring up to 5 ft (152 cm) in length were netted from the Basin Pond. In the latter year a number of small wels were released in the lake at Claydon Park, near Steeple Claydon, in Buckinghamshire: by 1961 they had become firmly established, and on 18 June of that year three fish were captured, the largest of which (then a national rod-caught record) measured 4 ft 8 in. (142 cm) in length and weighed 33 lb 12 oz (15·3 kg). During the coarse fishing season of 1969 at least 35 wels weighing between 7 lb and 25 lb (3–11 kg) were taken from Claydon Lake (where in January 1976 a 29 lb 4 oz (13·3 kg) specimen was taken), and several were caught in the lakes at Woburn, although Wheeler and Maitland report that 'since that year, one of the Woburn Abbey lakes – Lower Drakeloe Lake – has been netted in order to remove the wels and other fish before the release of sea lions in the water as part of the Game Reserve!' The same authors mention that other waters on the Bedfordshire/Buckinghamshire border – two of which are at Little Brickhill (between Bletchley and Leighton Buzzard) and at Tiddenfoot Pit – have been stocked with wels taken from the lakes at Woburn.

None of the wels released by Dr J. Berry in his ponds in Fife in 1925 survived the disastrous floods of 1934. In 1949 two specimens of *S. glanis* were released in the river Ousel – a tributary of the Great

*One report gives the weight of this fish as 'about 78 lbs.' (35 kg) and the length as 'more than six feet' (183 cm).

Ouse – but one was killed shortly afterwards before they had had time to breed.

Writing in 1951, Glegg quotes Williams that in

about 1906 Mr. Walter Rothschild (afterwards [the 2nd] Lord Rothschild) placed two or three large catfish – up to 30 lbs. [14 kg] in weight – in Marsworth Reservoir, at Tring. Nothing more was seen of them, but in 1928 a 3 lb. [1·4 kg] specimen was caught from this water. . . . There can be but little doubt that it was a descendant of the fish placed in the reservoir some twenty years before.

On 14 August 1947, Mr C. Double, a water-bailiff, found a dead 42 lb (19 kg) wels* in Marsworth, and later in the year a second dead fish of about 12 lb (5 kg) was recovered. In 1947 Dandy remarked that 'the species also found its way from the reservoir into the Grand Union Canal, but its present status there is unknown'. Wheeler and Maitland mention a 7 in. (18 cm) long 'catfish' caught in the canal near Alperton in Middlesex in 1968, although as this specimen possessed only four barbels it may have been a released aquarium *Ictalurus* sp.† The same authorities report that on 26 September, 1970 a 5 ft (152 cm) long wels weighing 43½ lb (19·7 kg) – the present British rod-caught record – was caught in Wilstone Reservoir (which is connected to Marsworth by a mile-long pipe) by Mr Richard Bray; this appears to be the first wels reported from this water, where a new colony may soon become established.

*In The Old Queen's Head public-house at Startops End is, or was, a photograph of two men with a wels of about 40 lb – which may possibly have been this specimen – as well as the skeletal heads of a number of small wels which have been caught in the canal or reservoirs.

†See below p. 475.

53. Channel Catfish
(Ictalurus punctatus)

The channel catfish is widely distributed in North America from the prairie Provinces of Canada and the Hudson Bay area to the Great Lakes and the St Lawrence basin, as far south as Florida and northern Mexico.

This species is silvery, olive, steely or slaty-blue dorsally and yellowish whitish ventrally, all specimens except adult males being liberally spotted. In the breeding season males have a blue-black head, a blackish-blue body dorsally and are whitish-blue ventrally. The channel catfish has eight or so barbels depending from around the mouth – from which it derives its name – and a deeply forked tail. Adults seldom exceed a length of about 2½ ft (76 cm) and a weight of 2–4 lb (1–2 kg). The channel catfish is a migratory species which in North America ascends small streams to spawn between April and June – when the hen lays from 4000–34000 eggs – but at other times prefers to live in deeper and larger rivers and in lakes with cool, clear water and clean sandy or gravelly bottoms with no thick vegetation. The young eat various species of small invertebrates, the adults preying largely on other fish.

The earliest documented importation of North American catfish to the British Isles appears to be one referred to by Harvie-Brown and Buckley who state enigmatically that they were introduced prior to 1892 to 'Loch Uisg: Lochbuie, island of Mull'. It was, perhaps, this introduction which a few years later caused *Punch* to publish the following effusion:

> Oh, do not bring the Catfish here!
> The Catfish is a name I fear.
> Oh, spare each stream and spring,
> The Kennet swift, the Wandle clear,
> The lake, the loch, the broad, the mere,
> From that detested thing!

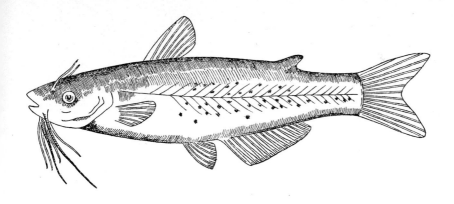

The Catfish is a hideous beast,
A bottom-feeder that doth feast
Upon unholy bait;
He's no addition to your meal,
He's rather richer than the eel;
And ranker than the skate.

His face is broad, and flat, and glum;
He's like some monstrous miller's thumb;
He's bearded like the pard,
Beholding him the grayling flee,
The trout take refuge in the sea,
The gudgeons go on guard.

He grows into a startling size,
The British matron 'twould surprise
And raise her burning blush
To see white catfish as large as man,
Through what the bards call 'water wan',
Come with an ugly rush!

They say the Catfish climbs the trees,
And robs the roosts, and down the breeze
Prolongs his caterwaul.
Oh, leave him in his western flood
Where the Mississippi churns the mud;
Don't bring him here at all!*

*Quoted by D. S. Jordan and B. W. Evermann in their *American Food and Game Fishes* (1902) – see Bibliography. I am indebted to Mr A. C. Wheeler for drawing these verses to my attention.

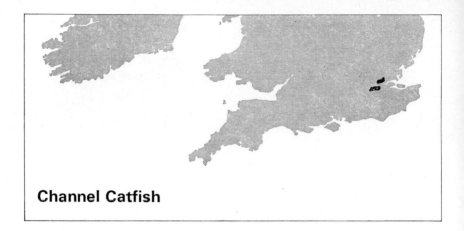

Channel Catfish

No further importations to Britain of North American catfish are recorded until the offer for sale in December 1968, for stocking purposes, of a shipment of channel catfish ostensibly imported as aquarium specimens from the United States. The only fish released into open waters appear to have been nine hundred 4 in. (10 cm) fingerlings aged about six months, purchased by the Hall & Co. Angling scheme. Mr J. Newby of Leisure Sport Ltd (successors to Hall & Co.) reports that these fish were introduced in January 1969 to four separate waters unconnected to any river or stream. These waters are the conjoined North and South Lakes (200 and 250 fish respectively) at Wraysbury, near Staines in Buckinghamshire; the Fleet Lake (250) at Thorpe, near Chertsey; the Sailing Club Lake (100) at Paper Court Farm, near Ripley; and the Car Park Lake (100) at Yateley, near Sandhurst – all in Surrey. As the stocking density was too low (the recommended figure is 1500–2000 per acre (0·4 hectare), and since channel catfish require warm and moving water devoid of predators and with a summer temperature of 80°–90°F (27°–32°C) for successful spawning, the likelihood of a long-term viable breeding population must be in some doubt.

Mr Wheeler reports that a colony of catfish or 'bullheads' has survived for a number of years in a lake near North Weald in Essex; their origin is unknown, although local pet-shops – in Epping and elsewhere – have from time to time sold black bullheads (*Ictalurus melas*) to aquarists, and they have occasionally been caught by anglers.

Wheeler and Maitland list the following records of 'American catfish' (some of which may have been *I. punctatus*) in British waters: 1. A 6 in. (15 cm) specimen caught at Kingsdown Road Flash, near Abram, Wigan, Lancashire, in July 1969; 2. Stock in a pond in Sussex reported by Mr D. F. Leney in December 1955; 3. A 7 in. (18 cm) specimen taken from the Grand Union Canal near Alperton (between Ealing and Willesden, in Middlesex) in September 1968; 4. Specimens of *Ameiurus nebulosus* introduced to ponds in Fife in 1925 by Dr J. Berry: none survived beyond 1934; 5. A specimen of *I. melas* found in a garden pond at Monkton, Ayrshire, in January 1970.

Three catfish (species unknown) were reported in the *Anglers' Mail* of 26 June 1975 to have been caught earlier in the year at Hill Top in the Bridgewater Canal, and another was taken at Norton: they had apparently been accidently introduced in a consignment of bream by the Warrington Angling Association in April 1974, and had thus successfully over-wintered in the canal.

54. Guppy
(Poecilia reticulata)

The guppy* is a native of north-eastern South America north of the Amazon and some of its offshore islands (the Leeward Islands, Trinidad, and Barbados in the Windward group) whence it has been widely introduced to many West Indian islands to control mosquitoes, and elsewhere for ornamental purposes. It was first imported to North America in about 1908.

The males of this tiny fish (which measure only about 1 in. (2·5 cm) in length) exhibit very considerable natural differences in both form and colouring; typical features are large dark 'eye' spots on the body and fins, the edges of which shine iridescently. The females (which are about twice as long as the males) are generally some shade of olive brown or yellowish green. As with the goldfish, numerous ornamental domesticated varieties have been obtained by selective breeding, two of the most popular of which are the 'Delta-tail' and the 'Veil-tail'. The food consists mainly of small aquatic insects and some aquatic algae. Guppies are ovoviviparous, the young hatching from the eggs before they emerge from the body of the female: broods average about twenty fish each, delivered over a period of some four weeks, and the female is capable of producing up to six broods a year from one fertilization by the male. A water temperature range of between about 68° and 86°F (20–30°C) is necessary for successful breeding.

There have, in recent years, been two well-established naturalized colonies of guppies at large in England. One of these, which was the subject of a study by B. S. Meadows between July 1966 and February 1968, is or was in the river Lee (or Lea) at Hackney in north-east

*Named after the nineteenth-century American naturalist and geologist, Robert J. L. Guppy, who settled in the island of Trinidad.

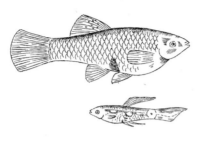

[TOP, FEMALE; BOTTOM, MALE]

London, where the temperature of the water was raised to an ideal height for the guppies by the cooling water discharge from the Central Electricity Generating Board's Hackney power station. This discharge was usually constant throughout the year, and the water temperature was therefore only subject to comparatively minor fluctuations, and was normally maintained even when the river was exceptionally full, since excess water was diverted over the Lee Bridge Weir some 500 yards (457 m) upstream of the power station's point of discharge. Throughout the year almost the entire flow in the Lee was diverted for use as cooling water; this, as Meadows points out, was important as the river water was thus constantly being recirculated, so ensuring that a ½ mile (804 m) dammed section of the river was perpetually at a higher temperature. The low quality of this stretch of water (where the guppies were established) was due to pollution in the late 1960s by the discharge of sewage effluent from Pymmes Brook which enters the Lee at Tottenham Hale, 1¼ miles (2 km) upstream of Hackney; this provided an effective barrier to many species of fish, and may thus have assisted the guppies to become established by keeping out competing species other than the three-spined stickleback (*Gasterosteus aculeatus*). Meadows reported that the young guppies appeared to feed principally in the vicinity of algal growths, while the adults frequented the more open water where they probably fed on mosquitoes and their larvae.

Guppy

In the early 1970s the pollution in the Lee decreased, and other native species began to re-colonize this stretch of the river. Following the recent closure of the generating station with a consequent lowering of the water temperature, and a further reduction of pollution in the river which will allow the return of more predating and competing indigenous species, this colony of guppies may now be extinct.

The second colony of naturalized guppies is established in the 400 yard (366 m) Church Street stretch of the St Helens Canal in Lancashire, the water of which is warmed by the discharge from the adjacent Pilkington Brothers glass factory: these guppies, which are said to have been established since 1963, when the stock of a defunct pet-shop was released in the canal, are sometimes used by local anglers as live-bait for perch.* In 1975 the Secretary of the St Helens Angling Association and other local anglers reported that the canal still contained a very large population of guppies, and that they were breeding.

*This stretch of water also contains a breeding population of the cichlid *Tilapia zillii* (see pages 497–9), as well as non-breeding stocks of angel-fish, catfish, chequered and rosy barbs, gouramies, and mollies.

This last population of guppies – which is without doubt the result of the liberation of pet aquarium fish – is almost certainly dependent for its survival on the continuing discharge of cooling water from the Pilkington glass works (as the colony in the river Lee owed its existence to the Hackney power station) and to the absence of competing or predatory species. It is, nevertheless, an extremely interesting example of a currently naturalized British fish.

55. Largemouth Bass

(Micropterus salmoides)

The largemouth bass is a native of Canada and the United States, where its range extends from Ontario and Quebec in the north through the Great Lakes and the valley of the Mississippi as far south as Texas. It has been introduced to other parts of North America and to continental Europe – to the latter first to Germany and Holland in 1883. Seven years later it was said to be breeding in a pond at Versailles, and it is now naturalized in many southern and eastern rivers of France.

The dorsal surfaces and the sides of the body and head of the largemouth bass vary in colour from a deep bronze-green to dull olive: the underside is a milky-yellow. The sides of the head are olive or golden-green, and along the sides runs a conspicuous solid black bar. The fins are a light olive-green – the dorsal fin being deeply divided. There are two indistinct dark bars on the gill cover, with a black spot on the extreme edge. In England adults seldom exceed 8–15 in. (20–38 cm) in length and 3 lb (1–2 kg) in weight, although in North America they may reach a length of up to 2 ft (61 cm) and a weight of 10 lb (4·5 kg). This species appears to favour warm and weedy lakes and slow-flowing lowland rivers with plenty of aquatic vegetation and soft bottoms: the young are usually found in the shallows, while the adults seem to prefer rather deeper waters. Largemouth bass feed when young principally on tadpoles, insect larvae, crustaceans and fish-eggs, and when adult on frogs and small fish. They spawn early, usually between March and May; the female prepares a shallow pit lined with aquatic vegetation on a sandy or pebbly bottom, in which she deposits 1000–4000 eggs, which she guards until they are hatched; the fry are also guarded for the first few days of their lives.

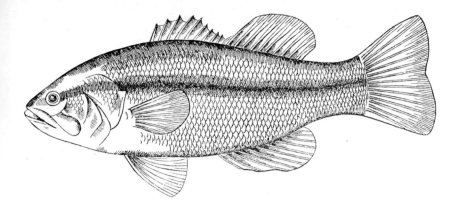

The earliest introduction of the largemouth bass to the British Isles appears to be that recorded by J. A. Harvie-Brown and T. E. Buckley who, in *A Fauna of Argyll and the Inner Hebrides* (1892), wrote:

These fish were imported by the Marquis of Exeter in 1879 – 'So far as I recollect, I imported through my pisciculturist, Mr. Silk, the Bass which were turned into Loch Baa by the Duke of Argyll in 1881 or 1882. These were, I think, mostly small-mouthed Bass, but there were a good many of the big-mouthed among them.'

In 1909, A. H. Patterson saw a specimen of the largemouth bass in the Wherry Hotel at Oulton Broad, near Lowestoft in east Suffolk, 'exhibited as the only survivor (!) captured out of a consignment from Austria that had been deposited in local waters; the others, it is believed, were all devoured by the Oulton Pike'. The same writer goes on to describe the largemouth as 'an introduced species, which did not flourish; had it done so, I think anglers would have very soon desired the extirpation of so voracious a fish'.

Writing in 1959 Fitter records that

about thirty years ago renewed attempts to establish black-bass fishing in Britain were definitely made with the large-mouthed species. . . . a German fish breeder had established a breeding stock of this form in Europe, and it was probably this acclimatized stock which was used for the second, more successful, wave of black-bass introductions. In 1927 the Norwich Angling Club put 250 large-mouthed yearlings into a three-acre [1·2 hectare] lake, but they disappeared in the following year. At about the same time fry from imported eggs were put into the River Ouse at Earith Bridge, but . . . some at least of these may actually have been pike-perch.

2L

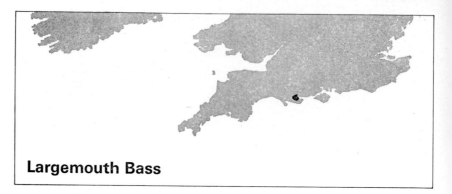

Largemouth Bass

In 1969 Mr D. F. Leney reported to Mr A. C. Wheeler and Dr P. S. Maitland that in 1934 or 1935 he had obtained some 4–7 in. (10–18 cm) specimens for the Surrey Trout Farm and United Fisheries at Shottermill near Haslemere in Surrey from an established colony in the river Loire near Bourges in central France. These fish were subsequently introduced to a gravel-pit at Send near Woking in Surrey, and to two pools near Wareham in Dorset and at Frome in Somerset; at Send a number of young fish were netted when the pit was cleared in 1937 or 1938, but Mr Leney considers that since then the stock has all died out. Fitter records that largemouth bass from Shottermill were also introduced to a pond in north-west Kent and to a lake in south Devon where they grew 'to a fair size'. Two further attempts to naturalize largemouth bass in Britain have been recorded; in 1935 a number were released in two reservoirs on the moors of Lancashire, but although a few were caught in June of the following year, the remainder apparently soon disappeared: a hundred yearlings liberated before the war in the lake at Coombe Abbey in Warwickshire, similarly failed to become established.

The current status of the largemouth bass in Britain is uncertain, and it may be that the colony in a disused clay-pit near Wareham, Isle of Purbeck is the only one presently extant. Fitter believes that the success of the fish (which have grown up to 3 lb (1–2 kg) in weight) at Wareham is due at least in part to the fact that the pit is sheltered and spring-fed and holds no pike but plenty of roach and perch. The British Museum collection contains a specimen taken from this water in 1965, and in 1969 an angler caught another weighing 2½ lb (1·1 kg), proving that the population is well established. This species is highly regarded by anglers as a sporting fish, but since when adult it is a piscivore and is known to be a direct competitor with trout, its introduction to waters holding other species could have serious consequences. It may be that the cold weather which frequently occurs in Britain during the breeding season of this early-spawning species, is one reason for its failure to become established in otherwise apparently suitable waters.

(In 1969 Dr P. S. Maitland and Mr C. E. Price reported the discovery of a North American monogenetic trematode (*Urocleidus principalis*) – a flattened leaf-shaped parasite (fluke) with adhesive suckers – on the gills of largemouth bass from Wareham. They suggest that this species (hitherto unrecorded in Britain) was probably introduced with the original stock of fish from the Loire, where the largemouth bass are presumably similarly infested).

56. Pumpkinseed

(*Lepomis gibbosus*)

The pumpkinseed, one of the family of sunfish, is a native of North America, where it ranges from southern Canada as far south as South Carolina, and eastwards from North and South Dakota to western Pennsylvania.

This species has a deep, compressed body which is greeny-bronze or olive on the back, and dull golden-green on the upper sides: the lower sides are golden, and are marked with irregular wavy inter-connecting blue-green lines. An indistinct dark band runs along the lateral line from head to tail, and the sides and head are flecked with olive, red or orange spots. The ventral surface varies from bronze to reddish orange. The two dorsal fins unite, and there is a distinguishing red spot edged with black on the rear corner of the gill cover. The pumpkinseed rarely exceeds a length of 9 in. (23 cm). It is found principally in small, cool, clear, shallow, weedy lakes and the lower reaches of slow-flowing rivers with alternate rapids and quiet pools. Here it feeds on worms, crustaceans and aquatic insects. Between May and August the female lays from 600 to 5000 yellow eggs in clumps in a hollow in the river- or lake-bed, where they are zealously guarded by the male until they hatch and the young disperse.

The pumpkinseed was much introduced to continental Europe between 1885 (when it was imported to France where it is now widely distributed throughout the lowland rivers) and about 1917, as a popular aquarium fish. It is now also well established in the Low Countries, Germany, western Russia, and in the basin of the Danube as far east as Rumania.

There appear to have been few attempts to naturalize the pumpkin-seed in Britain, although it has become established and bred in at least four separate waters. Mr D. F. Leney told Mr A. C. Wheeler and

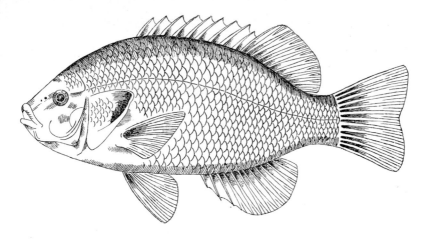

Dr P. S. Maitland that a breeding population existed at Groombridge (south-west of Tunbridge Wells) in east Sussex during the First World War. At about the same time Dr John Berry succeeded in establishing a small breeding population on his Tayfield estate at Newport-on-Tay in Fife. The British Museum collection contains two examples caught in a lake near Crawley (on the east Sussex/west Sussex border) in July 1938, and a third specimen caught in a pond near Bridgwater, Somerset in July 1953. The population at Crawley was still flourishing in 1976, although its future was thought to be threatened by encroaching development. In 1969 Mr Leney wrote to Mr Wheeler and Dr Maitland that he 'had bought a number of fine specimens which were caught locally, from a fishing tackle dealer in Bridgwater', and that 'this colony was established and breeding'. A single pumpkinseed was taken by an angler in the Hollow Pond, Whipp's Cross, Leytonstone, Essex, in 1974, and in the *Angling*

Pumpkinseed

Times of 24–25 September 1975 another angler, writing from Sanderstead, reported catching one in 'a Surrey pond'.

As both this and the following species are locally well established in southern or central Europe, their comparative failure to become more widely naturalized in Britain may be due to adverse climatic conditions.

57. Rock Bass

(Ambloplites rupestris)

Originally found only in the eastern central regions of North America west of the Appalachian Mountains, the rock bass has been introduced to other parts of the continent where it is now widely distributed, and has even penetrated, *via* man-made waterways, to the eastern seaboard.

The blotched and barred dorsal surface of the rock bass varies in colour from olive to a dull greeny or golden brown. The lower part of the head and the body below the pectoral fins and the belly are silvery white. The dorsal and caudal fins are freckled with brown, and there is an indistinct black spot on the upper edge of the gill cover. The eyes are orange or red. Adults reach a length of from 8–12 in. (20–30 cm) and weigh on average about 8 oz (226 g). This species is mainly found in cool, weedy lakes and in the lower stretches of deep and rocky streams and small rivers, where it feeds principally on small aquatic invertebrates. Spawning takes place between May and June when the female lays between 3000 and 11000 eggs in clumps in a nest on the sandy bottom of standing or slow-flowing waters.

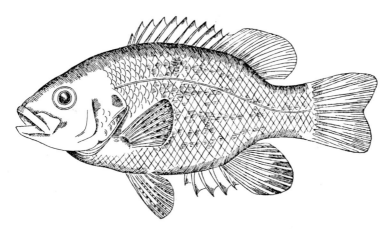

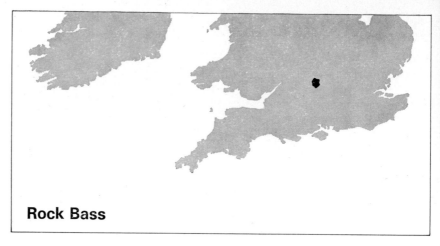

Rock Bass

The rock bass is known to be fully naturalized in only one water in the British Isles; for the past thirty-five years or more a colony has existed in a flooded three-acre (1·2 hectare) gravel-pit within 2 miles (3 km) of the northern outskirts of Oxford. How these fish came to be in this water is not known, but the first specimen – now in the collection of the British Museum – was caught in July 1937 by an angler who later captured a number of smaller specimens in the same water. In July 1956 the museum received a number of recently taken bass from Mr D. F. Leney, who reported that the pit was 'teeming' with the fish 'probably but few exceeding 6 or 7 in. [15–18 cm]'. In January 1969 Mr R. L. Manuel reported to Mr A. C. Wheeler and Dr P. S. Maitland that this colony was still in existence 'a few years ago', and so far as is known it continues to survive today.

58. Zander or Pike-Perch

(Stizostedion lucioperca)

The zander or pike-perch is widely distributed throughout north-western Europe as far north as Sweden and Finland, south to Yugoslavia and east to the Urals in Russia; a large and important commercial fishery is established in the lower reaches of the Volga river. In western Europe it has considerably extended its range by natural means following a number of artificial introductions – to many lakes in Schleswig-Holstein before the 1860s; to Lake Constance in 1883 and 1888, and at Huningue on the French bank of the Rhine in the early 1890s. By 1932 the zander had reached the valley of the Rhône and it later penetrated the Carmargue. Subsequent introductions have encouraged a further expansion of range.

The body-form of the zander or pike-perch is, like that of the pike (*Esox lucius*), long and narrow; like the common perch (*Perca fluviatilis*), however, the zander possesses two dorsal fins – one with spines, the other rayed – the former of which is marked with dark smudged spots which form a number of broken stripes. The back and sides – which are marked with dark vertical bars – vary from greyish to greeny-brown (the sides being of a lighter shade) and the underside is a dull milky white. The caudal fin is lightly scattered with dark spots. The narrow and elongated head encloses the typical aquatic predator's wide, flat, sharp-fanged mouth, the upper jaw of which extends well back past the pupil of the eye.

In the larger rivers of continental Europe (where it is widely exploited as a source of food and is known in Germany as '*zander*' or '*hechtbarsch*', meaning 'pike-perch'; in Hungary as '*fogosk*', and in Russia as '*berschik*') it may reach a weight of 25 lb (11 kg) or more and a length of about 3 ft (91 cm), although in lakes and smaller rivers it averages only about 5 lb (2·3 kg) and seldom exceeds 10 lb (4·5 kg) in weight.

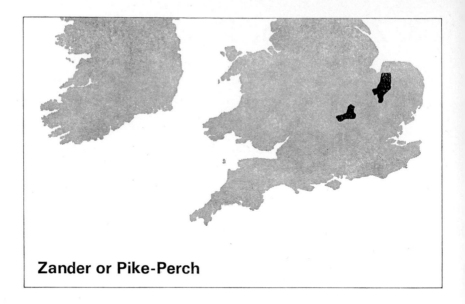

Zander or Pike-Perch

The zander prefers the warm and turbid parts of large and medium-sized lakes with plenty of aquatic vegetation and an abundant supply of oxygen, and the lower reaches of slow-flowing rivers. Here, when young, it feeds on invertebrates, and when adult preys on other fish.

In continental Europe breeding fish disperse from deep waters to shallows in the spring (when the ice melts in the northern part of their range); in lakes such movements are usually fairly localized, but may extend for up to 16 miles (25 km) or so. Populations living in brackish waters, as, for example, the Baltic Sea, migrate up in-flowing rivers for spawning, which takes place between April and June when the water has reached a temperature of at least 59°F (15°C). The female lays between 75 000 and 100 000 eggs per pound (453 g) of body-weight in clumps in shallow hollows on firm sandy or gravelly bottoms near the edges of lakes, where they adhere to growing water plants – usually *Typha* or *Phragmites* – or stones. Here they hatch, having been closely guarded by both parents, after about a week.

As early as 1861 the Acclimatisation Society was inveighing against the proposed introduction to the waters of the British Isles of so potentially dangerous a predator as the pike-perch, as it was then usually known. Nevertheless, exactly twenty years later, the in-

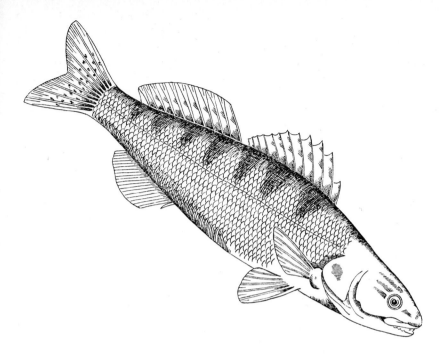

defatigable Frank Buckland – one of the founders of the Society –
wrote:

Many attempts have been made to transport this fish alive to England.
My friend, Mr. T. R. Sachs has taken immense pains in this matter, and
the following is his report to me of successful experiments to bring this
fish over alive for His Grace the Duke of Bedford.

'Mr. Dallmer, chief fishing master of Schleswig-Holstein, relates in the
German Fishery Circular of April, 1878, that he had the honour of being
requested by the President of the German Fishery Association to supply
His Grace the [9th] Duke of Bedford with about one hundred small
Zander, he having many opportunities of observing the peculiarities of
these fish in Schleswig-Holstein. It is well known that these fish are not
plentiful in the Elbe and Oder, abound in the Danube, and [are] not known
in the Rhine or Weser, although these two last are equally well suited for
this species. In many Schleswig lakes Zander are plentiful, in others not,
although connected with one another, and to all appearance the same
kind of water. . . . Sand, stone and clay ground they must have; they never
thrive on a muddy bottom.

'Mr. Dallmer came to the conclusion to get the Zander from shallow
lakes, as he knew such fish were better to transport, and in winter they
lived long in a fish box. He therefore wrote to the President that he thought

the undertaking possible, and would select fish of a large sort: first, because small fish were weak; second, they would more easily fall prey to larger fish; third, larger fish would soon after their arrival in England produce a family of English Zander.

'Mr. Dallmer selected 24 two-pounders [907 g] – twelve male, twelve female [from the lake at Bothkamper], so as to procure as many marriages as possible. . . . The next ship to London was the steamer *Capella*, and [due] to depart on Thursday night January 31, from Hamburg . . . arriving in London by Sunday mid-day. There they were met by servants of His Grace the Duke, who took charge of the carriers, and conveyed them to the railway station, about 3 English miles [5 km] distant. . . . On arrival [at Bedford] carriages were waiting to convey the fish to the estate, 4 English miles [6 km] more, where His Grace received them in person at the door of his mansion. "Are the fish alive?" he called out . . . twenty-three were very lively, the twenty-fourth being rather doubtful. His Grace had him cooked for dinner; four fish, two male, two female, were placed in one sheet of water, the others in a lake of about twenty-three acres [9 hectares], which was full of small fish but no pike, the gravelly bottom being eminently suited for Zander.

'This excellent result has succeeded . . . the fish were taken from the Schleswig lake on the 25th. January, and arrived at their destination on the 3rd. February, thus having been nine days on their journey. After this who will say Zander will not bear transporting? Henceforth, England possesses the celebrated aristocratic choice-eating fish, *Luceo perca zandra*, or pike-perch, thanks to His Grace the Duke of Bedford.

'Three previous attempts were made by my friend, Alexander Seydell of Stettin, and self to procure the Zander from Stettin by sea to London. On the first voyage all the fish died; the second produced three small live five-inch [13 cm] Zander. They died in Mr. Frank Buckland's museum at South Kensington, and are now there preserved in spirits. The third voyage nearly proved a shipwreck, as the steamer was thrown on her beam ends and all the gear on deck, the large Brighton Aquarium iron carrier included, were tossed into the North Sea.'

Thus ends Sachs's account of the earliest successful introduction of zander to the British Isles.*

In 1910 the 11th Duke of Bedford reinforced his stock of zander

*In 1959 Fitter wrote: 'on this occasion about a score of fish were sent from Schleswig-Holstein, having apparently been passed off as the North American black bass, which they somewhat resemble. The successful introduction was in 1910.' In fact exactly twenty-four fish were despatched from Schleswig-Holstein to Woburn, and as Wheeler and Maitland point out, the contemporary account by Sachs clearly indicates that the fish were pike-perch and not North American black bass. The 1910 introduction (*q.v.*) was no more than a re-inforcement of the already existing stock.

with a second consignment of fish from the continent to the lakes at Woburn Abbey, where they flourished and were angled for by members of the household. In 1947 the lakes were netted by the 12th Duke, who in a letter to *The Field* shortly afterwards wrote: 'A good stock of pike-perch, running up to 2½ lb [over 1 kg] were disclosed . . . I may add that the presence of pike-perch had not prevented an ample, and indeed excessive, stock of native fish from sharing the same water.' The Leighton Buzzard Angling Club were given about fifty zander, ranging in size from fingerlings up to fish of about 4 lb (nearly 2 kg) in weight; a small number were introduced into the river Ouzel and into an artificial watercourse which runs into the Grand Union Canal near Tring, but the majority were liberated in Firbank's Pit at Leighton Buzzard. No zander were reported from any of these waters until in 1950 a dead fish weighing 7 lb 2 oz (3·2 kg) was discovered at Firbank's Pit. The first fish, weighing 5½ lb (2·5 kg), to be caught on rod and line in Firbank's Pit was taken in 1951, and a number were caught in succeeding years until 1966 when the pit was filled in.

In 1950 the Leighton Buzzard anglers were presented with a further thirty young zander, each about 5 in. (13 cm) long, which they liberated in Claydon Lake near Steeple Claydon, 4 miles (6 km) south of Buckingham. Three years later these fish were breeding, and this colony is now considered to have become established, and has been the source of new stock for at least one other water nearby. In 1954 a few small zander were caught by anglers, and in the following year one of 4½ lb (2 kg) was taken as well as many smaller specimens. In the 1956–7 coarse fishing season six zander each weighing over 5 lb (2·3 kg) were caught at Claydon, the heaviest weighing 8 lb (3·6 kg). These figures suggest that the zander has a faster rate of growth in similar conditions than either the pike or the perch.

From the early 1960s the extension of the zander's distribution in England grew apace.* In the winter of 1959–60 the lakes at Woburn were once again netted, and ninety-seven zander were removed; either these fish were directly released into the Ouse Relief Channel of the Great Ouse river system at Stow Bridge, 3 miles (5 km) downstream of Downham Market by the water-bailiff, Mr Cawkwell, or, having been kept for some time in a stock-pond at Hengrave Hall,

*This has been well documented by Mr Alwyne Wheeler and Dr Peter Maitland, from whom much of the following information has been drawn.

near Bury St Edmunds, their progeny were liberated in the Channel in 1963 (two versions of this stocking exist). However that may be, this introduction appears to have been an immediate success; in July 1965 a 4½ lb (2 kg) specimen was caught in the Channel on rod and line at Downham Market, 11 miles (18 km) south of King's Lynn. In July 1969 Mr C. H. A. Fennell of the Great Ouse River Authority reported that 'zander are now prevalent in the Relief Channel between Denver [south of Downham Market] and King's Lynn and also in the cut-off channel upstream of Denver sluice'.

In July 1965 a water bailiff of the Great Ouse River Authority stated that, in order to reinforce the current stock, one hundred zander from Sweden had recently been released in the Relief Channel at Stow Bridge, some 2 miles (3 km) north of Downham Market. (Another source reported that five hundred fish had been liberated.) There is also evidence that an attempt was made in 1969 to introduce zander to 'a private fishery' in the Fenland district of Lincolnshire: this introduction may have been at Landbeach Lakes near Cambridge, where in 1965 it was reported that a specimen had been caught after 'a few pike-perch were put in as fingerlings three seasons ago'.

In the 1968 and 1969 coarse fishing season local anglers caught considerable numbers of zander in the Relief Channel where, according to Wheeler and Maitland, they 'are now numerous, breeding and firmly established . . . and have shown a rapid growth rate'.

In flood conditions the Relief Channel connects, through its head and tail sluices, with the Great Ouse river system, while a second channel runs from the head sluice into the rivers Little Ouse and Wissey. In this last river the *Angling Times* of 30 January 1969 reported that 'zander have now been seen . . . as far upriver as Stoke Ferry [7 miles (11 km) south-east of Downham Market]'. Wheeler and Maitland mention the following early reports in angling news-papers* of zander caught in waters other than the Relief Channel: in the river Cam (August 1969) and 'on the Dimmock's Cote stretch of the river' (July 1969); in 'the Old Bedford River just downstream of Welney' 8 miles (13 km) south-west of Downham Market (August 1969 and 1970); at Sutton Gault (August 1970); and in the River Nene at Milton Ferry and at Peterborough (August 1969). In October 1969 two zander weighing 7½ lb (3·4 kg) and 10 lb (4·5 kg) were caught in the cut-off Channel which runs into the Relief Channel, and a fish of 1½ lb (679 g) was taken at Brownshill Staunch

*See Bibliography.

– half-way between the confluence with the Hundred Foot Drain and St Ives. In October of the following year a 6 lb 13 oz (3·1 kg) zander was captured in the Great Ouse near St Ives, Huntingdonshire. Two zander were caught in the Cresswells reach of the Great Ouse near Ely in August 1970, and in December of that year a 2 oz (56 g) fingerling was netted in the Middle Level Drain. In September 1970, Dr Barrie Rickards furnished Mr Wheeler and Dr Maitland with particulars of a 7½ oz (213 g) zander caught in the Cam 600 yards (550 m) upriver of the confluence of the river and Swaffham Lode, and of a larger specimen from the Great Ouse near the A10 road-bridge at Littleport.

Since the publication of Wheeler's and Maitland's paper in 1973, references to zander in the angling press have included the following:

now showing up in good numbers throughout the whole length of the river [Delph] . . . the Old Bedford River . . . holds fair numbers . . . the Great Ouse itself holds zander . . . the advance is getting dangerously close to the Nene system too.

January, 1974.

Zander are now being taken in rapidly increasing numbers in the Forty Foot, Sixteen Foot and Pophams Eau. They have also appeared for the first time in The Twenty Foot and Old River Nene as they continue their spread in a westwards direction. Seven zander over 10 lb. [4·5 kg] have been taken – four from the Relief Channel (best 11 lb 14 oz [5·4 kg]), two from Roswell Pits (best 11 lb 8 oz [5·2 kg]), and one from the Wissey (13 lb 4 oz [6 kg]). Specimens around 8 lb (3·6 kg) have been reported from the Delph and Ouse at Littleport.

November, 1975.

In the same month it was reported that in the winter of 1973–4 a number of zander had been illegally released in a gravel-pit at Maxey (close to the Maxey Cut, a flood-relief channel running into the river Welland) 8 miles (13 km) north-west of Peterborough. A little under two years later the Anglian Water Authority removed six adults, 730 fry and fifty-five yearlings from this water.*

During preparation for Dr Maitland's mapping of the distribution of freshwater fish in the British Isles, Mr P. H. Langton supplied details of single zander caught in the Hundred Foot Drain, 4 miles (6 km) from Ely on 24 July 1967, and in the river Delph and the

*I am grateful to Mr Wheeler for drawing my attention to these references.

Old Bedford River, both the same distance from Downham Market. Additional reports were received from Mr W. Coyne of one or two zander caught some years ago in the river Mease, near Tamworth in Staffordshire: from Mr W. Pitt of some in a lake at Stoke Poges, near Slough in Buckinghamshire; and from Mr R. N. Campbell of the apparently hitherto unrecorded stocking in 1952 of a pond near Sevenoaks in Kent with fish from Woburn.

'There seems to be little doubt', Mr Wheeler and Dr Maitland conclude,

that the zander will spread through at least the rivers of East Anglia. Its spread has been considerably assisted by the stocking of fishing waters, and the presence of zander in numerous accessible waters will probably lead to misguided attempts to introduce the species elsewhere for angling purposes.* It is noticeable that not until it was released into a river system (the Ouse Relief Channel) did the expansion in its range become uncontrolled, and apparently uncontrollable.

In 1925 twenty fingerlings of the walleye (*S. vitreum*), which were believed to be black bass, were liberated in the river Ouse at Earith Bridge on the Cambridgeshire/Huntingdonshire border. In March 1934 an $11\frac{3}{4}$ lb (5·4 kg) specimen aged ten years was caught at Welney in the river Delph in the Ouse system. No further examples have been recorded in the Delph, and it is highly unlikely that any of this stock survives.

Wheeler and Maitland refer to evidence of another and previously unrecorded introduction of a species of *Stizostedion* which has recently come to light. On 25 March 1909 Mr Frank Batterson sent to Mr G. A. Boulenger at the British Museum a fish for identification which had been taken in 'the eel trap at Kings Mill, Skefford, Beds. on March 24th. Seven others of the same species having been caught in the trap during the last few months'. Mr Boulenger's answer is not extant, but on 31 March of the same year Mr Batterson wrote again in acknowledgement: 'I beg to thank you most sincerely for your kind and valuable answer to my enquiry re Pikeperch'. Unfortunately the specimen was not retained by the museum, and there is no way of knowing whether or not this specimen was *S. lucioperca*.

---

*In 1973 a $4\frac{1}{2}$ lb (2 kg) zander was caught in the river Severn below Tewkesbury Weir: previously, two small specimens were released in the river Lee in Hertfordshire. **Note:** See also Addenda.

59. 'Cichlid'
(Tilapia zillii)

This member of the family of *Cichlidae* – which has no English or native name – is widely distributed throughout northern and western Africa (where it is especially common in Lakes Victoria and Albert in Tanzania and Uganda) as far north as Israel.

Tilapia zillii has a blue-tinged olive-green body which is marked with six or seven dark vertical stripes. The lips are bright green, and the throat, which is normally a deep pink, becomes noticeably brighter and redder in the breeding season. The dorsal, anal and caudal fins are oliveaceous marked with yellow spots, and the edges of both the dorsal and anal fins are usually outlined with orange. The characteristic black '*Tilapia* spot' (which is rimmed with yellow) at the base of the rear half of the dorsal fin, is particularly distinct in *zillii*. This cichlid feeds exclusively on plants, and is therefore found principally in densely weeded waters, usually off exposed rocky or sandy beaches, but sometimes also in muddy coves. The females of some cichlids, having laid their eggs, take them into their mouth where they remain until the young fish have developed. The male of *Tilapia zillii*, however, constructs a simple basin-shaped nest in the sand among the reeds or tamarisk-bushes near the shore where the female deposits her eggs, which are subsequently zealously guarded by one or other of the parents, who constantly clean them and maintain a current of water over them. Adult *zillii* reach a maximum length of about 11 in. (28 cm).

The only breeding population of cichlids in the wild in Britain is found in the Church Street stretch of the St Helens Canal in Lancashire, where the water is heated by effluent discharged from the glassworks of Pilkington Brothers. This water also contains a breeding population of guppies and non-breeding stocks of other species of

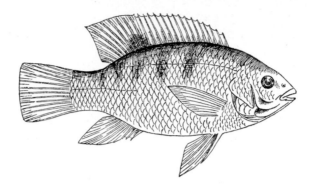

tropical fish, which became established in the canal in 1963 following the closure of a tropical-fish shop and the subsequent liberation of its stock.* In 1973 Mr A. C. Wheeler and Dr P. S. Maitland reported that Dr James Chubb of Liverpool University had made three cichlids (which had been caught in the spring of 1967 in the St Helens Canal and had subsequently been kept alive for some time in an aquarium) available for examination: these specimens, which were identified by Dr Ethelwynn Trewavas as *Tilapia zillii*, are now in the

*See pages 478–9.

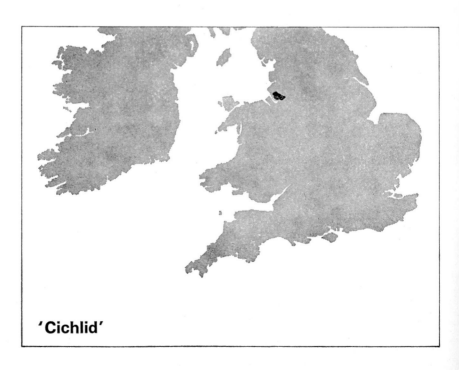

'Cichlid'

collection of the British Museum. In 1975 the Secretary of the St Helens Angling Association and other local anglers confirmed that this breeding population of cichlids – which, like that of the guppies, is presumably dependent for its existence and survival on the continued discharge into the canal of heated water – was flourishing, and that specimens weighing up to 12 oz (339 g) were being caught.

Epilogue

Two of the principal factors which today concern the introduction of animals from one part of the world to another are the effects such introductions have on the environment of the importing country and on its indigenous fauna and flora, and the ever present risk of transmitting contagious or infectious diseases. So far as the British Isles are concerned, the first of these problems has already been discussed in the foregoing chapters.

'Foot-and-mouth' disease, outbreaks of which occur from time to time in Britain, decimates stocks of man's domestic animals. Psittacosis or ornithosis, a respiratory infection transmitted principally by members of the parrot family, can, and sometimes does, prove fatal to man: in the winter of 1929–30 an outbreak in Europe, North Africa, and North America caused 750 cases which resulted in 143 deaths. Histoplasmosis, which causes high fever, lesions of the lungs and enlargement of the liver and spleen; cryptococcal meningitis; toxoplasmosis; encephalitis, which attacks human brain and nerve tissues, and infectious equine encephalomyelitis (or 'sleeping sickness') are just some of the other diseases with which animals can infect man. But the most virulent and feared of all such illnesses is rabies.

Rabies has been known to medical science and dreaded by man for over four thousand years; it is referred to in the code of Eshunna in about 2200 B.C., and the symptoms were clearly described by Aristotle (384–322 B.C.). It is still one of the most dangerous diseases known to man.

In Europe the rabies virus spread *via* the 'Polish Corridor' into Germany after the last war; from there it travelled south-westwards through Holland and Belgium to the Channel coast and the North Sea ports, first appearing in France in 1968. Between that year and 1975 a total of 8955 rabid animals were recorded in France, of which 7437 (83%) were foxes, 965 (11%) were cattle, 319 (3·5%) were cats and 234 ·(25%) were dogs. According to a map prepared by *Les Cahiers de Médecine Vétérinaire* in May 1970 for the World Health Organization, the disease had reached eastern Ardennes, and northern Meuse, Moselle and Bas-Rhin by the preceding month. Between 1970 and 1974 the number of known cases of rabies in France almost quintupled from a total of 482 in the former year to 2385 – of which 2151 (90%) were foxes – in the latter. It was projected that by 1976 rabies could have penetrated as far south as Aube and Haute Marne and by 1980 to Saône-et-Loire and Ain, advancing at the rate of between 20 and 30 miles (32 and 48 km) a year.

All warm-blooded mammals can catch and transmit rabies. In Europe, cats, dogs and, especially, foxes are the principal carriers of the virus: in Britain foxes are both increasing in numbers and are becoming more commensal, and it is therefore vital that they should not become infected. The W.H.O. emphasizes, however, that other animals can be, and often are, the origin of outbreaks of the disease: examples are bats in the United States and central America; skunks in Canada; monkeys in central Africa; jackals in Rhodesia; rats in Algeria; domestic cattle in South Africa, and mongooses in Puerto Rico.*

The rabies virus travels by two main routes – *via* wild or deliberately introduced animals: the natural water barrier of the English Channel provides a valuable defence for the British Isles against the transmission of the virus by wild animals, although the possibility always exists that it could gain entry through migrant bats, some species of which are known to cross the Channel, or by rats or feral cats stowing away on cross-Channel ships.

In order to combat the risk caused by animals imported by man, the Rabies Act of 1975, introduced following the Waterhouse Committee of Inquiry on Rabies, not only continued legislation under which dogs, cats and other mammals are subject both to an import licence and to a six months period of quarantine, but in addition, brought in three important new provisions. First, animals landed illegally may be destroyed; second, cats and dogs may now only enter at certain designated sea- and airports which possess quarantine quarters: third, fines of an unlimited amount (in place of the previous maximum of £400) as well as a prison sentence of up to one year may be imposed on indictment on serious offenders. The

*E. S. Tierkel *et al.*, 'Mongoose Rabies in Puerto Rico', *Publ. Health Reps*, 67(3), pp. 274–8, 1952.

authorities are also given power to control outbreaks by banning movement, by vaccinating, and by quarantining or destroying infected animals.

*

The expanding zoo and pet trade – witness the increasing number of wildlife and safari parks – is another reason for strictly controlling the transference of wild animals from one country to another;* firstly, because of the risk involved of denuding the exporting country of its native fauna and the threat thus caused to endangered species;† secondly, because of the ever-present danger of the escape from captivity in the importing country of potentially harmful and dangerous animals; and thirdly, because of the considerable amount of wanton cruelty entailed.

The Royal Society for the Protection of Birds in *All Heaven in a Rage: a Study of the Importation of Wild Birds into the United Kingdom*, compiled by Timothy Inskipp, and the International Society for the Protection of Animals, recently provided some statistics in connection with world wild life trade.‡

Six hundred thousand wild birds, with an estimated retail value of some £1 670 000, are believed to be imported into Britain every year, and the same number die after being caught before they even leave their native country: perhaps a fifth of those which reach Britain die soon after their arrival. The world trade may involve 5 500 000 birds *per annum*. Since 1970 birds from more than fifty-five endangered species have been imported to Britain, the principal offenders being Africa (with the exception of Kenya which permits trapping and exportation only under individual licence); Asia

---

*Some organizations have, however, been instrumental in saving from possible extinction a number of endangered species by breeding them in captivity; these include, among others, the Hawaiian Goose or né-né (*Branta sandvicensis*) and Pére David's deer (*Elaphurus davidianus*). Due to genetical and evolutional hazards which result from inbreeding, and on ethical grounds, this policy must be regarded as of doubtful validity.

†An additional and serious threat to some of the world's rarer animals – the discussion of which is outside the scope of this book – is that posed by the international and profitable trade in furs and skins.

‡Other references include: 'Agriculture and the Introduction of Non-Native Species', *Avic. Mag.* (Ed.), 75, pp. 70–1, 1969; K. Blackwell, 'Escapes', *North. and Soke of Peter. Bird Rep.*, 1971; C. L. Boyle, 'Control of the Importation of Wildlife', *Rep, to I.C.B.P. Confr.*, 1970; J. O. D'Eath, 'On Keeping Free-Winged Water-Fowl', *Avic. Mag.*, 79, pp. 70–3, 1973; M. D. England: 'Escapes', *Avic Mag.*, 76, pp. 150–2, 1970; D. Goodwin, 'The Problem of Birds Escaping from Captivity', *Brit. Birds*, 49, pp. 339–49, 1956.

(especially Bangkok and Singapore); Australia; India (especially Calcutta); Indonesia, and Papua New Guinea. North Korea and Thailand allow the export of native species in unlimited numbers; In 'The Plight of Thailand's Birdlife' (*Avic. Mag.*, 79, pp. 131–6, 1973) R. M. Martin quotes a statement from *Animal Exportations from Thailand in 1962–71* which provides an appalling example of that country's mercenary attitude to the problem:

Wild animals make up one of the natural resources of the country which are of value to the economy. . . . In the past Thailand has received millions of Baht income from the exportation of wild animals alone. Many species are in demand and therefore commercial business dealing with wild animals seems to be a real promising one.*

South American countries imposed restrictions on exports in 1973, and there is also restrictive legislation in force in North America: some North American breeding-species, however, are exported from their Central and South American winter-quarters. The United Kingdom is by no means blameless: in 1974, only ducks, geese, birds of prey (including owls), partridges, quails, pheasants, guinea- and domestic fowl and turkeys required an import licence – birds of prey and owls for reasons of conservation, and the remainder for health reasons. A ban on the importation of parrots had been withdrawn several years previously.† The R.S.P.B. also draws attention to the fact that the transportation of birds often involves considerable cruelty: it instances the case of birds captured in bird-lime; of those travelling in such cramped conditions that they suffer permanent injury or die; of live pigeons being placed in containers with birds of prey to provide food for the journey; and of the latter sometimes travelling with their eyelids sewn together so that they will not struggle. The R.S.P.C.A. hostel at London Airport (Heathrow) is called upon to deal with thousands of such cases annually.

Between 1965 and 1967, 750000 Greek tortoises (according to the International Union for the Conservation of Nature and Natural Resources now 'heavily depleted throughout its range' due to 'collection for the pet trade') were imported into Britain from

---

*See also: E. McClure and S. Chaiyaphum, 'The Sale of Birds at the Bangkok Sunday Market,' *Nat. Hist. Bull. Siam Soc.* 24 (1–2), pp. 41–78, 1971; J. A. Burton, 'Bangkok Bird Market,' *Birds Int.* 1(3), pp. 40–4, 1975.

†On 1 March 1976 a quarantine period of thirty-five days was re-introduced for birds mported into Britain.

Morocco. Large numbers of golden frogs from Panama and young spectacled caimans from South America are imported to Europe annually as pets, the majority of which soon die because they are not properly cared for, and in India lion-tailed macaque and capped langur females are sometimes shot so that their young can be more easily captured. In the two hundred and fifty years up to 1850, only two mammals and three birds became extinct in the United States and Puerto Rico: the number of endangered vertebrates throughout the world may now approach one thousand. Those at risk through the wildlife trade include – among many others – the golden lion marmoset from the maritime forests of Brazil; various species of spotted cats; the golden-shouldered paradise parakeet from Australia; the rainbow boa; the Mexican golden trout; various species of otters; many birds of prey, and almost all crocodiles, alligators, caimans and gavials.* The world trade in primates alone totals an estimated 160 000 to 200 000 animals annually.

In order to attempt to combat this trade in wild animals, the I.U.C.N. held a Convention in Washington in March 1973 on International Trade in Endangered Species of Wild Fauna and Flora. Fifty-seven out of eighty nations were signatories to the Treaty which came into force on 1 July 1975 following ratification by Canada, Chile, Cyprus, Ecuador, Mauritius, Nigeria, Sweden, Switzerland, Tunisia, the United States, Uruguay, and the United Arab Emirates who assented directly, a total of only two more than the minimum required but clearly insufficient to make the Treaty fully effective. Costa Rica, Nepal and Peru subsequently also ratified their decision, followed later by a further eight nations, making a total of twenty-five by the spring of 1976. The Treaty controls trade in wild species by: 1. Imposing strict control on the exportation of some 400 endangered species (many of which are included in *The Red Book* of wildlife in danger published in 1969 for the I.U.C.N.; 2. Placing a restriction on the trade in some slightly less rare species which will require an export permit; 3. Permitting any country to impose restrictions on any species not contained in the first two categories.

The United Kingdom implemented the Treaty on 1 January 1976 by introducing controls under the Import, Export and Customs

*Janet Coates Barber, 'Trading in Endangered Wildlife', *Country Life*, 159 (4117), pp. 1386–8, 1976.

Powers (Defence) Act (1939), and formally ratified its decision on 2 August 1976. The Endangered Species (Import and Export) Bill, which both confirms and restates already existing legislation under the 1939 Act and also extends its powers, was under consideration by both Houses of Parliament in the autumn of 1976.

Chronological Table
of Introductions

Date of Introduction of Present Stock	Species	Locality of Introduction	Introduced By	Natural Range	Remarks
? Late Stone Age or Early Bronze Age, c. 2000 B.C.	Orkney and Guernsey Voles	Orkney and Guernsey	? Neolithic or Early Bronze Age Man	Europe	
? Late Stone Age or Early Bronze Age	Wild Goat	—	? Neolithic or Early Bronze Age Man	Eastern Mediterranean and Middle East	
Iron Age c. 1000 B.C.	House-mouse	—	—	Europe	
?Late Bronze Age or Iron Age, c. 1000 B.C.	Fallow Deer	Southern England	? Phoenicians	Southern Europe	Well established by Roman period
?Pre-historic	White-toothed Shrews	Channel and Scilly Isles	—	Europe	
?Pre-historic	Yellow-necked Mouse	—	—	Europe	

Date of Introduction of Present Stock	Species	Locality of Introduction	Introduced By	Natural Range	Remarks
? Pre-historic	Soay Sheep	St Kilda	—	Scandinavia	May have been introduced in pre-Viking period (c. 300–400 A.D.) or by the Vikings c. A.D. 800
?c. A.D. 900	St Kilda Field mouse	St Kilda	? Norsemen	? Scandinavia	
Before the Middle Ages	Domestic Cat	—	—	North Africa	
Before 1176	Rabbit	Scilly Isles	? Plantagenets	Central Europe	Possibly introduced previously as a source of food by the Romans
Before 1177	Common Pheasant	—	? Saxons	Asia Minor	
? 12th century	Black Rat	Ireland	? Crusaders	South-east Asia	
Late 13th century (ferret)	Polecat-ferret	—	? Plantagenets	Europe	May have been originally introduced by the Romans to hunt rabbits
Before 1496	Common Carp	Southern England	—	Central Asia	Probably introduced to monastic fish-stews

Date	Species	Place	Introducer	Origin	Remarks
17th century	Canada Goose	—	—	Canada	First mentioned in collection of Charles II (reigned 1660–85) in St James's Park, London
17th century	Egyptian Goose	—	—	Africa	Illustrated (under 'Gambo-Goose or Spur-wing'd Goose') in Willughby and Ray (1678)
? c. 1691	Goldfish	London	? East India Company	Asia/China	Other dates suggested vary between 1603 and c. 1710
1728–9	Brown Rat	—	In ships from Russian ports	South-east Asia	—
By 1768	Chinese Ring-Necked Pheasant	—	—	China	Mentioned by Thomas Pennant in 1768
About 1770	Red-legged Partridge	Sudbourne Hall, Suffolk	Earl of Hertford	South-west Europe	First introduced to Windsor and Richmond Parks by Charles II in 1673: to Belvoir Castle by the Duke of Rutland in 1682, and to Wimbledon, Surrey, by the Duke of Leeds in 1712.
About 1770	Red-legged Partridge	Rendlesham Hall Suffolk	Lord Rendlesham	South-west Europe	

Date of Introduction of Present Stock	Species	Locality of Introduction	Introduced By	Natural Range	Remarks
About 1770	Red-legged Partridge	Alnwick Castle, Northumberland	Duke of Northumberland	South-west Europe	
About 1770	Red-legged Partridge	St Osyth, Essex	Earl of Rochford	South-west Europe	
1837–8	Capercaillie	Taymouth Castle Perthshire	Marquess of Breadalbane	Northern Europe	Originally a native species which became extinct c. 1785. Unsuccessfully reintroduced to Norfolk in c. 1823 by Mr (later Sir) Thomas Buxton et al.
1837	Edible Frog	Morton and Foulden Norfolk	Mr George D. Berney	Europe	Apparently established at Foulmire Fen, Cambridgeshire, by 1770s, but origin unknown.
c.1840	Japanese Pheasant	Knowsley Park, Lancashire	Earl of Derby	Japan	Mentioned by John Gould in 1857
1853	Wels	Morton Hall, Norfolk	Mr George D. Berney	Europe	Subsequently introduced by Sir Stephen Lakeman in 1864; by Sir Joshua Rowley in 1865; by Lord Odo Russell in 1880, et al.

1860	Japanese Sika Deer	Powerscourt, Co. Wicklow	Viscount Powerscourt	Japan	Escaped
	Japanese Sika Deer	Regent's Park, London	Zoological Society	Japan	Escaped
1868	American Brook Trout or Brook Charr	London	Mr Frank Buckland	Eastern North America	
1874	Little Owl	Stonewall Park, Chiddingstone, Kent	Lieut. Col. E. G. B. Meade-Waldo	Europe	Five from Italy released at Walton Hall, Yorkshire, by Charles Waterton, 1842–3
1888	Little Owl	About 40 at Lilford Hall, Northamptonshire	Lord Lilford	Europe	
1874	Orfe	Woburn Abbey, Bedfordshire	Lord Arthur Russell	Eastern and central Europe	
1876	Grey Squirrel	Henbury Park, Cheshire	Mr T. V. Brocklehurst	North America	In October 1828 grey squirrels were reported at Llandisilio Hall, Denbighshire
1878	Zander or Pike-Perch	Woburn Abbey, Bedfordshire	Duke of Bedford	North-western Europe	
1878 (or 1898)	Midwife Toad	Bedford	Horton & Smart Ltd	Western Europe	

Date of Introduction of Present Stock	Species	Locality of Introduction	Introduced By	Natural Range	Remarks
1884	Rainbow Trout	Delaford Park, Iver, Bucks	National Fish Culture Association	North America	Breeding at Sir James Maitland's hatchery at Howietoun, Stirlingshire, by 1887.
c. 1898	Mongolian Pheasant	Isle of Bute	Marquess of Bute	Kirgizskaya	
c. 1898	Prince of Wales's Pheasant	Isle of Bute; Kent; Norfolk	Marquess of Bute, et. al.	Afganistan/Turkestan	
Late 19th century	Lady Amherst's Pheasant	Woburn Abbey Bedfordshire, etc.	Duke of Bedford, et. al.	China	First introduced, according to G. C. Bompas, in 1828
Late 19th century	Silver or Kalij Pheasant	Woburn Abbey, Bedfordshire, etc.	Duke of Bedford, et al.	China	First introduced in 1740
From about 1900	Mandarin Duck	Woburn Abbey, Bedfordshire, etc.	Duke of Bedford. et al.	South-east Asia	First introduced to Richmond, Surrey, shortly before 1745 by Sir Matthew Decker, Bt
c. 1900	Chinese and Indian Muntiac Deer	Woburn Abbey, Bedfordshire	Duke of Bedford	Southern China; India	Escaped

Date	Species	Location	Introduced by	Origin	Notes
c. 1900	Chinese Water Deer	Woburn Abbey, Bedfordshire	Duke of Bedford.	China and Korea	Escaped
Since c. 1900	Fire-bellied Toad	Newdigate, Surrey; Woburn, Bedfordshire	Mr T. B. Rothwell; Duke of Bedford	Central and Eastern Europe and Asia	
Early 20th century	Pumpkinseed	Southern England	—	North America	Spread by release by anglers of surplus live-bait
Early 20th century	Bitterling	Cheshire and Lancashire	Surplus-stock from pet-fish shops	Europe	
20th century	Golden Pheasant	Elveden Hall, Suffolk, etc.	Earl of Iveagh, et al.	China	First recorded bird in Britain was in the collection of Eleazer Albin, about 1725
20th century	Golden Pheasant	Galloway	Duke of Bedford	China	
20th century	European Pond Tortoise	—	—	Europe	Earliest recorded introduction 1890–1
20th century	Southern or Italian Crested Newt	Newdigate, Surrey	Mr T. B. Rothwell	Southern Europe	
1902	Edible Dormouse	Tring Park Hertfordshire	Lord Rothschild	Europe	Possibly previously introduced as a source of food by the Romans

Date of Introduction of Present Stock	Species	Locality of	Introduced By	Natural Range	Remarks
About 1908	Red-necked Wallaby	Leonardslee Park, Sussex	Sir Edmund Loder, Bt	Tasmania	Three introduced in April 1865 by Acclimatisation Society. Present populations escaped in 1939
Mid-1930s	Red-necked Wallaby	Roaches House, Staffordshire	Captain H. C. Brocklehurst	Tasmania	
1900–10	European or Green Tree Frog	Beaulieu Abbey Estate, New Forest, Hampshire	Mr Jones	Europe	Earliest recorded introduction was made at St Lawrence, Isle of Wight, by Lord Walsingham in the 1840s
1929	North American Mink	—	—	North America	Escaped from fur-farms
1929	Coypu	—	—	South America	Escaped from (nutria) fur-farms
1920s–30s	Alpine Newt	Newdigate, Surrey	Mr T. B. Rothwell	Europe	
Before 1930	Pallas's Pheasant	Norfolk	—	Manchuria	

Date	Species	Place	Introducer	Origin	Remarks
1932	Wall Lizard	Farnham Castle, Surrey	—	Europe	
1954	Wall Lizard	Totnes, Devon	Viscount Chaplin	Europe	
1962	Wall Lizard	Isle of Wight	—	Europe	
1934–5	Marsh Frog	Stone-in-Oxney, Kent	Mr E. P. Smith (Edward Percy)	Europe	
1934 or 1935	Largemouth Bass	Wareham, Dorset	Surrey Trout Farm	North America	Said to have been first imported by the Marquis of Exeter in 1879
Before 1937	Rock Bass	North Oxford	—	North America	
1942	*Ph. c. satscheuensis*	Kent	—	Kansu	
1947	Painted Frog	Manchester area	Mr L. A. Lantz	Southern Europe and North-west Africa	
Early 1960s	Painted Frog	London area	—		
1950	Night Heron	Edinburgh Zoo	Royal Zoological Society of Scotland	North America	Imported by the Zoo in 1936. (A pair turned out by Lord Lilford in 1887 did not long survive)
c. 1954	Yellow-bellied Toad	Totnes, Devon	Viscount Chaplin	Europe	

Date of Introduction of Present Stock	Species	Locality of Introduction	Introduced By	Natural Range	Remarks
c. 1964	Yellow-bellied Toad	Bishopsteighton, Devon	Mr W. H. Coleridge	Europe	
1956	Bobwhite Quail	Dunwich, Suffolk	Major Pardoe	North America	Introduced to Ireland before 1813 by General Gabbit. Over a dozen other introductions later in the 19th century
1957	Bobwhite Quail	Helmingham Hall, Suffolk	Lord Tollemache	North America	
1964	Bobwhite Quail	Tresco, Isles of Scilly	Lieut.-Commander and Mrs T. Dorrien-Smith	North America	
1957	Ruddy Duck	Chew Valley Reservoir, Somerset	Wildfowl Trust, Slimbridge, Gloucestershire	North America	A pair in the collection of Mr Noel Stevens at Walcot Hall, Shropshire, bred in the 1930s. Present stock imported by the Wildfowl Trust in 1948: birds first escaped in 1952–3

1963	Guppy	St Helens Canal, Lancashire	Surplus stock from pet-fish shop	North-eastern South America	Survival presumed to be dependent on discharge of heated cooling water. R. Lee population may be extinct
c. 1963	Guppy	R. Lee, Hackney, north-east London	Surplus stock from pet-fish shop	North-eastern South America	
1963	'Cichlid' Tilapia zillii	St Helens Canal Lancashire	Surplus stock from pet-fish shop	North and west Africa	Survival presumed to be dependent on discharge of heated cooling water
1967	African Clawed Toad	Freshwater, Isle of Wight	Mr Frank Boyce	Africa	
1969	Channel Catfish	Four ponds in Buckinghamshire and Surrey	Hall & Co. Angling Scheme	North America	Earliest introduction may have been on Mull in 1802
1969	Himalayan or Hodgson's Porcupine	Okehampton, Devon	Pine Valley Wildlife Park	India	Escaped
1969	Ring-necked Parakeet	South London	—	Africa/India	Escaped and/or released
Late 1960s	Wood Duck	Between Guildford and Farnham, Surrey	—	North America	'Breeding freely' in Devon by the late 19th century, but soon died out

Date of Introduction of Present Stock	Species	Locality of Introduction	Introduced By	Natural Range	Remarks
1969	Budgerigar	Tresco, Isles of Scilly	Lieut.-Commander and Mrs T. Dorrien-Smith	Australia	First introduced to England by John Gould in about 1840
1970	Reeves's Pheasant	Kinveachy, Inverness-shire	—	China	Earliest British record is that of a bird presented to the Zoological Society by John Reeves in 1831
c. 1970	Tesselated Snake	—	—	Southern Europe, central Asia, Eastern North Africa	Bred first in Yorkshire in 1971.
1972	Crested Porcupine	Staffordshire. (near Stoke-on-Trent)	Alton Towers Botanical Gardens	North Africa	Escaped
1973	Mongolian Gerbil	Fishbourne, Isle-of-Wight	—	Mongolia	Escaped from or released by a film company

Chapter Bibliography

The Acclimatisation Society

BLUNT, Wilfrid, *The Ark in the Park*, Hamish Hamilton–Tryon Gallery, 1976.

BOMPAS, G. C., *Life of Frank Buckland*, Smith, Elder & Co., 1888.

BUCKLAND, F. T., in: *Journal of the Royal Society of Arts*, 9, pp.19–34, 1860.

—, Personal papers in the Library of the Royal College of Surgeons of England.

BURGESS, G. H. O., *The Curious World of Frank Buckland*, John Baker, 1967.

FITTER, R. S. R., *The Ark in Our Midst*, pp. 31–9, Collins, 1959.

GRAY, J. E., in: *Report of the British Association for the Advancement of Science* (1864–5), *The Field*, 12 and 19 July 1862; 4 and 18 July 1863.

MITCHELL, D. W., in: *Edinburgh Review*, 61, pp. 161–88, 1860.

—, *All the Year Round*, 5, pp. 492–6, 1861.

MITCHELL, P. Chambers, *Centenary History of the Zoological Society of London*, London Zoological Society, 1929.

OWEN, Rev. R., *The Life of Richard Owen*, John Murray, 1894.

SOCIETY FOR THE ACCLIMATISATION OF ANIMALS, BIRDS, FISHES, INSECTS AND VEGETABLES WITHIN THE UNITED KINGDOM, Annual Reports, 1861–5.

1. Red-Necked Wallaby

ANON., 'Britain's Wild Wallabies', *Animals*, 7, p. 524, 1965.

EDWARDS, K. C., *The Peak District*, Collins, 1962.

FITTER, R. S. R. *The Ark in Our Midst*, pp. 52–3, Collins, 1959.

GOULD, J., *The Mammals of Australia*, 1863.

MALLON, D., 'Britain's Wild Wallabies', *Animals*, 13, pp. 256–7, 1970.

MOLLISON, B. C., 'Food Regurgitation in Bennett's Wallaby', *C.S.I.R.O. Wildl. Res.*, 5, pp. 87–8, 1960.

PAGE, F. J. T., and TITTENSOR, A. M., *The Sussex Mammal Report, 1969, 1970, 1971*, Sussex Nat. Trust, Henfield, 1970–2.

RIDE, W. D. L., *Native Mammals of Australia*, O.U.P., 1970.

SHARMAN, G. B., and CALABY, J. H., 'Reproductive Behaviour in the Red Kangaroo in Captivity', *Int. Zoo. Yrbk*, 6, pp. 138–40, 1966.

TROUGHTON, Ellis, *Furred Animals of Australia*, Angus and Robertson, 1957.

YALDEN, D. W., and HOSEY, G. R., 'Feral Wallabies in the Peak District', *Jour. Zool. Lond.*, 165, pp. 513–20, 1971.

2. Greater and Lesser White-Toothed Shrews, 3. Orkney and Guernsey Voles, 4. Yellow-Necked Mouse and St Kilda Field Mouse

BEIRNE, B. P., 'The History of the British Land Mammals', *Ann. Mag. Nat. Hist. Ser.*, 11, 14, pp. 501–14, 1947.

—, *The Origin and History of the British Fauna*, London, 1952.

BERRY, R. J., 'History in the Evolution of *Apodemus Sylvaticus* at One End of its Range', *Jour. Zool. Lond.*, 159, pp. 311–28, 1969.

CORBET, G. B., 'Origin of the British Insular Races of Small Mammals and of the Lusitanian Fauna', *Nat. Lond.*, 191, pp. 1037–40, 1961.

—, 'The Identification of British Mammals', *Brit. Mus. Nat. Hist. Lond.*, 1969.

—, 'The Geological Significance of the Present Distribution of Mammals in Britain', *Bull. Mamm. Soc.*, 31, pp. 14–16, 1969.

—, *The Distribution of Mammals in Historic Times*, Systematics Association Special Vol. No. 6, 1974.

CORKE, D., 'The Local Distribution of the Yellow-necked Mouse', *Mamm. Rev.*, 1, pp. 62–6, 1970.

HINTON, M. A. C., 'On a New Species of *Crocidura* from Scilly', *Ann. Mag. Nat. Hist. Ser.*, 9, 14, pp. 509–10, 1924.

JEWELL, P. A., and FULLAGER, P., 'Fertility among Races of the Field Mouse and their Failure to form Hybrids with the Yellow-necked Mouse', *Evol.*, 19, pp. 175–81, 1965.

THURLOW, W. G., 'The Yellow-necked Mouse at Stowmarket', *Trans. Suff. Nat. Soc.*, 10, pp. 297–300, 1958.

WILLIAMSON, K., and BOYD, J. M., *St. Kilda Summer*, Hutchinson, 1970.

YALDEN, D. W., 'A Population of the Yellow-necked Mouse', *Jour. Zool. Lond.*, 164, pp. 244–50, 1971.

5. Mongolian Gerbil

AISTROP, J. B., *The Mongolian Gerbil*, Dobson, 1968.

ALLEN, G. M. (in Granger, W. ed.), *Natural History of Central Asia: The Mammals of China and Mongolia:* Vol. 2, Amer. Mus. Nat. Hist., 1940.

ANON., 'Gerbils' (12), *Zool. Soc. Lond.*, 1969.

GULOTTA, E. J., 'Mammalian Species', No. 3, *Meriones unguiculatus*, pp. 1–5, Amer. Socy. Mammalogists, 19 Jan. 1971.

KUEHN, R. E., and ZUCKER, I., 'Mating Behaviour of *M. unguiculatus*', *Amer. Zool.*, 6.

—, 'Reproductive Behaviour of the Mongolian Gerbil', *Jour. Comp. Physiol. Psychol*, 66 (3), pp. 747–52, 1968.

MARSTON, J. H., and CHANG, M. C., 'The Breeding, Management, and Reproductive Physiology of the Mongolian Gerbil', *Lab. Anim. Care*, 15 (1), pp. 34–48, U.S.A., 1965.

MONROE, B. N., *Gerbils in Color*, T.F.H. Publ. Inc., Jersey City, 1967.

NAKAI, K., 'Reproductive and Post-Natal Deveolpment of the Colony-Bred *M. unguiculatus*': *Bull. Exper. Anim*, 9 (5), Japan.

NAUMAN, D. J., 'Open Field Behaviour of the Mongolian Gerbil', *Psych. Sci.*, 10 (5), pp. 163–4, 1968.

ROBINSON, D. J., *How to Raise and Train Gerbils*, T.F.H. Publ. Inc., Jersey City, 1968.

ROBINSON, P. F., 'Metabolism of the Gerbil', *Sciences*, 130 (3374), pp. 502–3, 1959.

SCHNEIDER, E. (ed.), *Enjoy Your Gerbils*, Pet Library Ltd.

SCHWENTKER, V., 'The Gerbil – a New Laboratory Animal', *Illin. Veter.*, 6, pp. 5–9, 1963.

SOCOLOF, L., *Gerbils as Pets*, T.F.H. Publ. Inc., Jersey City, 1966.

WALKER, E., *Mammals of the World* (2 vols.), Johns Hopkins, 1964.

WALTERS, G., PEARL, J., and ROGERS, J. V., 'The Gerbil as a Subject in Behavioural Research', *Psych. Reps*, 12 (2), pp. 315–18, 1963.

6. Rabbit

ANDREWS, C. H., 'Myxomatosis in Britain', *Nature Lond.*, 174, pp. 529–30, 1954.

ANDREWS, C. H., THOMPSON, H. V., and MANSI, W., 'Myxomatosis', *Nature Lond.*, 184, pp. 1179–80, 1959.

ARMOUR, C. J., and THOMPSON, H. V., 'Spread of Myxomatosis in the first outbreak in Great Britain', *Ann. App. Biol.*, 43, pp. 511–18, 1955.

BACKHOUSE, K. M., and THOMPSON, H. V., 'Myxomatosis', *Nature Lond.*, 176, pp. 1155–6, 1955.

BRAMBELL, F. W. R., 'The Reproduction of the Wild Rabbit', *Proc. Zool. Soc. Lond.*, 114, pp. 1–45, 1944.

FENNER, F., and RATCLIFFE, F. N., *Myxomatosis*, C.U.P., 1965.

FITTER, R. S. R., *The Ark in Our Midst*, pp. 114–22, Collins 1959.

HAMILTON, J., *Reproduction in the Rabbit*, Oliver & Boyd, 1925.

HARTING, J. E., *The Rabbit*, Longmans Green, 1898.

LEVER, Christopher, 'The Invaders, and how they came here', *The Field*, p. 394, 28 August 1969.

LLOYD, H. G., and WALTON, K. C., 'Rabbit Survey in West Wales', *Agric.*, 75, pp. 32–6, 1969.

LOCKLEY, R. M., 'Some Experiments in Rabbit Control', *Nature Lond.*, 145, p. 767, 1940.

—, *The Private Life of the Rabbit*, André Deutsch, 1964.

MATHESON, C., 'The Rabbit and the Hare in Wales', *Antiquity*, 15, pp. 371–81, 1941.

PHILLIPS, W. M., STEPHENS, M. N., and WORDEN, A. N., 'Observations on the Rabbit in West Wales', *Nature Lond.*, 169, pp. 869–70, 1952.

RITCHIE, J., *The Influence of Man on Animal Life in Scotland*, 1920.

RITCHIE, J. N., HUDSON, J. R., and THOMPSON, H. V., 'Myxomatosis', *Vet. Rec.*, 66, pp. 796–804, 1954.

ROGERS, Thorold, *History of Agriculture and Prices in England* (Vol. 1).

SHEAIL, J., *Rabbits and their History*, David and Charles, 1971.

—, 'Historical Material on a Wild Animal – the rabbit', *Local Historian*, 9, pp. 59–64, 1970.

SPENCER, H. E. P., 'Rabbit', *Trans. Suff. Nat. Soc.*, 9, pp. 369–70, 1956.

STEPHENS, M. N., 'Seasonal Observations on the Wild Rabbit in West Wales', *Proc. Zool. Soc. Lond.*, 122, pp. 417–34, 1952.

THOMPSON, H. V., 'The Rabbit Disease: Myxomatosis', *Ann. App. Biol.*, 41, pp. 358–66, 1954.

THOMPSON, H. V., and WORDEN, A. N., *The Rabbit*, Collins, 1956.

VEALE, E. M., 'The Rabbit in England', *Agric. Hist. Rev.*, 5, pp. 85–90, 1957.

7. House Mouse, 8. Black Rat, 9. Brown Rat

BENTLEY, E. W., 'The Distribution and Status of *Rattus rattus L.* in the United Kingdom in 1951 and 1956', *Jour. Anim. Ecol.*, 28, pp. 299–308, 1959.

—, 'A further loss of ground by *Rattus rattus L.* in the United Kingdom during 1956–61', *Jour. Anim. Ecol.*, 33, pp. 371–3, 1964.

BERRY, R. J., 'The Natural History of the house mouse', *Field Study 3*, pp. 212–62, 1970.

BISHOP, I. R., and DELANEY, M. J., 'The Ecological Distribution of Small Mammals in the Channel Islands', *Mammalia*, 27, pp. 99–110, 1963.

CHITTY, D., and SHORTEN, M., 'Techniques for the Study of the Norway Rat', *Jour. Mamm.*, 27, pp. 63–78, 1946.

CHITTY, D., and SOUTHERN, H. N., (eds), *Control of Rats and Mice* (Vols. 1–3), Clarendon Press, 1954.

CORBET, G. B., *The Distribution of Mammals in Historic Times*, Systematics Association Special Volume No. 6, Academic Press, 1974.

COTTON, M. J., 'Fleas (*Siphonaptera*) Recorded from Lundy Island', *Ent. Mon. Mag.*, 96, p. 243, 1960.

CROWCROFT, P., *Mice All Over*, Foulis, 1966.

DEANE, C. D., 'The Black Rat in the North of Ireland', *Irish Nat. Jour.*, 10, pp. 296–8, 1952.

ELTON, C., *Voles, Mice and Lemmings*, Clarendon Press, 1942.

FAIRLEY, J. S., 'A Critical Re-appraisal of the Status in Ireland of the Eastern House Mouse . . .', *Irish Nat. Jour.*, 17, pp. 2–5, 1971.

FISHER, J., 'The Black Rat in London', *Lond. Nat.*, 29, p. 136, 1950.

—, 'St. Kilda, a natural experiment', *New Nat. Jour.*, 1, pp. 91–108, Collins, 1948.

FITTER, R. S. R., *The Ark in Our Midst*, pp. 107–14, Collins, 1959.

HINTON, M. A. C., 'Rats and Mice as Enemies of Mankind', *Brit. Mus. (Nat. Hist.)*, 1931.

LAURIE, E. M. O., 'The Reproduction of the House-mouse living in Different Environments', *Proc. Roy. Soc. Lond.*, 133B, pp. 248–81, 1946.

LEVER, Christopher, 'The Invaders, and how they came here', *The Field*, p. 394, 28 August 1969.

MATHESON, C., 'A Survey of the Status of *Rattus rattus* and its Sub-species in the Sea-ports of Great Britain and Ireland', *Jour. Anim. Ecol.*, 8, pp. 76–93, 1939.

—, *The Brown and the Black Rat in Wales*, The Nat. Mus. of Wales, Cardiff, 1931.

—, 'The Brown Rat', *Animals of Britain*, 16, Sunday Times Publications, London, 1962.

—, 'The Black-rat at British Sea-ports', *Pest Technol.*, 1, pp. 4–7, 1958.

M.A.F.F., *Rats and Mice in England and Wales, 1966–68*. Technical Circular No. 23.

NEWSOME, A. E., 'The Ecology of House-mice in Cereal Hay-stacks', *Jour. Anim. Ecol.*, 40, pp. 1–16, 1971.

ORR, R. T., 'Communal Nests of the House-mouse', *Wasmann Collector*, 6, pp. 35–7, 1944.

PALLAS, P. S., in STEINIGER, F., *Rattenbiologie und Rattenbekämpfung*, 1831; repr. Stuttgart, Ferdinand Enke Verlag, 1952.

RITCHIE, J., *The Influence of Man on Animal Life in Scotland*, 1920.

SCHWARZ, E., and H. K., 'The Wild and Commensal Stocks of the House-mouse', *Jour. Mammal*, 24, pp. 59–72, 1943.

SILVER, J., 'The Introduction and Spread of House Rats in the United States', *Jour. Mammal.* 8, pp. 59–60, 1927.

SOUTHERN, H. N., and LAURIE, E. M. O., 'The House-mouse in Corn-ricks', *Jour. Anim. Ecol.*, 15, pp. 134–49, 1946.

TWIGG, Graham, *The Brown Rat*, David and Charles, 1975.

WATSON, J. S., 'The Melanic form of *Rattus norvegicus* in London', *Nature*, 154, pp. 334–5, 1944.

10. Edible Dormouse

CARRINGTON, R. P., 'The Edible Dormouse', *Jour. Assoc. Sch. Nat. Hist. Soc.*, 3, pp. 27–9, 1950.

FITTER, R. S. R., *The Ark in Our Midst*, pp. 98–100, Collins, 1959.

FLOWER, Major F. S., 'Exhibition of Skin and Skull of Female Fat Dormouse', *Proc. Zool. Soc. Lond.*, p. 769, 1929.

LEUTSCHER, Alfred, 'The Edible Dormouse', *Discovery*, 15, pp. 70–1, 1954.

LEVER, Christopher 'The Invaders, and how they came here', *The Field*, p. 395, 28 August, 1969.

LLOYD, B., 'A List of the Vertebrates of Hertfordshire. 4. Mammals', *Trans. Herts. Nat. Hist. Soc.*, 22, p. 34, 1947.

MIDDLETON, A. D., 'Whipsnade Ecological Survey, 1936–7', *Proc. Zool. Soc. Lond. (A)*, 107, p. 480, 1937.

PALMER, A., *Countryside*, N.S. 16, 1, pp. 7–9, 1951.

POTTS, G., 'The Garden Dormouse in Shropshire', *Northwest Nat.*, 17, p. 246, 1942.

PROCTOR, Elsie, 'The Fat Dormouse in Captivity', *Beds. Nat.*, 3, p. 21, 1949.

STREET, Philip, 'The Edible Dormouse in England', *Country Life*, 188, pp. 1484–5, 1955.

THOMPSON, H. V., 'The Edible Dormouse in England 1902–51', *Proc. Zool. Soc. Lond.*, 122, pp. 1017–24, 1953.

THOMPSON, H. V., and PLATT, F. B., 'The Present Status of *Glis* in England', *Bull. Mamm. Soc. Brit. Isles*, 21, pp. 5–6, May 1964.

VESEY-FITZGERALD, B., 'Welcome or Unwelcome Guest?' *The Field*, 168, p. 1075, 1936.

—, 'Squirrel-tailed Dormouse', *The Field*, 171, p. 927, 1938.

VEVERS, G. M., 'The Fat Dormouse and other Wild Mammals at Whipsnade', *Bedford Nat.*, 2, pp. 42–4, 1947.

VIETINGHOFF-RIESCH, A. F. von., 'Der Siebenschläfer (*Glis Glis* L)', *Monogrn. Wildsäugetiere*, 16, pp. 1–196, 1960.

11. Grey Squirrel

BANKALOW, F. S., and SHORTEN, M., *The World of the Gray Squirrel*, Lippincott, 1973.

BARRINGTON, R. M., 'On the Distribution of Squirrels into Ireland', *Sci. Proc. Roy. Dub. Soc.*, 2, pp. 615–31, 1880.

BENHAM, E. M., 'The Distribution of Squirrels in Dorset', *Proc. Dor. Nat. Hist. and Arch. Soc.*, 74, pp. 121–32, 1953.

COLQUHOUN, M. K., 'The Habitat Distribution of the Grey Squirrel in Savernake Forest', *Jour. Anim. Ecol.*, 11, pp. 127–30, 1942.

FITTER, R. S. R., *The Ark in Our Midst*, pp. 91–8, Collins, 1959.

—, *Lond. Nat.*, pp. 6–19, 1938.

FORESTRY COMMISSION LEAFLET NO. 31, *The Grey Squirrel: A Woodland Pest*, H.M.S.O., 1953.

HARVIE BROWN, J. A., 'Squirrels in Great Britain', *Proc. Roy. Phys. Soc. Edin.*, 5, pp. 343–8, 6, pp. 31–63 and 115–83, 1880–1.

JOHNSTON, F. J., 'The Grey Squirrel in Epping Forest', *Lond. Nat.*, pp. 94–9, 1937.

LEVER, Christopher, 'The Invaders, and how they came here', *The Field*, p. 394, 28 August 1969.

LLOYD, H. G., 'The Distribution of Grey Squirrels in England and Wales, 1959', *Jour. Anim. Ecol.*, 31, pp. 157–66, 1962.

M.A.F.F., *Squirrels: Their Biology and Control*, Bull. No. 184, 1962.

MARSTRAND, P. K., 'Biologist', *Jour. Instit. Biol.*, May (2), pp. 21, 68, 1974.

MIDDLETON, A. D., 'The Ecology of the American Grey Squirrel in the British Isles', *Proc. Zool. Soc. Lond.*, pp. 809–43, 1930.

—, 'The Grey Squirrel in the British Isles in 1930–32', *Jour. Anim. Ecol.*, 1, pp. 166–7, 1932.

—, *The Grey Squirrel*, Sidgwick and Jackson, 1931.

—, 'The Distribution of the Grey Squirrel in Great Britain in 1935', *Jour. Anim. Ecol.*, 4, pp. 274–6, 1935.

MIDDLETON, A. D., and PARSON, B. T., 'The Distribution of the Grey Squirrel in Great Britain in 1937', *Jour. Anim. Ecol.*, 6, pp. 286–90, 1937.

RUTTLEDGE, R. F., 'Note on the Distribution of Squirrels in Ireland' *Irish Nat.*, 33, p. 73, 1924.

SHORTEN, Monica, 'A Survey of the Distribution of the American Grey Squirrel and the British Red Squirrel in England and Wales in 1944–45', *Jour. Anim. Ecol.*, 15, pp. 82–92, 1946.

—, 'Grey Squirrels in Britain', *The New Nat. Journal*, pp. 42–6, Collins, 1948.

—, 'Some Aspects of the Biology of the Grey Squirrel in Great Britain', *Proc. Zool. Soc. Lond.*, 121 (2), pp. 427–59, 1951.

—, 'Notes on the Distribution of the Grey Squirrel and the Red Squirrel in England and Wales from 1945–52', *Jour. Anim. Ecol.*, 22, pp. 134–40, 1953.

—, *Squirrels*, Collins, 1954.

—, 'Squirrels in England, Wales and Scotland in 1955', *Jour. Anim. Ecol.*, 26, pp. 287–94, 1957.

—, 'The Grey Squirrel' (*Animals of Britain, 5*), Sunday Times Publications, 1962.

—, 'Introduced Menace', *Nat. Hist. Mag.*, 10, pp. 43–8, 1964.

—, *Squirrels: their Biology and Control*, M.A.F.F. Bull. 184, 1962.

SHORTEN, M., and COURTIER, F. A., 'A Population Study of the Grey Squirrel in May 1954', *Ann. App. Biol.*, 43, pp. 492–510, 1955.

TAYLOR, J. C., 'Home Range and Agonistic Behaviour in the Grey Squirrel', *Symp. Zool. Soc. Lond.*, 18, pp. 229–35, 1966.

TAYLOR, K. D. *et. al.*, 'Movements of the Grey Squirrel as Revealed by Trapping', *Jour. App. Ecol.*, 8, pp. 123–46, 1971.

THOMPSON, H. V., and PEACE, T. R., 'The Grey Squirrel Problem', *Quart. Jour. For.*, 56 (1), pp. 33–42, 1962.

THOMPSON, H. V., and PLATT, F. B., *Bull. Mamm. Soc. Brit. Isles*, 21, pp. 7–8, May 1964.

VIZOSO, M., 'Squirrel Populations and their Control', *Forestry, Suppl.*, pp. 15–20, 1967.

12. Coypu

BOURDELLE, E., 'American Mammals Introduced into France in the Contemporary Period, Especially Myocastor and Ondatra', *Jour. Mammal.*, 20, pp. 287–91, 1939.

CARILL-WORSLEY, P. E. T., 'A Nutria Fur-farm in Norfolk', *Trans. Norf. Nor. Nat. Soc.*, 13 (2), pp. 105–115, 1930–1.

COTTON, K. E., 'The Coypu', *River Bds. Ass. Yb.*, 31, 1963.

DAVIS, R. A., 'The Coypu', *Discovery*, June 1955.

—, 'The Coypu', *Jour. Ag. Min. Ag.*, 63, pp. 127–9, 1956.

—, 'Feral Coypus in Britain', *Ann. Appl. Biol.*, 5, pp. 345–8, 1963.

—, *Bulletin of the Mammal Society of the British Isles*, 6, pp. 17–18, 1956.

DAVIS, R. A., and JENSON, A. G., 'A Note on the Distribution of the Coypu (*Myocastor coypus*) in Great Britain', *Jour. Anim. Ecol.*, 29, p. 397, 1960.

ELLIS, E. A., 'Some Effects of the Selective Feeding by the Coypu (*Myocastor coypus*) on the Vegetation of Broadland', *Trans. Norf. Nor. Nat. Soc.*, 20, pp. 32–5, 1963.

—, *The Broads*, Collins, 1965.

FITTER, R. S. R., *The Ark in Our Midst*, pp. 103–4, Collins, 1959.

GOSLING, L. M., 'The Coypu in East Anglia', *Trans. Norf. Nor. Nat. Soc.* (*Nor. Bird and Mamm. Rep.*), 23 (1), pp. 49–59, 1974.

LAURIE, E. M. O., 'The Coypu in Great Britain', *Jour. Anim. Ecol.*, 15, pp. 22–34, 1946.

LEVER, Christopher, 'The Invaders, and how they came here', *The Field*, p. 394, 28 August 1969.

M.A.F.F., *The Coypu*, Advisory Leaflet No. 479.

NEWSON, R. M., 'Reproduction in Feral Coypu', *Symposium of Zool. Soc. Lond.*, 15, pp. 323–4, 1966.

NORRIS, J. D., 'A Campaign against Feral Coypu in Great Britain', *Jour. Appl. Biol.*, 4, pp. 191–9, 1967.

—, 'A Campaign against the Coypus in East Anglia', *New Scien.*, 331, pp. 625–6, 21 March 1963.

—, 'The Control of Coypus (*Myocastor coypus* Molina) by Cage Trapping', *Jour. Appl. Ecol.*, 4, pp. 167–89, 1967.

PARRY, E. A., 'The Beaver-rat – neither Beaver nor Rat', *Nature*, 23, pp. 256–8, 1939.

THOMPSON, H. V., and PLATT, F. B., *Bull. Mamm. Socy Brit. Isles*, 21, pp. 6–7, May 1964.

WARWICK, T., 'Some Escapes of Coypus from Nutria Farms in Great Britain', *Jour. Anim. Ecol.*, 4, pp. 146–7, 1935.

WILLIAMS, C. T., *Modern Fur-farming*, London, 1934.

13. Himalayan or Hodgson's Porcupine and Crested Porcupine

BAKER, J., 'New Menace to Forests – Bark-eating Himalayan Porcupines', *Eastern Daily Press*, 24 October 1974.

BUTCHER, A. J., 'Porcupine Threat in Devon', *Western Morning News*, 3 April 1974.

COMFORT, N., 'Farmers Facing Threat from Pest Porcupines', *Daily Telegraph*, 25 October 1974.

CORBET, G. B., and JONES, L. A., 'The Specific Characters of the Crested Porcupines, Subgenus *Hystrix*', *Proc. Zool. Soc. Lond.*, 144, pp. 285–300, 1965.

DORST, Jean, and DANDELOT, Pierre, *A Field Guild to Larger Mammals of Africa*, Collins, 1970.

GARDNER, V., 'A Trapped Porcupine', *The Field* (letter), 31 January 1974.

GOSLING, L. M., M.A.F.F., *Current Topics* (21 October 1974); (with WRIGHT, M.) *Feral Porcupines* (Hystrix hodgsoni) in Devon (unpublished, 3 July 1973); *Feral Porcupines in Staffordshire* (unpublished, 11 June 1974); *Wiltshire Porcupines* (unpublished, 19 March 1975).

HADDON, Celia, 'British Wildlife', *Sunday Times*, 4 August 1974.

MILBURN, F. M. H., 'Porcupines Defended', *Western Morning News*, 13 April 1974.

PRATER, S. H., *The Book of Indian Animals*, India, 1965.

WALKER, E. P., *Mammals of the World*, Baltimore, 1965.

14. Polecat-Ferret

FITTER, R. S. R., *The Ark in Our Midst*, pp. 87–8, Collins, 1959.

GORDON, S., *A Highland Year*, 1944.

MACINTYRE, D., 'A New British Mammal', *The Field*, 183, p. 408, 1944.

MILLER, G. S., 'The Origin of the Ferret', *Scot. Nat.*, pp. 153–4, 1933.

POCOCK, R. I., 'Ferrets and Polecats', *Scot. Nat.*, pp. 97–108, 1932.

TETLEY, H. 'On the British Polecats', *Proc. Zool. Soc. Lond.*, 109B, pp. 37–9, 1939.

—, 'Notes on British Polecats and Ferrets', *Proc. Zool. Soc. Lond.*, 115, pp. 212–17, 1945.

YEAMAN, J. A., 'Polecat-Ferrets in Mull', *Scot. Nat.*, p. 66, 1932.

20

15. North American Mink

CLARK, S. P., 'Field Experience of Feral Mink in Yorkshire and Lancashire', *Mamm. Rev.*, I, pp. 41–7, 1970.

DEANE, C. D., and O'GORMAN, F., 'The Spread of Feral Mink in Ireland', *Irish Nat. Jour.*, 16, pp. 198–202, 1969.

FITTER, R. S. R., 'Man's Additions to the British Fauna', *Discovery*, II, pp. 58–62, 1950.

—, 'When Mink is just a Pest', *Sunday Observer*, 2 February 1958.

—, *The Ark in Our Midst*, pp. 84–7, Collins, 1959.

JONES, J. L., 'From Fur to Fugitive: the spread of wild mink', *Country Life*, 159 (4103), pp. 420–2, 19 February 1976.

THOMPSON, H. V., 'Wild Mink in Britain', *New Scien.*, 13 (220), pp. 130–2, 18 January 1962.

—, 'Mink at Large in Britain', *Country Life*, 131 (3394), p. 645, 1962.

—, 'Wild Mink', *Agriculture*, 71 (12), pp. 564–7, 1964.

—, 'Feral Mink in Britain', *Animals*, 6 (9), pp. 240–4, 1965.

—, 'Control of Wild Mink', *Agriculture*, 74 (3), pp. 114–16, 1967.

—, 'British Wild Mink', *Ann. Appl. Biol.*, 61 (2), pp. 345–9, 1968.

—, 'Mink, the New British Beast of Prey', *The Field*, 232 (6046), pp. 1094–5, 5 December 1968.

—, 'Animal Immigrants', *Geog. Mag.*, pp. 762–72, February, 1966.

—, 'Wild Mink in Britain', *Country Landowner*, 20 (2), p. 100, 1969.

—, 'British Wild Mink – a Challenge to Naturalists', *Agriculture*, 78 (10), pp. 421–5, 1971.

THOMPSON, H. V., and PLATT, F. B., *Bull. Mamm. Soc. Brit. Isles*, 21, p. 7, May 1964.

16. Domestic Cat

FITTER, R. S. R., *The Ark in Our Midst*, pp. 88–90, Collins, 1959.

HUDSON, W. H., *Birds in London*, 1898.

MATHESON, C., 'The Domestic Cat as a Factor in Urban Ecology', *Jour. Anim. Ecol.*, 13, pp. 130–3, 1944.

MIVART, St George, *The Cat*, John Murray, 1881.

TEGNER, H., 'Wild Cats: Feral Cats', *Wildlife*, 18 (2), pp. 78–9, 1976.

17. Chinese or Reeves's Muntjac and Indian Muntjac,
18. Fallow Deer, 19. Japanese Sika Deer, 20. Chinese Water Deer

ARMSTRONG, N. *et. al.*, 'Observations on the Reproduction of Female Wild and Park Fallow Deer in Southern England', *Jour. Zool. Lond.*, 158, pp. 27–37, 1969.

BUCKHURST, E. A. J., COPLAND, W. O., and ROBERTS, E. A., 'The Status of

Sika Deer in the Poole Basin', *Proc. Dorset Nat. Hist. & Archae. Soc.*, 86, pp. 96–101, 1964.

CADMAN, W. A., 'The Management and Control of Fallow Deer in the New Forest', *Forestry Suppl.*, pp. 59–63, 1967.

CARNE, P. H., 'The Deer of Ashdown Forest', *Sussex Co. Mag.*, 27 (2), pp. 82–4, February 1953.

CARTER, D., 'The Haldon Fallow Deer', *Jour. Brit. Deer Soc.*, February, 1967.

CHAPLIN, R., *Reproduction in British Deer*, 1966.

—,'The Antler Cycle of Muntjac Deer in Britain', *Jour. Brit. Deer Soc.*, November 1972.

—, 'Some Observations on the Mating, Vocal and Territorial Behaviour of Wild Muntjac Deer', *Jour. Brit. Deer Soc.*, November 1971.

CHAPMAN, D. I., and N., 'Observations on the Biology of Fallow Deer in Epping Forest, England', *Biol. Conserv.*, 2, pp. 55–62, 1969.

—, 'Preliminary Observations on the Reproductive Cycle of Male Fallow Deer', *Jour. Reprod. Fert.*, 21, pp. 1–8, 1970.

—, 'Muntjac Deer Hybrids'. *Jour. Brit. Deer Soc.*, March 1972.

—, *Fallow Deer*, 1, British Deer Society, 1970.

CHAPMAN, D. I., and CHAPLIN, R. E., 'Research on Fallow Deer Biology', *Jour. Brit. Deer Soc.*, July 1967.

CHAPMAN, D. I., and DANSIE, O., 'Reproduction and Foetal Development in Female Muntjac', *Mammalia*, 34, pp. 303–19, 1970.

CLARK, M., 'Fraying by Muntjac', *Jour. Brit. Deer Soc.*, November 1968.

—, 'Notes on Muntjacs in South Hertfordshire', *Jour. Brit. Deer Soc.*, November 1971.

CLAY, Theresa, 'Lice on the Muntjac', *Jour. Brit. Deer Soc.*, November 1966.

DANSIE, O., 'Are Muntjac Deer?', *Jour. Brit. Deer Soc.*, June 1969.

—, 'A Policy for Muntjac?' *Jour. Brit. Deer Soc.*, November 1969.

—, *Muntjac*, 2, Brit. Deer Soc., 1970.

—, 'The Spread of the Remarkable Muntjac', *Jour. Brit. Deer Soc.*, February 1971.

DEANE, C. Douglas, 'Deer in Ireland'. *Jour. Brit. Deer Soc.*. November 1972.

DELAP, P., 'Deer in Co. Wicklow', *Irish Nat. Jour.*, 6, pp. 82–8, January 1936–November 1937.

—, 'Hybridization of Red and Sika Deer in Rural Western England', *Jour. Brit. Deer Soc.*, 1, p. 131, 1967.

—, 'Observations on Deer in North-west England', *Jour. Zool. Lond.*, 156, 4, pp. 531–3, 1968.

DE NAHLIK, A. J., *Wild Deer*, Faber and Faber, 1959.

—, *Deer Management*, David and Charles, 1974.

EASTCOTT, B., 'Deer in Essex', *Essex Nat.*, 32 (5), pp. 313–16, 1971.

EDLIN, H. L., *Changing Wildlife of Britain*, B. T. Batsford, 1952.

FITTER, R. S. R., *The Ark in Our Midst*, pp. 41–75, Collins, 1959.

HARRIS, R. A., and DUFF, K. R., *Wild Deer in Britain*, David and Charles, 1970.

HARTING, J. E., *Essays on Sport and Natural History*, 1883.

—, 'The Deer of Epping Forest', *Essex Nat.*, 1, pp. 46–62, 1882.

—, 'Hertfordshire Deer Parks', *Trans. Herts. Nat. Hist. Soc.*, 11 (3), pp. 97–110, 1883.

HORWOOD, M. T., 'Sika Deer Studies in the Poole Basin', *Progress Reports* (unpub.), 1966–9.

—, 'The World of Wareham Sika', *Jour. Brit. Deer Soc.*, February 1973.

HORWOOD, M. T., and MASTERS, E. H., *Sika Deer*, 3, Brit. Deer Soc., 1970.

IDLE, E. T., and MITCHELL, J., 'The Fallow Deer of Loch Lomondside', *Jour. Brit. Deer Soc.*, November 1968.

JOHNSTONE, G., *The Field*, 187, pp. 180–2, 1946; 195, pp. 48–9, 14 January 1950.

—, *Sussex Co. Mag.*, 24 (11), pp. 472–8, November 1950.

—, *Countryman*, 40, pp. 250–5, 1949.

KING, R. J., 'Fallow Deer in Savernake Forest', *Jour. Brit. Deer Soc*, November 1958.

LEEKE, C. J., 'Notes on Feral Barking Deer in the Reading Area', *Read. Nat.*, 22, pp. 25–9, 1970.

LEFTWICH, A. W., 'Deer in Northamptonshire', *Jour. Northants. Nat. Hist. Soc.*, 1958.

LEVER, Christopher, 'The Invaders, and how they came here', *The Field*, p. 395, 28 August 1969.

LOWE, V. P. W., and GARDINER, A., 'Red Deer – Sika Deer Hybridisation', *Jour. Zool. Lond.*, 177, pp. 553–66, 1975.

MARSHALL, F., 'The Future for Fallow Deer', *Jour. Brit. Deer Soc.*, July 1974.

MATHESON, C., *Changes in the Fauna of Wales Within Historic Times*, Cardiff, 1932.

MITCHELL, W. R., and ROBINSON, J., 'The Bowland Sika: Some Aspects of the Rut', *Jour. Brit. Deer Soc.*, July 1973.

—, 'The Bowland Sika: Some Notes on Vocal Activity', *Jour. Brit. Deer Soc.*, February 1974.

—, 'The Bowland Sika: Some Notes on Antlers', *Jour. Brit. Deer Soc.*, November 1974.

—, 'The Bowland Sika: Their History, Status and Distribution', *Jour. Brit. Deer Soc.*, November 1971.

—, 'The Bowland Sika: Some Notes on Calving', *Jour. Brit. Deer Soc.*, July 1975.

MOFFAT, C. B., 'The Mammals of Ireland', *Proc. Roy. Irish Acad.*, 44 (B.6), 1938.

MOONEY, O. V., 'Irish Deer and Forest Relations', *Irish Forestry*, 9 (1), pp. 11–27, 1952.

MULLOY, F., 'A Note on the Occurrence of Deer in Ireland', *Jour. Brit. Deer Soc.*, 2, pp. 502–4, 1970.

NATURE CONSERVANCY, *Sika Deer Research: Second Progress Report*, 1971.

NELSON, G. S., 'A Note on the Internal Parasites of the Muntjac', *Jour. Brit. Deer Soc.*, November 1966.

PAGE, F. J. T., 'The Wild Deer of East Anglia', *Trans. Norf. Nor. Nat. Soc.*, 17 (5), pp. 316–21, 1954.

—, *Field Guild to British Deer*, 1957.

PICKVANCE, T. J., and CHARD, J. S. R., 'Feral Muntjac (*Muntiacus spp.*) in the West Midlands, with special reference to Warwickshire', *Proc. Birm. Nat. Hist. and Phil. Soc.*, 19 (1), pp. 1–8, 1960.

POWERSCOURT, Viscount, 'On the Acclimatization of the Japanese Deer of Powerscourt', *Proc. Zool. Soc. Lond.*, pp. 207–9, 1884.

PRIOR, R., *Living with Deer*, André Deutsch, 1956.

RITCHIE, J., *The Influence of Man on Animal Life in Scotland*, 1920.

ROBINSON, J., 'The World of Bowland Sika', *Jour. Brit. Deer Soc.*, February 1973.

SCOTT, W. A., 'Fallow Deer in the Teign Valley', *Jour. Brit. Deer Soc.*, February 1967.

SOPER, Eileen A., *Muntjac*, Longmans, 1969.

SPRINGTHORPE, G., 'Longhaired Fallow Deer', *Jour. Brit. Deer Soc.*, November 1966.

SPRINGTHORPE, G., and VOYSEY, J., 'The Fallow Deer of Mortimer Forest', *Jour. Brit. Deer Soc.*, November 1969.

SUTCLIFFE, A., 'Joint Mitnor Cave, Buckfastleigh', *Trans. Torquay Nat. Hist. Soc.*, 13, (Pt. 1), pp. 1–26, 1960.

—, 'Excavations in the Torbryan Caves, Devonshire', *Proc. Devon Archae. Explor. Soc.*, 5, pp. 5–6, 1962.

TAYLOR, Sir W. L., 'The Distribution of Wild Deer in Scotland', *Jour. Anim. Ecol.*, 18 (2), pp. 187–92, November 1949.

—, 'The Distribution of Wild Deer in England and Wales', *Jour. Anim. Ecol.*, 8 (1), pp. 6–9, May 1939, and 17 (2), pp. 151–4, November 1948.

VENNER, B. G., 'Fallow Deer in the Forest of Dean', *Jour. Brit. Deer Soc.*, February 1970.

WARNER, L. J., 'A Muntjac in Windsor Forest', *Jour. Brit. Deer Soc.*, July 1972.

WHITEHEAD, G. K., *Deer and their Management in the Deer Parks of Great Britain and Ireland*, Country Life, 1950.

—, *The Deer of Great Britain and Ireland*, Routledge & Kegan Paul, 1964.

—, 'Wild Deer in Scotland', *The Field*, 201 (5219), p. 83 (fallow deer), 15 January 1953; 201 (5223), p. 239 (Sika deer), 12 February 1953.

WHITEHEAD, G. K., 'Deer from Asia at Home in England, *The Field*, 203 (5284), p. 663,15 April 1954.

—, 'Black Sika at Whipsnade', *Country Life*, 116 (3002), p. 371, 29 July 1954.

—, 'Future of Scotland's Wild Deer', *Country Life*, 132 (3416) pp. 424–6, 23 August 1962.

—, 'Muntjac Deer in England', *Gamekeeper and Countryside*, 781, pp. 12–13, October 1962.

WILLET, J. A., and MULLOY, F., 'Wild Deer, their Status and Distribution', *Jour. Brit. Deer Soc.*, 2 (2), pp. 498–504, June 1970.

21. Wild Goat

BAKER, A. L. L., 'Wild Goats in Wales', *The Times*, 3 September 1973; *The Field*, 2 September 1954.

BARROW, G., *Wild Wales* (1854).

'B.B'., *Summer Road to Wales*, Nicholas Kaye, 1964.

BRITISH GOAT SOCIETY, 'British Wild Goats', Monthly Circular 12 No. 8, pp. 2–3, 1919; 13 No. 9, p. 3, 1920; 13 No. 11, pp. 5–7, 1920; 15 No. 11, pp. 210–12, 1922; Year Book, pp. 52–5, 1923; Monthly Journal, 24 No. 1, p. 11, 1931; 26 No. 11, p. 275, 1933.

BUCHANAN SMITH, A. D., 'Goats for the Island of Eigg', *British Goat Society Year Book*, pp. 57–8, 1927.

—, 'Wild Goats in Scotland', *British Goat Society Year Book*, pp. 120–5, 1932.

CAMPBELL, B., *Snowdonia*, Collins, 1949.

CASEBY, J. A., 'The Welsh Goat', *British Goat Society Year Book*, pp. 120–3, 1936.

CONDRY, W. M., *Snowdonia National Park*, Collins, 1966.

CROOK, I., 'Feral Goats in North Wales', *Animals*, 12 No. 1, pp. 13–15, 1969.

DARLING, F. Fraser, *Natural History in the Highlands and Islands*, pp. 75, 126, Collins, 1947.

FITTER, R. S. R., *The Ark in Our Midst*, pp. 53–9, Collins, 1959.

GLASGOW AND ANDERSONIAN NATURAL HISTORY SOCIETY Report, 1936.

GORDON, Seton. *Highways and Byways in the West Highlands*, pp. 167, 258, 280 and 376, 1935.

HARRIS, D. R., 'The Distribution and Ancestry of the Domestic Goat', *Proc. Linn. Soc. Lond.*, 173, pp. 79–81, 1962.

HARVIE-BROWN, J. A., and BUCKLEY, T. E., *A Vertebrate Fauna of the Outer Hebrides*, pp. 41–2, 1888.

—, *A Vertebrate Fauna of Argyll and the Inner Hebrides*, p. 45, 1892.

MACKENZIE, D., *Goat Husbandry*, 1971.

MANWOOD, J., *A Treatise of the Laws of the Forest*, 1665.

MARTIN, M., *A Description of the Western Islands of Scotland*, 1703.

MATHESON, C., *Changes in the Fauna of Wales within Historic Times*, Cardiff, 1933.

—, *Annual Report W. Wales Field Society*, pp. 13–16, 1953–4.

MCDOUGALL, P., 'Feral Goats of Kielderhead Moor', *Jour. Zool.*, 176 (2), pp. 215–46, 1975.

MCTAGGART, H. S., 'Feral Goats, Holy Island, Arran' *U.F.A.W. Annual Report*, pp. 22–4, 1968–9.

MILNER, C., GOODIER, R., and CROOK, I. G., 'Feral Goats', *Nat. Wales*, 11, pp. 3–11, 1968.

MONTROSE, Duke of, *The Countryman*, 1932.

NIALL, I., 'Wild Goats', *Country Life*, 24 January 1957.

PENNANT, T., *British Zoology*, pp. 13–15, 1766.

—, *A Tour of Scotland and a Voyage to the Hebrides, 1772*, 1, 1776.

RICHMOND, W. K., 'Wild Goats', *Scottish Field*, May 1955.

RUTTY, John, *An Essay Towards a Natural History of the County of Dublin*, 1772.

ST JOHN, Charles, *Sketches of the Wild Sports and Natural History of the Highlands*, John Murray, 1878.

SCHWARZ, E., 'On Ibex and Wild Goat', *Ann. Mag. Nat. Hist.*, 16, pp. 433–7, 1935.

STEWART, A. E., 'Wild Goats in Scotland', *The Field*, 16 December 1933.

TEGNER, H., 'Wild Goats of the Border', *Country Life*, 29 February 1952.

—, *A Border Country*, 1955.

—, 'Wild Goats of Britain', *Scottish Field*, 1965.

—, 'Goat Stalking in Scotland', *Shooting Times & Country Magazine*, 1 April 1967.

'W', 'Wild Goats of Cheviot', *Country Life*, 16 May 1908.

WATKINS-PITCHFORD, D., 'Wild Goats of the Welsh Mountains', *Country Life*, 21 November 1963.

WATT, H. B., and DARLING, F. Fraser, 'On the Wild Goat in Scotland' and 'Habits of Wild Goats in Scotland', *Jour. Anim. Ecol.*, 6, pp. 15–22, 1937.

WENTWORTH DAY, J., 'The Cashmere Goats of Windsor', *Country Life*, 14 October 1954.

WHITEHEAD, G. K., 'Wild Goats of Scotland', *The Field*, 10 March 1945.

—, 'The Horned Game of Great Britain', *Country Life*, 19 September 1947.

—, 'An Historic Herd of 'Wild' Goats', *Country Life Annual*, 1952.

—, 'Wild Goats of Wales', *Country Life*, 26 September 1957.

—, *The Wild Goats of Great Britain and Ireland*, David and Charles, 1972.

22. Soay Sheep

BOYD, J. M., 'The Sheep Population of Hirta, St. Kilda', *Scot. Nat.*, 65, pp. 25–8, 1953; 68, pp. 10–13, 1955.

BOYD, J. M., *St. Kilda National Nature Reserve Management Plan*, Nature Conservancy, 1964.

BOYD, J. M., *et al.*, 'The Soay Sheep on the Island of Hirta, St. Kilda', *Proc. Zool. Soc. Lond*, 142, pp. 129–63, 1964.

COCKBURN, A. M., 'The Geology of St. Kilda', *Trans. Roy. Soc. Edin.*, 58, pp. 511–48, 1935.

DARLING, F. Fraser, *The Natural History of the Highlands and Islands*, Collins, 1947.

ELWES, H. J., 'Primitive Breeds of Sheep in Scotland', *Scot. Nat.*, pp, 25–32, 1912.

EWART, J. C., 'Domestic Sheep and their Wild Ancestors', *Trans. High., and Agric. Soc. Scot.* (5), 25, pp. 160–91, 1913.

FISHER, J., 'St. Kilda, a Natural Experiment', *New. Nat. Jour.*, 1, pp. 91–108, Collins, 1948.

FITTER, R. S. R., *The Ark in Our Midst*, pp. 59–61, Collins, 1959.

FRASER, A., *et. al.*, 'Blackface Sheep of Boreray', *Jour. Blackface Breeders Assoc.*, 10, pp. 11–13, 1957.

GRUBB, P., and JEWELL, P. A., 'Social Grouping and Home Range in Feral Soay Sheep', *Symp. Zool. Soc. Lond.*, No. 18, pp. 179–210, 1966.

JEWELL, P. A. (ed.), *et al.*, *Island Survivors: The Ecology of the Soay Sheep on St. Kilda*, Athlone Press, University of London, 1974.

LOCKLEY, R. M., 'Wild Sheep in Wales', *Nature in Wales*, 6, pp. 75–8, 1960.

—, *Letters from Skokholm*, 1947.

POORE, M. E. D., and ROBERTSON, V. C., 'The Vegetation of St. Kilda in 1948', *Jour. Ecol.*, 37, pp. 82–99, 1949.

RYDER, M. L., 'The Evolution of Scottish Breeds of Sheep', *Scot. Studies*, 12, pp. 127–67, 1968.

SHILLITO, Elizabeth E, and HOYLAND, Valerie J., 'Observations on Parturition and Maternal Care in Soay Sheep', *Jour. Zool. Soc. Lond.*, 165, pp. 509–12, 1971.

STEEL, T., 'The Life and Death of St. Kilda', *Nat. Trust. Scot.*, 1965. (Reprinted by Fontana, 1975.)

TAYLOR, A. B., 'The Norsemen in St. Kilda', *Saga Book of the Viking Society*, 17; 2, pp. 116–44; 3, pp. 106–14, 1967–8.

WILLIAMSON, K., and BOYD, J. M., *St. Kilda Summer*, Hutchinson, 1960.

23. Ring-Necked, Rose-Ringed or Green Parakeet

CARPENTER, C. P., 'The Status of the Ring-necked Parakeet in the Bromley area of Kent . . . and elsewhere in south-east England' (unpublished, February 1976).

DERBYSHIRE ORNITHOLOGICAL SOCIETY, Bulletin No. 221, August 1975.

ENGLAND, M. D., 'Escapes', *Avic. Mag.*, 76, pp. 150–2, 1970.

ENGLAND, M. D., 'Feral Populations of Parakeets', (letter), *Brit. Birds*, 67 (9), pp. 393–4, September 1974.

—, 'A Further Review of the Problem of Escapes', *Brit. Birds*, 67 (5), pp. 177–97, May 1974.

FITTER, R. S. R., *The Ark in Our Midst*, p. 225, Collins 1959.

FORSHAW, J. M., *Parrots of the World*, Lansdowne, 1973.

HADDON, Celia, 'British Wildlife', *Sunday Times*, 4 August 1974.

HAWKES, B., 'The Invasion of the Asians', *Surrey Life*, 5 (6), p. 30, May 1976.

HUDSON, R., 'Feral Parakeets near London', 'Parakeets in the London Area', *Brit. Birds*, 67 (9), pp. 33, 174, April 1974.

KING, B., *et al.*, *A Field Guide to the Birds of South-East Asia*, Collins, 1975.

NATURE-TIMES NEWS SERVICE: Science Report: Ornithology: *Problem of Escaping Birds*, 1974.

REPORTS FOR 1970–3, London Natural History Society, Kent Ornithological Society, Surrey Bird Club.

24. Budgerigar

BLUNT, Wilfrid., *The Ark in the Park*, Hamish Hamilton – Tryon Gallery, 1976.

DORRIEN-SMITH, Mrs T., 'Budgerigars on Tresco' (unpublished memorandum, 1974).

FITTER, R. S. R., *The Ark in our Midst*, pp. 225–6, Collins, 1959.

FORSHAW, J. M., *Parrots of the World*, Lansdowne, 1973.

GOULD, J., *The Birds of Australia*, 1865.

ROWLEY, Ian, *Bird Life*, Australian Naturalist Library, Collins, 1973.

25. Little Owl

AINSLIE, D., 'The Little Owl in Bedfordshire', *The Zool.*, p. 353, 1907.

BALSTON, R. J., *et. al.*, *Notes on the Birds of Kent*, 1907.

BLATHWAYT, F. L., 'The Little Owl in Lincolnshire', *The Zool.*, p. 112, 1902; p. 74, 1904.

BRADSHAW, C. J., 'A Little Owl at Henley', *The Zool.*, p. 476, 1901.

BURTON, J. A. (ed.)., *Owls of the World*, Peter Lowe, 1973.

BUXTON, P. A., 'The Spread of The Little Owl in Hertfordshire', *The Zool.*, p. 430, 1907.

DAVIS, W. J., *Birds of Kent*, 1907.

ELLISON, A., 'Little Owls Breeding in Hertfordshire', *The Zool.*, p. 430, 1907.

FIELDING, C. H., *Memories of Malling and its Valley*, 1893.

FITTER, R. S. R., *The Ark in Our Midst*, pp. 215–19, Collins, 1959.

FROHAWK, F. W., *The Field*, 12 (10), 1907.

HAINES, C. Reginald, *Notes on the Birds of Rutland*, 1907.

HARTERT, E., and ROTHSCHILD, L. W., *Victoria History of Buckinghamshire*.

HIBBERT-WARE, A., 'Report of the Little Owl Food Inquiry, 1936–7', *Brit. Birds*, 31, pp. 162–87, 205–29, 249–64, 1937.

HOSKING, E., and NEWBERRY, C. W., *Birds of the Night*, Collins, 1945.

JOURDAIN, F. C. R., *Derb. Arch. and Nat. Hist. Soc. Jour.*, 1907.

LILFORD, Lord, *Lord Lilford on Birds*, Hutchinson, 1903.

—, *Notes on the Birds of Northamptonshire and Neighbourhood*, Porter, 1895.

MAPLES, S., 'The Little Owl in Hertfordshire', *The Zool.*, p. 353, 1907.

SPARKS, J., and SOPER, T., *Owls*, David and Charles, 1970.

STEELE-ELLIOT, J., 'The Little Owl in Bedfordshire', *The Zool.*, p. 384, 1907.

STERLAND, W. J., and WHITAKER, J., *A Descriptive List of the Birds of Nottinghamshire*, 1879.

WATERTON, Charles, *Essays on Natural History*, Warne, 1871.

WELLS BLADEN, W., *Trans. North Staffs. Field Club*, 1906–7.

WITHERBY, H. F., and TICEHURST, N. F., 'Spread of the Little Owl in Britain', *Brit. Birds*, 1, pp. 335–42, 1908.

26. Night-Heron

BRITISH BIRDS: *Reports and Editorial*, 41, p.24; 43, p.302; 47, pp.351–4, 1954.

DORWARD, D. F., 'The Night-Heron Colony in the Edinburgh Zoo', *Scot. Nat.*, 69, pp. 32–6, 1957.

FITTER, R. S. R., *The Ark in our Midst*, pp. 228–9, Collins, 1959.

LEBRET, T., 'Een Kleine broedkolonie van de Kwak, *Nycticorax nycticorax* (L) in Nederland in 1946 en 1947', *Ardea*, 35, pp. 149–56, 1947.

LILFORD, Lord, *Notes on the Birds of Northamptonshire and Neighbourhood*, Porter, 1895.

—, *Lord Lilford on Birds*, Hutchinson, 1903.

ROYAL ZOOLOGICAL SOCIETY OF SCOTLAND: Annual Reports, 1936–75.

27. Canada Goose

ATKINSON-WILLES, G. L. (ed.), *Wildfowl in Great Britain*, Nature Conservancy Monograph, H.M.S.O., 1963.

BAXTER, E. V., and RINTOUL, L. J., *The Birds of Scotland*, Edinburgh, 1953.

BERRY. J., 'The Status and Distribution of Wild Geese and Wild Duck in Scotland', *Wildfowl Inquiry Comm. Rep.* 2, Cambridge, 1939.

BLURTON-JONES, N. G., 'Census of Breeding Canada Geese 1953', *Bird Study*, 3, pp. 153–70, 1956.

DENNIS, R., 'Capture of Moulting Canada Geese on the Beauly Firth', *Wildfowl Trust Ann. Rep.*, 15, pp. 71–4, 1964.

FITTER, R. S. R., *The Ark in Our Midst*, pp. 184–8, Collins, 1959.

GLEGG, W. E., *A History of the Birds of Essex*, 1929.

GUTHRIE, D., 'Canada Geese in the Outer Hebrides', *Ann. Scot. Nat. Hist.*, p. 119, 1903.

HANSON, H. C., *The Giant Canada Goose*, Carbondale, Ill.: Southern Illinois University Press, 1965.

HANSON, H. C., and SMITH, R. H., 'Canada Geese of the Mississippi Flyway', *Bull. Ill. Nat. Hist. Surv.*, 25, pp. 67–210.

HINE, R. L., and SCHOENFELD, C. (eds.), *Canada Goose Management*, Madison, Wisc.: Dembar Educational Research Services Inc., 1968.

KEAR, J., 'The Assessment of Goose Damage by Grazing Trials', *Trans. 6th. Int. Union Game Biol.*, pp. 333–9, 1965.

—, A review of 'The Giant Canada Goose' by H. C. Hanson (*vide supra*), *Ibis*, 108, pp. 144–5, 1966.

KENNEDY, P. G. *et. al.*, *The Birds of Ireland*, Oliver and Boyd, 1954.

LIMENTANI, J. D., 'Changes in Wintering Habits by Canada Geese', *Camb. Bird Rep.*, 48, 1975.

MERNE, O. J., 'The Status of the Canada Goose in Ireland', *Irish Bird Rep.*, 17, pp. 12–17, 1970.

OATES, C., *The Birds of Lancashire*, 1953.

RUTTLEDGE, R. F., *Ireland's Birds*, 1966.

SALOMONSEN, F., 'The Moult Migration', *Wildfowl*, 19, pp. 5–24, 1968.

SCOTT, P., and BOYD, H., *Wildfowl of the British Isles*, Country Life, 1957.

STERLING, T., and DZUBIN, A., 'Canada Goose Molt Migrations to the North-West Territories', *Trans. North Amer. Wildf. Conf.*, 32, pp. 355–73, 1967.

TAVERNER, P. A., 'Study of *Branta canadensis*', *Ann. Rep. Nat. Mus. Can.*, 67, 1931.

WALKER, A. F. G., 'The Moult Migration of Yorkshire Canada Geese', *Wildfowl*, 21, pp. 99–104, 1970.

WOOD, J. S., 'Normal Development and Causes of Reproductive Failure in Canada Geese', *Jour. Wildl. Manag.*, 28, pp. 197–208, 1964.

28. Egyptian Goose

ATKINSON-WILLES, G. L. (ed.), *Wildfowl in Great Britain*, Nature Conservancy Monograph, H.M.S.O., 1963.

COUNTY BIRD REPORTS for Norfolk, Suffolk and Staffordshire.

FITTER, R. S. R., *The Ark in Our Midst*, pp. 197–8, Collins, 1959.

HAWKER, Colonel Peter, *Diaries*, published by Sir Ralph Payne-Gallwey, 1893.

SCOTT, P., and BOYD, H., *Wildfowl of the British Isles*, Country Life, 1957.

29. Mandarin Duck

CAMPBELL, B., and FERGUSON-LEES, I. J., *A Field Guide to Birds' Nests*, 1972.
FITTER, R. S. R., *The Ark in Our Midst*, pp. 188–93, Collins, 1959.
GORDON, Seton, *Edward Grey and his Birds*, Country Life, 1937.
GREY, Viscount of Fallodon, *The Charm of Birds*, Hodder and Stoughton, 1927.
LEVER, Christopher, 'The Mandarin Duck in Britain', *Country Life*, pp. 829–31, 17 October 1957.
OGILVIE, M. A., *Ducks of Britain and Europe*, 1975.
PHILLIPS, John C., *The Natural History of the Ducks*, 3, Houghton Mifflin & Co., Boston, 1925.
SAVAGE, Christopher, *The Mandarin Duck*, A. & C. Black, 1952.
SCOTT, P., and BOYD, H. *Wildfowl of the British Isles*, Country Life, 1957.
TOMLINSON, D., 'Surrey's Chinese Duck', *Country Life*, 159 (4115), pp. 1248–9, 13 May 1976.

30. Wood Duck or Carolina Duck

BRITISH ORNITHOLOGISTS' UNION RECORDS COMMITTEE REPORT, *Ibis*, 114, p. 447, 1972.
BRITISH ORNITHOLOGISTS' UNION: *The Status of Birds in Britain and Ireland*, Blackwell, 1971.
FITTER, R. S. R., *The Ark in Our Midst*, pp. 193–4, Collins, 1959.
FITTER, R. S. R., and RICHARDSON, R. A., *Collins Guide to Nests and Eggs*, Collins, 1968.
LEVER, Christopher, 'The Mandarin Duck in Britain', *Country Life*, pp. 829–31, 17 October 1957.
PARR, D. (ed.), *Birds in Surrey 1900–1970*, Batsford, 1972.
SAVAGE, Christopher, *The Mandarin Duck*, A. & C. Black, 1952.
WHEATLEY, J. J., *Surrey County Bird Report*, 20, pp. 64–5, 1972.
—, 'Status of Carolina Duck (*Aix sponsa*) in Surrey', *Surrey Bird Club Quarterley Bulletin*, 55, p. 15, 1970.

31. Ruddy Duck

ATKINSON-WILLES, G. L., *Wildfowl in Great Britain*, Nature Conservancy Monograph, 1963.
BRITISH ORNITHOLOGISTS' UNION RECORDS COMMITTEE REPORT, *Ibis*, 113, pp. 420–3, February, 1971.
CAMPBELL, B., and FERGUSON-LEES, I. J., *A Field Guide to Birds' Nests*, 1972
COUNTY BIRD REPORTS of Bristol; Cheshire; Gloucestershire; Shropshire; Somerset; Staffordshire; Warwickshire; West Midlands; Worcestershire.

DAWSON, L. R., 'Rearing the North American Ruddy Duck', *Avic Mag.* 80, p. 327, 1974.

DICKENS, R. F., and Pickup, J. D., *Fairburn and its Nature Reserve . . .,* Clapham, 1973.

FERGUSON-LEES, I. J., 'The Identification of White-headed and Ruddy Ducks', *Brit. Birds*, 51, pp. 239–41, 1958.

HUDSON, Robert, 'Ruddy Ducks in Britain', *Brit. Birds,* 69 (4), pp. 132–43, April 1976. (see also 'Ruddy Immigrant', *Country Life*, 59 (4116), p. 1351, 20 May 1976).

JOHNSTONE, S. T., 'Slimbridge 1947–57 . . .', *Wildfowl Trust Ann. Rep.*, 12: pp. 167–8 (1961).

KING, B., 'Feral North-American Ruddy Ducks in Somerset', *Wildfowl Trust Ann. Rep.*, pp. 167–8, 1959–60.

—, 'Association Between Male North-American Ruddy Ducks and Stray Ducklings', *Brit. Birds*, 69 (3), p. 34, 1976.

LORD, J., and MUNNS, D. J., *Atlas of Breeding Birds of the West Midlands 1966–68*, 1970.

MARTIN, R. M., 'The North American Ruddy Duck in Captivity', *Avic. Mag.* 80, pp. 132–5, 1974.

OGILVIE, M. A., *Ducks of Britain and Europe,* 1975.

PARSLOW, J. L. F., *Breeding Birds of Britain and Ireland*, 1973.

SEIGFRIED, W. R., 'Summer Foods and Feeding of the Ruddy Duck in Manitoba', *Can. Jour. Zool.*, 51. pp. 1293–7, 1973.

WENT, R., 'Rare Birds: the North American Ruddy Duck', *Bird Life 1975*, pp. 18–19, 1975.

WILDFOWL TRUST ANNUAL REPORTS: 1959–75.

32. Capercaillie

ANGUS, W. C., 'The Capercaillie', *Proc. Nat. Hist. Soc. Glas.*, New Ser. 1, p. 380, 1886.

BARTHOLOMEW, J., 'Capercaillie in West Stirling', *Scot. Nat.*, p. 61, 1929.

BONAR, H. N., 'Capercaillie in Midlothian', *Ann. Scot. Nat. Hist.*, pp. 51–2, 1907.

—, 'Capercaillie in East Lothian', *Ann. Scot. Nat. Hist.*, p. 120, 1910.

CAMPBELL, J. M., 'Capercaillies in Ayrshire', *Ann. Scot. Nat. Hist.*, p. 186, 1906.

DAVIDSON, J., 'Capercaillies in Moray', *Ann. Scot. Nat. Hist.*, p. 52, 1907.

FITTER, R. S. R., *The Ark in Our Midst*, pp. 164–8, Collins, 1959.

GLADSTONE, H. S., 'Capercaillie in Ayrshire', *Ann. Scot. Nat. Hist.*, p. 116, 1906.

—, 'The Last of the Indigenous Scottish Capercaillies', *Scot. Nat.*, pp. 169–77, 1921.

GRANT, W., and CUBBY, J., 'The Capercaillie Reintroduction Experiment at Grizedale', *W.A.G.B.I. Rep. and Year Book*, pp. 96–8, 1972–3.

HARVIE-BROWN, J. A., *The Capercailzie in Scotland*, Edinburgh, 1879.

—, 'The Capercaillie in Scotland', *Scot. Nat.*, pp. 289–94, 1880.

—, 'Capercaillie in S.E. Lanarkshire', *Ann. Scot. Nat. Hist.*, p. 118, 1898.

—, 'Extension of the Capercaillie in Moray', *Ann. Scot. Nat. Hist.*, p. 184, 1911.

INGRAM, C., 'A Few Notes on *Tetrao urogallus* and its Allies, *Ibis*, Ser. 10, 3, pp. 128–33, 1915.

KAY, F. C. Lister, 'Capercaillie in Argyll', *Ann. Scot. Nat. Hist.*, p. 189, 1904.

LECKIE, N., 'Capercaillie in Linlithgowshire', *Ann. Scot. Nat. Hist.*, p. 44, 1897.

LLOYD, L., *The Gamebirds and Wildfowl of Sweden and Norway*, London, 1867.

MACKEITH, T. T., 'The Capercaillie in Renfrewshire', *Scot. Nat.*, p. 270, 1916.

MACKENZIE, W. D., 'Capercailzie in Strathnairn', *Ann. Scot. Nat. Hist.*, p. 51, 1900.

MARSHALL, H. B., 'Capercaillie in Peebleshire', *Ann. Scot. Nat. Hist.*, p. 244, 1907.

MAXWELL, H., 'Capercaillie in the South of Scotland', *Ann. Scot. Nat. Hist.*, p. 116, 1907.

MENZIES, W. Stewart, 'Capercaillie and Willow Grouse in Moray', *Ann. Scot. Nat. Hist.*, pp. 116–17, 1907.

MILLAIS, J. G., *The Natural History of British Game Birds*, London, 1909.

PENNIE, Ian D., 'The History and Distribution of the Capercaillie in Scotland', *Scot. Nat.*, 62, pp. 65–87; 157–78; 63, pp. 4–18, 135; 1950–1.

POLLARD, Hugh, *Game-Birds and Game-Bird Shooting*, 1936.

PRICHARD, H. Hesketh, *Sport in Wildest Britain*.

REID, D. N., 'Spread of Capercaillie in Ross-shire', *Scot. Nat.*, p. 26, 1930.

RITCHIE, J., 'Northward Extension of Capercaillie to Sutherland', *Scot. Nat.*, p. 126, 1929.

—, *The Influence of Man on Animal Life in Scotland*, 1920.

ROSS, D. M., 'Capercailzie in the Mid-Deveron District', *Ann. Scot. Nat.*, p. 254, 1897.

TAYLOR, W. L., 'The Capercaillie in Scotland', *Jour. Anim. Ecol.*, 17, pp. 155–7, 1948.

TEMPLEWOOD, Lord, *The Unbroken Thread*, 1949.

THOMPSON, J. Vaughan, *Birds of Ireland*, 1900

THOMPSON, W., *Natural History of Ireland*, 1850.

VESEY-FITZGERALD, B., *British Game,* Collins, 1949.

33. Pheasant, 34. Golden Pheasant,

35. Lady Amherst's Pheasant, 36. Reeves's Pheasant

DAWKINS, W. B. (Letter), *Ibis*, (2), 5, p. 358, 1869.

DELACOUR, J., *The Pheasants of the World*, Country Life, 1951.

ELLIOT, D. G., *Monograph of the Phasianidae*, 1870–2.

FITTER, R. S. R., *The Ark in Our Midst*, pp. 136–49, Collins 1959.

GLADSTONE, H. S., 'The Introduction of the Ring-Necked Pheasant to Great Britain', *Brit. Birds*, 17, pp. 36–7; 18, p. 84, 1923–4.

—, A Sixteenth-century Portrait of the Pheasant', *Brit. Birds*, 15, pp. 67–9, 1921.

HACHISUKA, Hon. M. U., 'Nomenclature and Aberration of Pheasant (*tenebrosus*)', *Bull. B.O.C.*, 46, pp. 101–2; 47, pp. 50–2, 1926.

KENNEDY, P. G. *et. al.*, *The Birds of Ireland*, Oliver and Boyd, 1954.

LOWE, P. R., 'Some Remarks on . . . *Phasianus colchicus,* mut. tenebrosus', *Ibis*, (12) 6, pp. 314–20, 1930.

—, 'The Introduction of the Pheasant into the British Isles', *Ibis*, (13) 3, pp. 332–43, 1933.

MATHESON, C., 'Game Records in Wales', *Nat. Lib. Wales. Jour.*, 9, pp. 288–94; 10, pp. 205–14, 1956–7.

MAXWELL, H., 'Naturalization of the Golden Pheasant', *Ann. Scot. Nat. Hist.*, pp. 53–4, 1905.

PARKER, E., *Game Birds, Beasts and Fishes*, 1935.

RAWSTORNE, Lawrence, *Gamonia, or the Art of Preserving Game*, Rudolph Ackermann, 1837; reprinted with Introduction by Eric Parker, Herbert Jenkins, 1929.

SETH-SMITH, D., 'The Origin of the Melanistic Pheasant', *Ibis*, (13) 2, pp. 438–41, 1932.

TEGETMEIER, W. B., *Pheasants: Their Natural History and Practical Management*, 1904.

VESEY-FITZGERALD, B., *British Game*, Collins, 1949.

WAYRE, Philip, *A Guide to the Pheasants of the World*, Country Life, 1969.

37. Red-Legged or French Partridge

EDITORIAL, *British Birds*, 66 (4), pp. 133–5, 1974.

FITTER, R. S. R., *The Ark in Our Midst*, pp. 150–4, Collins, 1959.

HARTING, J. E., 'The Local Distribution of the Red-legged Partridge', *The Field*, 61, pp. 130–1, 1883. (Reprinted in *The Recreations of a Naturalist*, 1906.)

HUDSON, R., *British Birds*, 65 (9), pp. 404–5, 1972.

LEICESTER, Earl of, 'Date of the Introduction into England of the Red-legged Partridge', *The Field*, 137, p. 372, 1921.

PARKER, E., *Game Birds, Beasts and Fishes*, 1935.
POLWHELE, R., *History of Cornwall*, 1816.
VESEY-FITZGERALD, B., *British Game*, Collins, 1949.

38. Bobwhite Quail

BRITISH ORNITHOLOGISTS' UNION RECORDS COMMITTEE REPORT, *Ibis*, 116, p. 579, 1974.
CORNWALL BIRD REPORT: 38, p. 58, 1968.
DENNY, H., 'Bobwhite Quail in Norfolk', *Ann. Mag. Nat. Hist.*, 13, pp. 405–6, 1844.
FITTER, R. S. R., *The Ark in Our Midst*, pp. 155–7, Collins, 1959.
NORFOLK BIRD REPORT, 1974.
PERTHSHIRE BIRD REPORT, p. 6, 1974.

39. Alpine Newt

ARNOLD, E. N., BURTON, J. A., and OVENDEN, D. W., *A Field Guide to the Reptiles and Amphibians of Europe*, Collins, 1977.
FITTER, R. S. R., *The Ark in Our Midst*, p. 275, Collins, 1959.
FRAZER, J. F. D., 'Introduced Species of Amphibians and Reptiles in Mainland Britain', *Brit. Jour. Herp.*, 3, pp. 145–50, 1964.

40. Midwife Toad

ARNOLD, E. N., BURTON, J. A., and OVENDEN, D. W., *A Field Guide to the Reptiles and Amphibians of Europe*, Collins, 1977.
FITTER, R. S. R., *The Ark in Our Midst*, pp. 273–4, Collins, 1959.
FRAZER, J. F. D., 'Introduced Species of Amphibians and Reptiles in Mainland Britain', *Brit. Jour. Herp.*, 3, pp. 145–50, 1964.
SMITH, M., 'The Midwife Toad in Britain', *Brit. Jour. Herp.*, 1, pp. 55–6, pp. 89–91, 1949–50.
—, *The British Amphibians and Reptiles*, Collins, 1951.
TAYLOR, R. H. R., 'The Distribution of Reptiles and Amphibia in the British Isles . . .', *Brit. Jour. Herp.*, 1, pp. 1–25, 1948.
—, 'The Distribution of Amphibians and Reptiles . . .', *Brit. Jour. Herp.*, pp. 95–101, 1963.

41. African Clawed Toad

FRAZER, J. F. D., 'Introduced Species of Amphibians and Reptiles in Mainland Britain', *Brit. Jour. Herp.*, 3, pp. 145–50, 1964.

42. European or Green Tree Frog

ARNOLD, E. N., BURTON, J. A., and OVENDEN, D. W., *A Field Guide to the Reptiles and Amphibians of Europe*, Collins, 1977.

DALTON, R. F., 'Distribution of the Dorset Amphibia and Reptilia', *Proc. Dor. Nat. Hist. and Arch. Soc.*, 72, pp. 135–43, 1950.

FITTER, R. S. R., *The Ark in Our Midst*, p. 274, Collins, 1959.

FRAZER, J. F. D., 'Introduced Species of Amphibians and Reptiles in Mainland Britain', *Brit. Jour. Herp.*, 3, pp. 145–50, 1964.

TAYLOR, R. H. R., 'The Distribution of Reptiles and Amphibia in the British Isles . . .', *Brit. Jour. Herp.*, 1, pp. 1–25, 1948.

WADHAM, P., in Marley's *Natural History of the Isle of Wight*, 1909.

YALDEN, D. W., 'Distribution of Reptiles and Amphibians in the London Area', *Lond. Nat.*, 44, pp. 58–69, 1965.

43. Edible Frog

ARNOLD, E. N., BURTON, J. A., and OVENDEN, D. W., *A Field Guide to the Reptiles and Amphibians of Europe*, Collins, 1977.

ARNOLD, H. R., *Provisional Atlas of the Amphibians and Reptiles of the British Isles*, Nat. Conserv., 1973.

BELL, T., *History of British Reptiles*, 2nd ed., 1849.

—, 'The Edible Frog, Long a Native of Foulmire Fen', *Zool.*, 17, p. 6565, 1859.

BOND, F., 'Note on the Occurrence of the Edible Frog in Cambridgeshire', *Zool.*, 2. 393, p. 677, 1843.

BOULENGER, G. A., 'On the Origin of the Edible Frog in England', *Zool.*, 42, pp. 265–69, 1884.

—, 'On the Races and Variation of the Edible Frog', *Ann. Mag. Nat. Hist.*, 2, pp. 241–57, 1918.

—, *The Tailless Batrachians of Europe*, Ray Soc. Lond. 2 Vols., 1897–8.

BUNTING, W., 'Animals and Plants Introduced to Thorne District of Yorkshire', *Brit. Jour. Herp.*, 1, p. 70, 1957.

COLLENETTE, C. L., *A History of Richmond Park . . .*, 1938.

FISHWICK, J. L., *On the Reptiles and Batrachians*, Vict. Co. Hist. Bed., 1904.

FITTER, R. S. R., *The Ark in Our Midst*, pp. 262–70, Collins, 1959.

FORREST, H. E., *Caradoc Record*, 1891–1914.

FRAZER, J. F. D., 'Introduced Species of Amphibians and Reptiles in Mainland Britain', *Brit. Jour. Herp.*, 3, pp. 145–50, 1964.

GADOW, H., 'Reptilia and Amphibia of Cambridgeshire'. In *Handbook of the Natural History of Cambridgeshire*, ed. J. E. Marr and A. E. Shipley, C.U.P., 1904.

IMMS, A. D., *A Scientific Survey of the Cambridge District*, Rep. Brit. Assoc. Adv. Sci., 1938.

2P

KENSALL, J. E., 'The Reptiles of Hampshire and the Isle of Wight', *Proc. Hamp. Field Club*, 1897.

LEUTSCHER, A., 'An Invader's Decline and Fall', *Country Life*, 20 February 1975.

MILLER, S. H., and SKERTCHLY, S. B. J., *Fenland Past and Present*, 1878.

NEWTON, A., 'The Naturalization of the Edible Frog in England', *Zool.*, 17, pp. 6538–40, 1859.

RINTOUL, L. J., and BAXTER, E. V., *A Vertebrate Fauna of Forth*, 1935.

ROPE, E. J., 'The Reptiles of Suffolk', *Trans. Suff. Nat. Soc.*, 2, 1934.

RUSSELL, H., *Zool.*, pp. 352, 390; 1904.

SMITH, M., *The British Amphibians and Reptiles*, Collins, 1951.

SOUTHWELL, T., *On the Reptiles and Batrachians*. Vict. Co. Hist. Norf., 1901.

—, 'Mammalia and Reptilia of Norfolk', *Zool.*, pp. 2751–60, 1871.

SWANTON, E. W., 'The Mammals . . . of Haslemere and District', *Has. Nat. Hist. Soc.*, 10, 1928.

TAYLOR, R. H. R., 'The Distribution of Reptiles and Amphibia in the British Isles . . .', *Brit. Jour. Herp.*, 1, pp. 1–25, 1948.

—, 'The Distribution of Amphibians and Reptiles . . .', *Brit. Jour. Herp.*, pp. 95–101, 1963.

YALDEN, D. W., 'Distribution of Reptiles and Amphibians in the London Area', *Lond. Nat.*, 44, pp. 58–69, 1965.

44. Marsh Frog

ARNOLD, E. N., Burton, J. A., and OVENDEN, D. W., *A Field Guide to the Reptiles and Amphibians of Europe*, Collins, 1977.

ARNOLD, H. R., *Provisional Atlas of the Amphibians and Reptiles of the British Isles*, Nat. Conserv., 1973.

BUNTING, W., 'Animals and Plants Introduced to Thorne District of Yorkshire', *Brit. Jour. Herp.*, 1, p. 70, 1957.

BURTON, J., 'The Laughing Frogs of Romney Marsh', *Country Life*, 20 December 1973.

COLERIDGE, W. L., 'Laughing Frogs', letter to *Country Life*, p. 798, 4 April 1974.

FITTER, R. S. R., *The Ark in Our Midst*, pp. 270–4, Collins, 1959.

FRAZER, J. F. D., 'Introduced Species of Amphibians and Reptiles in Mainland Britain', *Brit. Jour. Herp.*, 3, pp. 145–50, 1964.

KNIGHT, M., 'Mystery of the Marsh Frog', *Country Life*, 1948.

MENZIES, J. I., 'The Marsh Frog in England', *Brit. Jour. Herp.*, pp. 43–54, 1962.

SMITH, E. P., 'On the Introduction and Distribution of *Rana esculenta* in East Kent', *Jour. Anim. Ecol.*, 8, pp. 168–70, 1939.

SMITH, M., 'The Feeding Habits of the Marsh Frog', *Brit. Jour. Herp.*, I, pp. 170–2, 1953.

—, *The British Amphibians and Reptiles*. Collins, 1951.

TAYLOR, R. H. R., 'The Distribution of Reptiles and Amphibia in the British Isles . . .', *Brit. Jour. Herp.*, I, pp. 1–25, 1948.

—, 'The Distribution of Amphibians and Reptiles . . .', *Brit. Jour. Herp.*, pp. 95–101, 1963.

YALDEN, D. W., 'Distribution of Reptiles and Amphibians in the London Area', *Lond. Nat.*, 44, pp. 58–69, 1965.

45. Wall Lizard

ARNOLD, E. N., BURTON, J. A., and OVENDEN, D. W., *A Field Guide to the Reptiles and Amphibians of Europe*, Collins, 1977.

FITTER, R. S. R., *The Ark in Our Midst*, p. 258, Collins, 1959.

FRAZER, J. F. D., 'The Reptiles and Amphibians of the Channel Islands . . .', *Brit. Jour. Herp.*, I (2), pp. 51–3, 1949.

—, 'Introduced Species of Amphibians and Reptiles in Mainland Britain', *Brit. Jour. Herp.*, 3, pp. 145–50, 1964.

SINEL, J., *Rep. and Trans. Guernsey Soc. Nat. Sci.*, 1905–08.

SMITH, M., 'The Wall Lizard in England', *Brit. Jour. Herp*, I., pp. 99–100 1951.

SPELLERBERG, I. F., 'Britain's Reptile Immigrants', *Country Life*, 20 February 1975.

TAYLOR, R. H. R., 'The Distribution of the Reptiles and Amphibians in the British Isles . . .', *Brit. Jour. Herp.*, I, pp. 1–25, 1948.

—, 'The Distribution of Amphibians and Reptiles . . .', *Brit. Jour. Herp.*, pp. 95–101, 1963.

OTHER AMPHIBIANS AND REPTILES

ARNOLD, E. N., BURTON, J. A., and OVENDEN, D. W., *A Field Guide to the Reptiles and Amphibians of Europe*, Collins, 1977.

COLERIDGE, W. L., 'Laughing Frogs', letter to *Country Life*, p. 798, 4 April 1974.

COLLENETTE, C. L., *A History of Richmond Park . . .*, 1938.

FITTER, R. S. R., *The Ark in Our Midst*, pp. 254–5, Collins, 1959.

FRAZER, J. F. D., 'Introduced Species of Amphibians and Reptiles in Mainland Britain', *Brit. Jour. Herp.*, 3, pp. 145–50, 1964.

LANTZ, L. A., Note in *Proc. Roy. Soc. Lond.*, B.134, pp. 52–6, 1947.

ROPE, E. J., 'The Reptiles of Suffolk', *Trans. Suff. Nat. Soc.*, 2, 1934.

ROPE, G. T., *On the Reptiles and Batrachians*, Vic. Co. Hist. Suff., 1911.

SPELLERBERG, I. F., 'Britain's Reptile Immigrants', *Country Life*, 20 February 1975.

SWANTON, E. W., 'The Mammals . . . of Haslemere and District', *Has. Nat. Hist. Soc.*, 10, 1928.

TAYLOR, R. H. R., 'The Distribution of Reptiles and Amphibia in the British Isles . . .', *Brit. Jour. Herp.*, 1, pp. 1–25, 1948.

—, 'The Distribution of Amphibians and Reptiles . . .', *Brit. Jour. Herp.*, pp. 95–101, 1963.

WADHAM, P., in Morley's *Natural History of the Isle of Wight*, 1909.

46. Rainbow Trout

BEILBY, G. H., *An Attempt to Establish Rainbow Trout in Cornish Streams 1968–69*, Corn. Riv. Auth., 1971.

CARR, D., 'Blagdon Reservoir', *Salm. Trout Mag.*, 45, pp. 368–70, 1926.

—, 'Thirty years of Blagdon', *The Field*, p. 1427, 2 December 1933.

CLARKE, B., 'IPN – the Virus Now Devastating Our Trout Fry', *Sunday Times*, 4 April 1976. (See also letters 19 April (LATTA, J. P.) and 25 April (CLARK, B. B.)).

DAY, F., *British and Irish Salmonidae*, Williams and Norgate, 1887.

EDITORIAL, 'What is a Rainbow Trout?' *Salm. Trout. Mag.*, 68, p. 199, 1932.

ELLISON, N. F., and CHUBB, J. C., 'The Marine and Freshwater Fishes', *Fauna Lancs. and Chesh.*, 44, pp. 1–8, 1963.

FITTER, R. S. R., *The Ark in Our Midst*, pp. 280–4, Collins, 1959.

FROST, W. E., 'Rainbows . . . of a Peat Lough on Arranmore', *Salm. Trout Mag.*, 100, pp. 234–40, 1940.

—, 'A Survey of the Rainbow Trout in Britain and Ireland', *Salm. Trout Assoc.*, July 1974.

FROST, W. E., and BROWN, M. E., *The Trout*, Collins, 1967.

GREY, Viscount of Fallodon, *Fly Fishing*, 1899.

HEWITT, E. R., *Telling on the Trout*, 1926.

HUNT, P. C., 'A Brief Assessment of the Rainbow Trout in Britain', *Fish Mgmt.*, 2, pp. 52–5, 1972.

JORDON, D. S., and EVERMANN, B. W., *American Food and Game Fishes*, Hutchinson, 1902.

LINFIELD, R. S. J., ANGLIAN WATER AUTHORITY REPORTS on Infectious Pancreatic Necrosis (IPN), (unpublished, 1975–6).

MCCASKIE, H. B., 'Changes in the Derbyshire Wye', *Fish. Gaz.*, 118, pp. 667–8, 1939.

MacCRIMMON, H. R., 'World Distribution of Rainbow Trout', *Jour. Fish. Res. Board Can.*, 28, pp. 663–704, 1971.

MAITLAND, J. R. G., *The History of Howietoun*, Guy, 1887.

MAITLAND, P. S., 'Rainbow Trout, *Salmo irideus* Gibbons, in Loch Lomond', *Glas. Nat.*, 18, pp. 421–3, 1966.

MAITLAND P. S., '. . . the distribution of Freshwater fish in the British Isles', *Jour. Fish. Biol.*, 1, pp. 45–58, 1969.

—, *Key to British Freshwater Fishes*, Scien. Publs., Freshw. Biol. Assn, 1972.

M.A.F.F., *Reports on Hertfordshire Rivers*, Fresh Fish. Pap., 6, 1928; 13, 1933.

MUUS, B. J. (ed. A. Wheeler), *The Freshwater Fish of Britain and Europe*, Collins, 1971.

RICHMOND, F. G., 'About Rainbow Trout', *Salm. Trout Mag.*, 20, pp. 63–73, 1919.

—, 'Future of Rainbow Trout in England', *The Field*, 30 April 1932.

STONE, L., *Domesticated Trout: How to Breed and Grow Them*, Charleston, 1877.

STUART, T. A., '. . . trout fishery of the Lake of Menteith', *Rep. Fish. Scot.*, pp. 155–7, 1967.

—, 'Studies on the Lake of Menteith', *Dir. Fish. Res.*, *Rep.* for 1968, pp. 132–4, 1969; for 1970, pp. 114–15, 1971.

TAVERNER, E., *Trout Fishing from all Angles*.

TEW, W. E., 'The Rainbows of the Derbyshire Wye', *Salm. Trout Mag.*, 61, pp. 362–4, 1930.

VARLEY, M. E., 'British Freshwater Fishes . . .', *Fishing News*, 1967.

WALKER, C. E., and PETTERSON, C. S., *The Rainbow Trout*, London, 1898.

WHEELER, A. C., *The Fishes of the British Isles and North-West Europe*, Macmillan, 1969.

—, 'The Fishes of the London Area', *Lond. Nat.*, pp. 80–101, 1958.

—, *Changes in the Freshwater Fish Fauna of Britain*, Sys. Ass. Sp. Vol. 6, Acad. Press, 1974.

WHEELER, A. C., and MAITLAND, P. S., 'The Scarcer Freshwater Fishes of the British Isles'. 1. Introduced species, *Jour. Fish. Biol.*, 5, pp. 49–68, 1973.

WORTHINGTON, E. B., 'Rainbows. A Report on Attempts to Acclimatize Rainbow Trout in Britain', *Salm. Trout Mag.*, 100, pp. 241–60; 101, pp. 62–99; 1941.

47. American Brook Trout or Brook Charr

ANON, 'A New Lakeland Fishery', *The Field*, 146 (3789), p. 238, 6 August 1925.

ARMISTEAD, J. J., *An Angler's Paradise, and How to Obtain It*, London, 1895.

AYTON, W. J., 'Llyn Tarw – American Brook Charr Investigations 1973–74' (unpublished, 15 July 1975).

BRASCH, J.; MCFADDEN, J., and KMIOTEK, S., *The Eastern Brook Trout: its Life History, Ecology and Management*, Wiscon. Conserv. Dept. Publ. 226, 1966.

BUCKLAND, F., *Land and Water*, 1871; 1886.

—, *Natural History of British Freshwater Fishes*, 1880.

CARRINGTON, J. T., 'Fish Culture for the Thames', *Zoologist*, 2, pp. 5110–13, 1876.

DAY, F., *British and Irish Salmonidae*, Williams and Norgate, 1887.

DE BUNSEN, J. M., 'The American Brook Trout in Britain' (unpublished, October 1962).

FISHERY BOARD FOR SCOTLAND REPORTS, 1884–5; 1888–90.

FITTER, R. S. R., *The Ark in Our Midst*, pp. 284–5, Collins, 1959.

FROST, W. F., 'Fish', *Freshw. Biol. Assoc. Ann. Rep.*, 36, p. 42, 1968.

GILLMORE, P., *The Times* (letter), 28 October 1885.

GREY, Viscount of Fallodon, *Fly Fishing*, 1899.

HARVIE-BROWN, J. A., and BUCKLEY, T. E., *A Fauna of Argyll and the Inner Hebrides,* Douglas, 1892.

HEWITT, E. R., *Telling on the Trout*, 1926.

JENKINS, J. T., *The Fishes of the British Isles*, 1925.

JORDAN, D. S., and EVERMANN, B. W., *American Food and Game Fishes*, Hutchinson, 1902.

LAMOND, H., *Loch Lomond*, Jackson, Wylie, 1931.

MacCRIMMON, H. R., and CAMPBELL, J. S., 'The World Distribution of Brook Trout', *Jour. Fish. Res. Board. Canada*, 26 (7), pp. 1699–1725, 1969; with GOTTS, B., 28 (3), pp. 452–6, 1971.

MACLEAN, J. P., *History of the Isle of Mull*, Jobes & Sons, Ohio, 1923.

MAITLAND, J. R. G., *The History of Howietoun*, Guy, 1887.

MAITLAND, P. S., *Key to British Freshwater Fishes*, Scien. Publs. Freshw. Biol. Assn, 1972.

MUUS, B. J. (ed. A. Wheeler), *The Freshwater Fish of Britain and Europe*, Collins, 1971.

ORKIN, P. A. (ed.), *Freshwater Fishes*, Thames and Hudson, 1957.

Salmon and Trout Magazine, 1919, 1932, 1940–1.

SCHINDLER, O., *Guide to Freshwater Fishes*.

SCOTT, T., and BROWN, A., 'The Marine and Freshwater Fishes', *Hand. Brit. Assn*, pp. 173–80, Maclehose, 1901.

SCOTT, W. B., and CROSSMAN, E. J., *Freshwater Fishes of Canada*, Fish. Res. Bd. of Can. Ottawa, 1973.

STONE, L., *Domesticated Trout: How to Breed and Grow Them*, Charleston, 1877.

TATE, R., *Freshwater Fish of the British Isles*, 1911.

TAVERNER, E., *Trout Fishing from all Angles*.

TORBETT, H., *The Angler's Freshwater Fishes*, Putnam, 1961.

WALKER, A. F., 'The American Brook Trout in Scotland', *Rod and Line*, 16 (2), pp. 24–6, February 1976.

WHEELER, A. C., *Changes in the Freshwater Fish Fauna of Britain*, Sys. Ass. Sp. Vol. 6, Acad. Press, 1974.

—, *The Fishes of the British Isles and North-West Europe*, Macmillan, 1969.

WHEELER, A. C., and MAITLAND, P. S., 'The Scarcer Freshwater Fishes of the British Isles', 1. Introduced Species. *Jour. Fish. Biol.*, 5, pp. 49–68, 1973.

48. Common Carp

BALON, Eugene K., 'Domestication of the Carp', *Cyprinus Carpio L,* Royal Ontario Museum, Canada, 29 April 1974.

—, 'Studies on the Wild Carp . . . New Opinions Concerning the Origin of the Carp.' Práce Laboratória rybárstva, 1969.

'B.B' *Confessions of a Carp Fisher*, Eyre and Spottiswoode, 1950.

DAY, F., *The Fishes of Great Britain and Ireland*, 1880.

FITTER, R. S. R., *The Ark in Our Midst*, pp. 292–4, Collins, 1959.

MAITLAND, P. S., 'A Population of Common Carp in the Loch Lomond District', *Glasg. Nat.*, 18, pp. 349–50, 1964.

—, *Key to British Freshwater Fishes*, Scien. Publs. Freshw. Biol. Assoc., 1972.

MUUS, B. J. (ed. A. Wheeler), *The Freshwater Fish of Britain and Europe*, Collins, 1971.

WALTON, Izaak, *The Compleat Angler*, 1653.

WENT, A. E. J., 'Notes on the Introduction of some Freshwater Fish into Ireland', *Jour. Dept Agric. Dublin*, 47, pp. 119–24, 1950.

WHEELER, A. C., *The Fishes of the British Isles and North-West Europe*, Macmillan, 1969.

—, *Changes in the Freshwater Fish Fauna of Britain*, Sys. Assoc. Sp. Vol. 6., Acad. Press, 1974.

YARRELL, W., *A History of British Fishes*, 1859.

49. Goldfish

ATKINS, E. M., *Goldfish and Other Cold-Water Fishes*, Atkins, 1936.

AXELROD, H. R., *Gold Fish Book*, TFH Publications, New Jersey, 1954.

BETTS, L. C., *The Goldfish*, Marshall Press, 1939.

BILLARDON DE SAUVIGNY, L. E., *Histoire Naturelle des Dorades de la Chine*, Paris 1780.

BLOCH, M. E., *Oeconomische Naturgeschichte der Fische Deutschlands*, Berlin, 1784.

EDWARDS, Arthur M., *Life Beneath the Waters, or the Aquarium in America*, New York, 1859.

EVANS, A., *Goldfish*, Muller, 1954.

FITTER, R. S. R., *The Ark in Our Midst*, p. 294, Collins, 1959.

HERVEY, G. F., 'The Goldfish of China in the XVIII Century', *China Society, Sinological Series,* 3, 1950

HERVEY, G. F., and HEMS, J., *The Goldfish,* Faber and Faber, 1968.

KISKINOUYE, K., 'The Goldfish and Other Ornamental Fish of Japan', *Natural Science,* 13, 1898.

KOH, Ting-pong, 'Notes on the Evolution of Goldfish', *China Journal of Science and Arts,* Shanghai, 1934.

MATSUBARA, S., 'Goldfish and their Culture in Japan', *Proc. 4th Int. Cong., Wash.,* 1908.

MULERTT, H., *The Goldfish and its Systematic Culture,* Cincinnati, 1883.

MUUS, B. J. (ed. A. Wheeler), *The Freshwater Fish of Britain and Europe,* Collins, 1971.

PEREIRA, R. A., 'The Goldfish', *Hong Kong Nat.,* 8 1937.

ROUGHLEY, T. C., *The Cult of the Goldfish,* Angus and Robertson, Sydney, 1936.

SCHAEK, M. de 'Histoire du Poisson Doré (*Carassius auratus*)', *Rev. Sci. Nat. Appliq.,* Paris, 1893.

SMITH, H. M., *Japanese Goldfish: Their Varieties and Cultivation,* Roberts, Washington, 1909.

50. Bitterling

ELLISON, N. F., 'A New Cheshire Fish', *Cheshire Life,* 85, November 1959.

ELLISON, N. F., and CHUBB, J. C., 'The Marine and Freshwater Fishes', *Fauna Lancs. and Chesh.,* 44: pp. 1–8, 1963.

FITTER, R. S. R., *The Ark in Our Midst,* pp. 298–9, Collins, 1959.

HARDY, E., 'The Bitterling in Lancashire', *Salm. Trout Mag.,* 142, pp. 548–53, 1954.

LANSBURY, I., 'Some Notes on . . . Fauna and Flora in southern Hertfordshire and north-eastern Middlesex', *Entom. Gaz.,* 7, pp. 97–111, 1956.

MAITLAND, P. S., *Key to British Freshwater Fishes.* Scien. Publs. Freshw. Biol. Assn, 1972.

MUUS, B. J. (ed. A. Wheeler), *The Freshwater Fish of Britain and Europe,* Collins, 1971.

ORKIN, P. A. (ed.), *Freshwater Fishes,* Thames and Hudson, 1957.

PERRING, F. H., 'Distribution Data Relating to Freshwater Fish', *Proc. Brit. Coarse Fish Conf.,* 3, pp. 57–61, 1967.

TORBETT, H. D., *The Angler's Freshwater Fishes,* Putnam, 1961.

WHEELER, A. C., *Changes in the Freshwater Fish Fauna of Britain,* Sys. Assoc. Sp. Vol. 6, Acad. Press, 1974.

—, 'The Fishes of the London Area', *Lond. Nat.,* pp. 80–101, 1958.

—, *The Fishes of the British Isles and North-West Europe,* Macmillan, 1969.

WHEELER, A. C., and MAITLAND, P. S., 'The Scarcer Freshwater Fishes of the British Isles', 1. Introduced species. *Jour. Fish. Biol.,* 5, pp. 49–68, 1973.

51. Orfe

BUCKLAND, F., *Natural History of British Fishes*, S.P.C.K., 1881.

FITTER, R. S. R., *The Ark in Our Midst*, p. 294, Collins, 1959.

MAITLAND, P. S., *Key to British Freshwater Fishes*, Scien. Publs. Freshw. Biol. Assn, 1972.

MUUS, B. J. (ed. A. Wheeler), *The Freshwater Fish of Britain and Europe*, Collins, 1971.

PERRING, F. H., 'Distribution Data Relating to Freshwater Fish', *Proc. Brit. Coarse Fish Conf.*, 3, pp. 57–61, 1967.

WHEELER, A. C., *Changes in the Freshwater Fish Fauna of Britain*, Sys. Assoc. Sp. Vol. 6, Acad. Press, 1974.

—, *The Fishes of the British Isles and North-West Europe*, Macmillan, 1969.

WHEELER, A. C., and MAITLAND, P. S., 'The Scarcer Freshwater Fishes of the British Isles', 1. Introduced species. *Jour. Fish. Biol.*, 5, pp. 49–68, 1973.

52. Wels or European Catfish

Anglers' Mail, 6 September 1968, 9 October 1969, 12 September and 3 October 1970, 28 January 1976.

Angling Times, 12 April and 13 May 1957.

BUCKLAND, F., *Natural History of British Fishes*, S.P.C.K., 1881.

COUCH, J., *A History of the Fishes of the British Islands*, 1865.

DANDY, J. E., 'A List of the Vertebrates of Hertfordshire', *Trans. Herts. Nat. Hist. Soc. Field Club*, 22, p. 168, 1947.

DAY, F., *The Fishes of Great Britain and Ireland*, 1880.

FITTER, R. S. R., *The Ark in Our Midst*, p. 298, Collins, 1959.

GLEGG, W. E., 'Introduction of Wels or Catfish into Hertfordshire', *Trans. Herts. Nat. Hist. Soc. Field Club*, 23, p. 131, 1951.

GUNTHER, A. (in Laver, H.), 'Mammals, Reptiles and Fishes of Essex', *Essex Field Club Spec. Mem.*, 3, 1898 (Reprinted from *The Field*, p. 411, 8 September 1894).

LLOYD, L., *Scandinavian Adventures*, 2 vols., 1854.

LOWE, J., 'The Arrival of the *Silurus glanis* in England', *The Field*, 24, p. 199, 17 September 1864.

MAITLAND, P. S., *Key to British Freshwater Fishes*, Scien. Publs. Freshw. Biol. Assn, 1972.

MUUS, B. J. (ed. A. Wheeler). *The Freshwater Fish of Britain and Europe*, Collins, 1971.

PERRING, F. H., 'Distribution Data Relating to Freshwater Fish', *Proc. Brit. Coarse Fish Conf.* 3, pp. 57–61, 1967.

SACHS, T. R., *Land and Water*, 20 January 1875.

WHEELER, A. C., *The Fishes of the British Isles and North-West Europe*, Macmillan, 1969.

WHEELER, A. C., and MAITLAND, P. S., 'The Scarcer Freshwater Fishes of the British Isles', 1. Introduced species. *Jour. Fish. Biol.* 5, pp. 49–68, 1973.
WILLIAMS, A. C., *Angling Diversions*, 1946.
YARRELL, W., *A History of British Fishes*, 1859.

53. Channel Catfish

Anglers' Mail, 2 January 1969.
JORDAN, D. S., and EVERMANN, B. W., *American Food and Game Fishes*, Hutchinson, 1902.
SCOTT, W. B., and CROSSMAN, E. J., *Freshwater Fishes of Canada*, Fish. Res. Bd. Can., Ottawa, 1973.
WHEELER, A. C., and MAITLAND, P. S., 'The Scarcer Freshwater Fishes of the British Isles', 1. Introduced species. *Jour. Fish. Biol.*, 5, pp. 49–68, 1973.
WHEELER, A. C., *Changes in the Freshwater Fish Fauna of Britain*, Sys. Assoc. Sp. Vol. 6, Acad. Press, 1974.

54. Guppy

DUMBILL, P., *Brit. Ich. Soc. Monthly Newsletter*, 1, March 1967.
LEE CONSERVANCY CATCHMENT BOARD, *Annual Report, 1965–66*, 1967.
MEADOWS, B. S., '. . . guppy . . . in the River Lee', *Essex Nat.*, 32, pp. 186–9 1968.
WHEELER, A. C., *Changes in the Freshwater Fish Fauna of Britain*, Sys. Assoc. Sp. Vol., 6, Acad. Press, 1974.
WHEELER, A. C., and MAITLAND, P. S., 'The Scarcer Freshwater Fishes of the British Isles', 1. Introduced species. *Jour. Fish. Biol.*, 5, pp. 49–68, 1973.

55. Largemouth Bass

Angling Times, 7 August 1969.
FITTER, R. S. R., *The Ark in Our Midst*, pp. 296–7, Collins, 1959.
HARVIE-BROWN, J. A., and BUCKLEY, T. E., *A Fauna of Argyll and the Inner Hebrides,* Douglas, 1892.
JORDAN, D. S., and EVERMANN, B. W., *American Food and Game Fishes*, Hutchinson, 1902.
MAITLAND, P. S., *Key to British Freshwater Fishes*, Scien. Publs. Freshw. Biol. Assn., 1972.
MAITLAND, P. S., and Price, C. E., '. . . a North American . . . trematode . . . probably introduced with the largemouth bass', *Jour. Fish. Biol.*, 1, pp. 17–18, 1969.
MUUS, B. J. (ed. A Wheeler), *The Freshwater Fish of Britain and Europe*, Collins, 1971.

PATTERSON, A. H., '. . . Fish and Fisheries of East Suffolk', *Zool.*, (4) 13, pp. 361–92, 414–21, 1909.

PERRING, F. H., 'Distribution Data Relating to Freshwater Fish', *Proc. Brit. Coarse Fish Conf.*, 3, pp. 57–61, 1967.

SCOTT, W. B., and CROSSMAN, E. J., *Freshwater Fishes of Canada*, Fish. Res. Bd. of Can. Ottawa, 1973.

WHEELER, A. C., *Changes in the Freshwater Fish Fauna of Britain*, Sys. Assoc. Sp. Vol., 6, Acad. Press, 1974.

WHEELER, A. C., and MAITLAND, P. S., 'The Scarcer Freshwater Fishes of the British Isles', 1. Introduced species. *Jour. Fish. Biol.*, 5, pp. 49–68, 1973.

56. Pumpkinseed

JORDAN, D. S., and EVERMANN, B. W., *American Food and Game Fishes*, Hutchinson, 1902.

MAITLAND, P. S., *Key to British Freshwater Fishes*, Scien. Publs. Freshw. Biol. Assn, 1972.

MUUS, B. S. (ed. A Wheeler), *The Freshwater Fish of Britain and Europe*, Collins, 1971.

PERRING, F. H., 'Distribution Data Relating to Freshwater Fish', *Proc. Brit. Coarse Fish Conf.*, 3, pp. 57–61, 1967.

SCOTT, W. B., and CROSSMAN, E. J., *Freshwater Fishes of Canada*, Fish. Res. Bd. of Can., Ottawa, 1973.

TORBETT, H. D., *The Angler's Freshwater Fishes*, Putnam, 1961.

TORTONESE, E., 'I pesci gatto', *Riv. It. Pisc. Itt–A*, 11, pp. 46–7, 1967.

WHEELER, A. C., *Changes in the Freshwater Fish Fauna of Britain*, Sys. Assoc. Sp. Vol. 6, Acad. Press, 1974.

—, *The Fishes of the British Isles and North-West Europe*, Macmillan, 1969.

WHEELER, A. C., and MAITLAND, P. S., 'The Scarcer Freshwater Fishes of the British Isles', 1. Introduced species. *Jour. Fish. Biol.*, 5, pp. 49–68, 1973.

57. Rock Bass

FITTER, R. S. R., *The Ark in Our Midst*, p. 297, Collins, 1959.

HAY, M., *Fish. Gaz. Lond.*, 115 (3245), p. 163, 1937.

JORDAN, D. S., and EVERMANN, B. W., *American Food and Game Fishes*, Hutchinson, 1902.

MAITLAND, P. S., *Key to British Freshwater Fishes*, Scien. Publs. Freshw. Biol. Assn, 1972.

PERRING, F. H., 'Distribution Data Relating to Freshwater Fish', *Proc. Brit. Coarse Fish Conf.*, 3, pp. 57–61, 1967.

SCOTT, W. B., and CROSSMAN, E. J., *Freshwater Fishes of Canada*, Fish. Res. Bd. of Can. Ottawa, 1973.

WHEELER, A. C. *Changes in the Freshwater Fish Fauna of Britain*, Sys. Assoc. Sp. Vol. 6, Acad. Press, 1974.

WHEELER, A. C., and MAITLAND, P. S., 'The Scarcer Freshwater Fish of the British Isles', 1. Introduced species. *Jour. Fish. Biol.*, 5, pp. 49–68, 1973.

58. Zander or Pike-Perch

Anglers' Mail, 6 February 1969; 12 December 1970; 14 February 1973; 10 April; and 19 June 1974; 19 November 1975.

Angling Times, 16 July and 24 December 1965; 11 November 1966; 28 November 1968; 30 January, 31 July, 7, 14, 21 August, 9, 23 October, 25 December 1969; 13, 27 August, 22 October 1970; 30–31 January 1974; 5 June, 7–8 and 20–21 August, 26 November 1975.

BARR, D., 'In Pursuit of Zander', *Country Life*, 159 (4104), p. 486, 26 February 1976.

BUCKLAND, F., *Natural History of British Fishes*, S.P.C.K., 1881.

FITTER, R. S. R., *The Ark in Our Midst*, p. 297, Collins, 1959.

MAITLAND, P. S., '. . . the distribution of Freshwater Fish in the British Isles', *Jour. Fish. Biol.*, 1, pp. 45–58, 1969.

—, *Key to British Freshwater Fishes*, Scien. Publs. Freshw. Biol. Assn, 1972.

MANSFIELD, K., 'Pike-Perch in England', *Salm. Trout Mag.*, 153, pp. 94–8, 1958.

MUUS, B. J. (ed. A. Wheeler), *The Freshwater Fish of Britain and Europe*, Collins, 1971.

PERRING, F. H., 'Distribution Data Relating to Freshwater Fish', *Proc. Brit. Coarse Fish Conf.* 3, pp. 57–61, 1967.

SACHS, T. R., 'Transportation of Live Pike-Perch . . .', *Land and Water*, p. 476, 25 May 1878.

WHEELER, A. C., *Changes in the Freshwater Fish Fauna of Britain*, Sys. Assoc. Sp. Vol., 6, Acad. Press, 1974.

—, *The Fishes of the British Isles and North-West Europe*, Macmillan, 1969.

WHEELER, A. C., and MAITLAND, P. S., 'The Scarcer Freshwater Fish of the British Isles', 1. Introduced species. *Jour. Fish. Biol.*, 5, pp. 49–68, 1973.

59. 'Cichlid'

Angling Times, 18 November 1966, 29 February 1968, 8 January 1970.

GREENWOOD, P. H., *Fishes of Uganda*, Ug. Soc. Kampala, 1966.

HOLDEN, M., and REED, W., *West African Freshwater Fish*, West African Nature Handbooks, Longman, 1972.

LOWE, R. H., 'Species of *Tilapia* in East African Dams . . .', *East Afr. Ag. Jour.*, 20 (4), April, 1955.

WHEELER, A. C., *Changes in the Freshwater Fish Fauna of Britain*, Sys. Assoc. Sp. Vol. 6., Acad. Press, 1974.

WHEELER, A. C., and MAITLAND, P. S., 'The Scarcer Freshwater Fishes of the British Isles', 1. Introduced species. *Jour. Fish Biol.*, 5, pp. 49–68, 1973.

General Bibliography

ARNOLD, E. N., BURTON, J. A., and OVENDEN, D. W., *A Field Guide to the Reptiles and Amphibians of Europe*, Collins, 1977.

BARRETT-HAMILTON, G. E. H., and HINTON, M. A. C., *A History of British Mammals,* Gurney and Jackson, 1910–21.

BRINK, E. H. van den, *A Field Guide to the Mammals of Britain and Europe*, Collins, 1967.

CORBET, G. B., *The Identification of British Mammals*, British Museum, 1964.

—, *The Terrestrial Mammals of Western Europe*, Foulis, 1966.

COUCH, J., *A History of the Fishes of the British Islands*, 1865.

DAY, F., *The Fishes of Great Britain and Ireland*, 1880.

DELACOUR, J., and SCOTT, P., *The Waterfowl of the World*, Country Life, 1954.

EDLIN, H. L., *The Changing Wildlife of Britain*, 1952.

ELTON, C., *Animal Ecology*, Sidgwick and Jackson, 1927.

FITTER, R. S. R., *The Ark in Our Midst*, Collins, 1959.

—, (ed.), *Book of British Birds*, Reader's Digest, 1974.

JENKINS, J. T., *The Fishes of the British Isles*, Warne, 1925.

LAWRENCE, M. J., and BROWN, R. N., *Mammals of Britain, Their Tracks, Trails and Signs*, Blandford, 1973.

LAYCOCK, G., *The Alien Animals: the Story of Imported Wildlife*, Ballantine Books Inc., N.Y., 1970.

MATTHEWS, L. Harrison, *British Mammals*, Collins, 1952.

MAXWELL, H., *British Freshwater Fishes*, Hutchinson, 1904.

MILLAIS, J. G., *The Mammals of Great Britain and Ireland*, Longmans, 1904–06.

MORRIS, F. O., *A History of British Birds*, Groombridge, 1856.

MUUS, B. J. (ed. A. Wheeler), *The Freshwater Fish of Britain and Europe*, Collins, 1971.

NIETHAMMER, G. and J., and SZIJJ, J., *Die Einbürgerung von Säugetieren und Vögeln in Europa*, Paul Parey, 1963.

PAGE, F. J. T., *Field Guide to British Deer*, Mammal Society, 1957.

PENNANT, T., *The British Zoology*, 1761–6.

REGAN, C. T., *The Freshwater Fishes of the British Isles*, 1911.

RITCHIE, J., *The Influence of Man on Animal Life in Scotland*, C.U.P., 1920.

ROOTS, C., *Animal Invaders*, David and Charles, 1976.

SCHINDLER, O., *Guide to Freshwater Fishes*, Thames and Hudson, 1957.

SHARROCK, J. T. R. (ed.), *Atlas of Breeding Birds in Britain and Ireland*, British Trust for Ornithology, 1976.

SMITH, M., *The British Amphibians and Reptiles*, Collins, 1951.

SOUTHERN, H. N. (ed.), *The Handbook of British Mammals*, Blackwell, 1964.

STREET, P., *Mammals in the British Isles*, Hale, 1961.

TAVERNER, E., *Anglers' Fishes and their Natural History*, 1957.

VESEY-FITZGERALD, B., *British Game*, Collins, 1949.

WAYRE, Philip, *A Guide to the Pheasants of the World*, Country Life, 1969.

WELLS, L., *Freshwater Fishes*, Warne, 1957.

WHEELER, A. C., *The Fishes of the British Isles and North-West Europe*, Macmillan, 1969.

WILLIAMS, A. C., *Angling Diversions*, 1946.

WITHERBY, H. F. *et. al., The Handbook of British Birds*, (5 vols.), Witherby, 1938-41.

YARRELL, W., *A History of British Fishes*, 1859.

—, *A History of British Birds*. 1837-43.

Addenda

23. Ring-Necked, Rose-Ringed or Green Parakeet

Ring-necked parakeets were reported to have bred in 1976 at Thundersley, north-west of Southend in Essex, and at Stockport in the Greater Manchester area – the latter an entirely new district. In both places they have been present since 1975 or 1974. (I am indebted to Mr Robert Hudson for drawing my attention to these two new records.)

24. Budgerigar

Only about a dozen pairs of budgerigars were reported by Mr David Hunt to have nested successfully on Tresco in the Scilly Isles in 1976, compared with an estimated 35–45 pairs in 1975. This suggests that the birds may be more dependent on artificial feeding, which was reduced during the winter of 1975–6, than had previously been supposed.

25. Little Owl

The 1976 population of little owls in the British Isles was estimated at between 7000 and 14000 pairs. (J. T. R. Sharrock, *The Atlas of Breeding Birds in Britain and Ireland*, British Trust for Ornithology, 1976.)

28. Egyptian Goose

Egyptian geese were reported to have colonized Thorpeness in east Suffolk during 1975, and to have bred there in 1976.

29. Mandarin Duck

In 1976 there were free-flying flocks of mandarins at Leeds Castle in Kent and at Salhouse, north-east of Norwich in Norfolk.

31. Ruddy Duck

In 1974 a pair of ruddy ducks are reported to have nested successfully on the shores of Lough Neagh in Northern Ireland.

Waterfowl collections at Apethorpe in the Soke of Peterborough, at St Neots in Huntingdonshire and near Grimsby in Lincolnshire – all of which rear ruddy ducks – are potential sources of further expansion of the species in eastern England. On 3 December 1976 the author saw a single male ruddy duck on the Berkshire end of Virginia Water – an exceptional south-easterly record.

32. Capercaillie

Three counts of capercaillie in the Black Wood of Rannoch in Perthshire, one of the finest remnants of the ancient Caledonian forest (see p. 325), in 1965–6 revealed totals of 63, 52, and 71 birds in an area of approximately 920 acres (372 hectares), a population density of about one to 15 acres (6 hectares). (G. W. Johnstone and F. C. Zwickel, 'Numbers of Capercaillie in the Black Wood of Rannoch', *Brit. Birds*, 59: pp. 498–9, 1966.)

In 1973 successful nesting is reported to have taken place on the Forestry Commission property at Grizedale, north Lancashire (see p. 323).

The total Scottish population in 1976 may have been as many as 10 000 pairs. (J. T. R. Sharrock, *The Atlas of Breeding Birds in Britain and Ireland*, British Trust for Ornithology, 1976.)

(See also C. E. Palmar, 'The Capercaillie', *Forestry Commission Publication*, No. 37.)

33. Pheasant

George Owen in his *Description of Pembrokeshire* (1603), wrote:

As for the Pheasant, in my memorie there was none breedinge within the shire untill about XVJ yeares past Sr Thomas Perrot Knight procured certaine hens and cockes to be transported out of Ireland,which he purposinge to endenize in a pleasant grove of his owne plantinge adioyninge to his house of Haroldston [see p. 337] gave them libertie therine, wherein they partly stayed and bredd there, and neere at hand, but afterwardes chose other landlordes in other places, and, as I heare of no great multiplyeinge, so are they not altogether destroyed, but some few are yet to be found in some places of the shire, thoughe but thinne.

G. Jones in *Welsh Folklore and Folk Customs* (1930), records that a cock pheasant observed in the Vale of Edeyrnion in 1812 was an unusual enough sight to be mistaken by the inhabitants for a winged and brightly scaled viper.

J. Garsed in the *Records of the Glamorganshire Agricultural Society*, 1890, reported that in 1781 Thomas Mansel Talbot had been awarded the

Society's gold medal 'for his spirited endeavours to introduce the English Pheasant into this county'. (All quoted by C. Matheson in 'The Pheasant in Wales'; *Brit. Birds,* 56: pp. 452–6, 1963.)

In 1976 the total population of pheasants in the British Isles may have been as many as half a million pairs. (J. T. R. Sharrock, *The Atlas of Breeding Birds in Britain and Ireland,* British Trust for Ornithology, 1976.)

34. Golden Pheasant

Only one or two pairs of golden pheasants were reported by Mr David Hunt to have bred successfully on Tresco in the Scilly Isles in 1976.

The total British population in that year was estimated at 500–1000 pairs. (J. T. R. Sharrock, *The Atlas of Breeding Birds in Britain and Ireland,* British Trust for Ornithology, 1976.)

35. Lady Amherst's Pheasant

In 1971 some Lady Amherst's pheasants were reported to be hybridizing with golden pheasants in Galloway. It is not known whether these resulted from a recent introduction or were the descendants of the Duke of Bedford's *amherstiae* x *pictus* introduction of about 1895 (see p. 343).

The total British population of Lady Amherst's pheasant in 1976 was estimated at 100–200 pairs. (J. T. R. Sharrock, *The Atlas of Breeding Birds in Britain and Ireland,* British Trust for Ornithology, 1976.)

36. Reeves's Pheasant

Robert Gray in 'The Introduction of Reeves's Pheasant into Scottish Game Preserves' (*Proc. Roy. Soc. Edin.,* 7, p. 239, 1882) reported that

During the past eight years Reeves's pheasants had been flying about wild in the woods of Guisachan in . . . Inverness, the property of Sir Dudley Marjoribanks, now Lord Tweedmouth. More than one hundred had been shot there in the course of a single season, and the birds were found to be as hardy as (the young indeed more so than) the commoner varieties . . . many found their way to Balmacara. It was also stated that forty male birds had been turned loose in the woods of Tulliallan, Clackmannanshire. Two birds . . . had lived in a wild state along with others of the same species for six years at Elvedon [*sic*] estate in Suffolk belonging to the Maharajah Duleep Singh, and were shot there by Lord Balfour [of Burleigh] in October last. Several hybrids had appeared during these years in the Elvedon [*sic*] grounds. They were dark-coloured birds, heavier than either parent, and with a white head and neck.

On 2 March 1972 forty-one wing-clipped Reeves's pheasants (thirty-two hens and nine cocks, of which twenty-six had been hatched in 1971 and

2Q

fifteen in 1970) were turned out by Mr Frederick Courtier of the Forestry Commission in a 30 acre (12 hectare) fox-proof enclosure at Denny Lodge near Lyndhurst in the New Forest. In 1972 between 120 and 130 eggs were laid, many of which were sucked by magpies or crows. As the birds' feathers grew some left the enclosure and dispersed over the Forest, where at least one pair is believed to have reared young in 1975. In 1976 two cocks and several hens remained inside the enclosure.

A total of fifty-two birds (mostly cocks) was liberated in 1970 at Kinveachy, from where individuals have wandered as far south and west as Alvie and Kinrara. In 1975 several were reported in the woods above Aviemore and around the Highland Wildlife Park near Kincraig, and one (possibly a hybrid) was killed during a capercaillie shoot near Carrbridge. The failure of the birds to become more firmly established at Kinveachy is attributed at least in part to the lack of subsequent reinforcing introductions, due to a ban on the importation of eggs or birds to Scotland from England following an outbreak of fowl-pest.

Reeves's pheasants were reported still to be present at Elveden in Suffolk at least until 1972.

The British Trust for Ornithology's *Atlas* (1976) includes reports of seventy Reeves's pheasants released in Morayshire and others in Cumberland in 1969. Feral populations are also said to have existed in Ross-shire and in Dorset until at least the mid-1950s.

(See also: Anon; 'Rearing Trials: Reeves Pheasants and Prairie Chickens' *Eley Game Advis. Stn. Ann. Rev. 1965–6*; pp. 33–7, 1966; T. H. Blank, Reeves Pheasant', *Game Conserv. Ann. Rev. 1969–70,* pp. 81–2, 1970.)

37. Red-Legged or French Partridge

The 1976 population of red-legged partridges in the British Isles was estimated at between 100 000 and 200 000 pairs. (J. T. R. Sharrock, *The Atlas of Breeding Birds in Britain and Ireland,* British Trust for Ornithology, 1976.)

(See also: Anon. 'National Game Census', *Game Conserv. Ann. Rev. 1970,* pp. 19–22, 1971; T. H. Blank and R. P. Bray, 'The National Game Census', *Game Conserve. Ann. Rev. 1969–70,* pp. 30–7, 1970; G. Howells, 'The Status of the Red-legged Partridge in Britain,' *Game Res. Assoc. Ann. Rep.* 2; pp. 46–51, 1963;

38. Bobwhite Quail

A covey of twenty-four bobwhite quail (including several young birds) was observed by Mr David Hunt on Tresco in the Scilly Isles in October 1976. The birds nested late that year, due, it is believed, to the severe summer drought. The total population appeared to be about the same as in 1975, when it was estimated to number 45–50 birds, with a preponderance of males. Bobwhite quail have been recorded from, but have not so far bred on, Bryher.

49. Goldfish

Dr C. W. Coates of the New York Zoological Society has suggested that the fish seen by Samuel Pepys at the house of Lady Pen (see p. 451) were in fact specimens of the labyrinth-gilled paradise fish (*Macropodus opercularis*) which lives in the paddy fields and drainage ditches of Korea, Vietnam, China and Taiwan. This species is very finely marked, and by using laminated accessory gill-like breathing organs situated in chambers on either side of its head, is able to live and breed in very poorly oxygenated waters and in small containers. It has for long been kept as a pet in captivity in China.

50. Bitterling

In August 1976 Mr Mark Ridgway, Fisheries Inspector for the Severn-Trent Water Authority, discovered a new population of bitterling in Hanley Park Pool, a 3–4 acre (1·5 hectare) artificial water averaging about 4 feet (120 cm) in depth, in the centre of Stoke-on-Trent. This colony, estimated to number several thousand individuals, ranging in age from five years to a few months, is assumed to originate from the disposal of surplus live-bait or unwanted cold-water aquarium stock. (I am indebted to Mr Alwyne Wheeler for drawing my attention to this new colony which was reported to him by Mr Keith Easton of the Severn–Trent Water Authortiy.)

58. Zander or Pike-Perch

In the summer of 1976 a new colony of zander was discovered in Coombe Abbey Lake, Nottingham. By October a total of eighty one-year-old fish had been captured, their small size suggesting that they had been illegally transferred as spawn or fry from the Fens. As the outflow debouches from Coombe Abbey into the river Sowe, which in turn flows into the Warwickshire Avon, a tributary of the Severn, there is considerable potential for the further spread of zander from this source.

Also in the summer of 1976 a zander reported as weighing 16 lb 6 oz (7·5 kg) and measuring 35½ in. (90 cm) long with a girth of 20½ in. (51 cm), was caught in the Cut-off Channel of the Great Ouse river system by Mr S. Smith of Daventry.

Index of People

Index of Species

NOTE: Species in CAPITALS are discussed in individual chapters.

Index of Places

Reptiles, Amphibians and Fish

Rainbow Trout

American Brook Trout
or Brook Charr
Rock Bass
Pumpkinseed
Largemouth Bass
Channel Catfish

Guppy

Painted Frog

Afric
Claw
'Cich
Til

Alpine Newt
Midwife Toad
European or Green Tree
Edible Frog
Marsh Frog
Wall Lizard
European Pond Tortois
Tesselated Snake
Southern or Italian Cre
Yellow-Bellied Toad
Fire-Bellied Toad
Green Lizard
Bitterling
Orfe
Zander or Pike-Perch

Note: Dual entries for a single species mark the
approximate extremities of its natural range.